# california's changing landscapes

GORDON B. OAKESHOTT

Former Deputy Chief and Chief
Division of Mines and Geology
State of California



McGRAW-HILL BOOK COMPANY

New York St. Louis San Francisco Auckland Bogotá Düsseldorf
Johannesburg London Madrid Mexico Montreal New Delhi
Panama Paris São Paulo Singapore Sydney Tokyo Toronto

Second Edition

# california's changing landscapes

A GUIDE TO THE GEOLOGY OF THE STATE

**CALIFORNIA'S CHANGING LANDSCAPES**
A GUIDE TO THE GEOLOGY OF THE STATE



567890 DODO 85

This book was set in Zenith by York Graphic Services, Inc.
The editors were Donald C. Jackson
and Anne T. Vinnicombe;
the production supervisor was Charles Hess.
R. R. Donnelley & Sons Company was printer and binder.

*To*

**OLAF P. JENKINS**

*Who launched me on my career with the state*
*and*
*who has been my companion on so many field trips*
*exploring California geology*

**Library of Congress Cataloging in Publication Data**

Oakeshott, Gordon B date
California's changing landscapes.

Bibliography: p.
Includes index.
1. Geology—California. I. Title.
QE89.02 1978 557.94 77-22045
ISBN 0-07-047584-9

# contents

## PART II CALIFORNIA THROUGH THE GEOLOGIC AGES

## PART III GEOLOGIC VIEWS AND JOURNEYS IN THE NATURAL PROVINCES

# list of illustrations

## PHOTOGRAPHS

## MAPS AND DIAGRAMS

# preface to the second edition

■ Since *California's Changing Landscapes* was published in 1971, I have had the opportunity of using the book in two different courses at two of California's great state universities. At the request of the department chairman at California State University, Sacramento, I taught two classes of elementary physical geology, supplementing the lectures with reading assignments and intensive study of the illustrated California examples. Later, at California State University, San Francisco, I used the book as primarily intended—for a one-semester course in geology of California.

Both ventures were quite successful, but the instructor learned something, too! The great majority wanted (1) a glossary of geological terms and (2) more on plate tectonics.

In this new edition, I have tried to supply both these needs as well as pick up those inevitable, inadvertent errors that creep into most manuscripts. Geologic facts and references have been thoroughly updated.

Unfortunately, the cost of stock and printing have continued to rise so as to challenge the skills of composers, printers, and author to maintain reasonable prices. This is why we have omitted the state geologic map as in the back pocket of the first edition. Students are strongly advised to get the similar, highly colored, 1:2,500,000-scale state geologic map by the U.S. Geological Survey and the California Division of Mines and Geology available at nominal costs.

My thanks to Margaret W. Turner for typing and editorial assistance.

**Gordon B. Oakeshott**

# preface to the first edition

■ This book was really written because, after living California geology as student, teacher, and professional geologist, I feel that I must impart some of my own enjoyment and appreciation of the fascinating phenomena of the state's long geologic history to all those—young and old, amateur and professional—who are eager to learn about their geologic environment.

It probably all began (for me, that is) when my mother, as a very young woman, tutored John Muir's daughters in the old Muir home in Martinez (now a National Historic Site) and listened to him tell of his great and lonely adventures in the Yosemite. Thus stimulated, her love of California "geography" was handed down to her school children in the 1890s and later to my brothers and me. My career was fully committed in my freshman year at Berkeley by the enthusiastic lectures of the late Professor Andrew C. Lawson, greatest of early day California geologists.

I have not written for my professional colleagues—I am not able to add materially to their knowledge and dedication—but rather for those young people in our secondary schools, colleges, and universities, who are studying the earth sciences, and for the thousands of amateur earth scientists who would like to know more of what they see in the minerals, rocks, and landscapes of the state. There is presently no single book available on the geology of California. This book has been written and illustrated to fill this gap.

Specific objectives of the book are to:

**1.** Appeal to all those many people who are interested in California's rocks, minerals, and geologic history, including the "rockhounds" and amateur geologists
**2.** Furnish a reliable background for earth science teachers at all levels, as far as basic geology applied to California is concerned
**3.** Outline the principles of geology, and features and history of the California landscape, as necessary background knowledge for engineers, planners, local and regional government officials, and others who must deal with the geological environment in making their judgments on the wisest and safest management of the resources
**4.** Supply a useful volume on the geology of California to serve as a text for the considerable number of courses given in junior colleges, state colleges, universities, and extensions that are called "Geology of California" and are generally given without prerequisites

The book contains all materials needed for a beginning study of California geology—geologic and road maps, relief map, charts and diagrams, and photographs especially chosen to illustrate California's most striking and interesting geologic features. For those wishing to study a particular area in more detail, the text is coordinated with the 27 sheets of the *Geologic Map of California*. These maps are drawn to a scale of 1 inch equals 4 miles, and are available from the California Division of Mines and Geology.

The text is organized in three parts. Part I, Rocks, Principles, and Processes—planned to be self-explanatory and easy reading for the upper high-school to college student and amateur geologist—uses California examples to discuss geologic principles, landscape-making processes, and the important rocks and minerals.

Part II, California through the Geologic Ages, is a chronological account of geologic formations and history, starting with the state's oldest rocks—nearly two billion years old—and following through the great changes in landscapes. The strange and different life forms recorded as fossils in the rocks of past ages are discussed.

Part III, Geologic Views and Journeys in the Natural Provinces, guides the interested viewer and visitor along the state's "geologic" highways—through the deserts, along the coasts, and into the High Sierra. This is a series of geologic views and journeys throughout the state.

An extensive list of further readings and other sources of information on California geology follows Part III.

This book was planned and written in my study at home beneath a forest of tall Monterey pines, high on the crest of the Berkeley Hills, resting safely on Upper Cretaceous shale bedrock, albeit less than a mile from the active Hayward fault!

**Gordon B. Oakeshott**

# acknowledgments

■ The manuscript text was greatly improved by the thoughtful and constructive criticism of many readers. Foremost among these was my editor and friend, Robert L. Heller, who made point-by-point criticisms and suggestions as the manuscript was written and rewritten. I am especially grateful also to A. O. Woodford who reviewed the bulk of the manuscript, and to James B. Koenig who was most helpful on Parts I and II because of his experience in teaching a university extension course on the geology of California. My state colleagues, Charles W. Chesterman, Earl W. Hart, Douglas M. Morton, Salem J. Rice, Bennie W. Troxel, and F. Harold Weber carefully reviewed areas within their special knowledge. University of California professors Robert F. Heizer and Bruce A. Bolt checked parts dealing with the antiquity of man, and with earthquakes, respectively. Lauren A. Wright and Bennie W. Troxel furnished data for the Death Valley road guide.

The many beautiful photographs by Mary Hill and the unique scratchboard drawings by my brother, Peter H. Oakeshott, truly embellish the text and help to make it more complete. The text is also enhanced by the many excellent photographs of Ernest S. Carter, Charles W. Chesterman, Charles W. Jennings, the Los Angeles County Museum, the National Park Service, John S. Shelton, the Smithsonian Institution, G Dallas Hanna, geologists and photo library of the U.S. Geological Survey, and F. Harold Weber. Drafting of maps and diagrams was done by R. Merl Smith.

Most of the typing was done by Susanna J. Bruff, wife of the late professor of geology at the College of Marin, Stephen C. Bruff, and Sandra J. Shrock. Additional typing was done by Dorris Campbell, Lucy O'Brien, Robyn Roles, and Margaret Turner.

The patience and encouragement of my wife, Bee, made it possible to write and compose this book during many weekends and holidays.

Finally, I am deeply grateful to my old friend to whom this book is dedicated, and to all my geology teachers, professional friends, and colleagues, who have contributed so much to a lifetime of work and enjoyment of the geology of California.

**Gordon B. Oakeshott**

☐ Time, rocks, and fossils are the ingredients of geologic history, in which time is measured, not in years, but in millions and hundreds of millions of years. Given enough time, soft mud and loose sand can become firm rock. Given enough time, a great mountain range can be lowered to rounded hills. Given a few billion years, primitive single-celled life of the seas evolved to man and his primate relatives.

We look at the mountains and valleys of California through the eyes of the geologist and know that this is but the landscape of this instant of geologic time. This is not the landscape of the 10 million years ago that was yesterday; nor is it the landscape of tomorrow—10 million years hence. The seas have advanced and retreated; mountains have been built, eroded away, and rebuilt; strange creatures have populated the seas and roamed the lands, constantly being replaced by evolving life forms. Change—evolution—has been the keynote; only geologic principles and basic processes have stayed the same throughout the aeons.

# I
# ROCKS PRINCIPLES AND PROCESSES

# 1 california's mountains and valleys

*California's mountains and valleys stretch from the coast to the Sierra Nevada. The far-distant Sierra viewed across San Joaquin Valley from Mount Diablo.*

■ California's 100 million acres (158,300 square miles) make it one of the three largest states in the United States; only Alaska and Texas are larger. No state approaching its size can equal California in diversity of landforms, topography, climate, and landscapes. An overview of California as seen from space, but denuded of its highly varied vegetation and unobscured by clouds and haze, gives an impression of high and rugged mountains trending predominantly in a north-northwesterly direction, with a great central trough or valley oriented in about the same way (see relief map, inside front cover). The impression of high topographic relief is no delusion, for Mount Whitney in the southern Sierra Nevada has an elevation of 14,495 feet and the bottom of Death Valley, only a few tens of miles away, is 282 feet below sea level. These are the highest and lowest points in the United States (exclusive of Alaska and Hawaii).

Only two of the boundaries of California are natural: on the west the waters of the Pacific Ocean lap onto a thousand miles of beaches and sea cliffs, and in the southeastern desert the state shares something like 200 miles of the Colorado River with the state of Arizona. On the north, the Oregon boundary is the 42nd parallel of latitude; the southern boundary against Mexico is also a westerly trending straight line, south of the 33rd parallel. The boundary with Nevada on the east, like those with Oregon and Mexico, consists of straight lines that cut sharply across all natural features of the land surface. The state has a roughly uniform "width" of about 200 miles.

Even rather casual examination of the relief map (inside cover), the 1:2,500,000-scale colored geologic map of California (available from the U.S. Geological Survey or the California Division of Mines and Geology), and other maps showing precipitation (Figure 1-3), drainage basins (Figure 1-4), and the like, shows that the state can be readily divided into natural regions or provinces (inside cover). Each of these provinces has certain broad and more or less characteristic features—relief, landforms, geology, and landscapes—that distinguish it. We shall find that the natural provinces are fundamentally geologic regions and that they have developed their distinctive characteristics of trend, relief, drainage, and landscape largely as the result of natural geologic processes acting on diverse rock types and structures through millions of years of geologic time.

In order to develop an initial and basic acquaintance with the geography and geology of California, we shall make a brief overall examination of each of the natural provinces. The predominant northwesterly trend of the state's natural features is accentuated by two mountain ranges and the valley between them. We shall therefore discuss, in order, the distinguishing features of the Coast Ranges, the Sierra Nevada, and the Great Valley, then continue with the adjoining provinces.

## COAST RANGES

The Coast Ranges natural region, or geomorphic province, trends north-northwesterly, roughly parallel to the Sierra Nevada, but lies between the Great Valley and the Pacific Ocean. It extends for about 400 miles along the coast into Oregon (narrowing markedly at the extreme north), and measures from a few to about 70 miles in the east-west dimension. At the north, this province is separated from the Klamath Mountains by the South Fork Mountain fault zone [see the 1:2,500,000-scale colored geologic map of California available from the U.S. Geological Survey or the California Division of Mines and Geology (see details p. 341)], but topographically the mountains are continuous and the separation is not a sharp one. The southern end of the Coast Ranges butts against the east-west trending Santa Ynez Mountains in the Transverse Ranges province, but the Coast Ranges also swing eastward near their southern end into the east-trending San Emigdio Range and Tehachapi Mountains with complete closure around the southern end of the Great Valley into the Sierra Nevada.

Since the geologic structures of the Coast Ranges trend 30 to 40 degrees west of north, the irregular, but more northerly trending coastline cuts obliquely across basic lineation of the ranges. Thus, geologic features that characterize the Coast Ranges extend out onto the continental shelf under the Pacific Ocean. The coast itself is extremely rugged and is marked by nearly vertical sea cliffs, sea stacks (near-shore islands separated from the mainland by wave and current erosion), and steep mountains at the shore line, in many places with gently sloping, narrow terraces a few feet to a few hundred feet above the sea.

Relief and topography of the Coast Ranges are entirely different from those of the Sierra Nevada. Highest peaks are seldom more than 6,000 feet and crests average between 2,000 and 4,000 feet in elevation. Most crests are rounded by erosion, while the jagged summit of the Sierra Nevada reflects recent glaciation in the High Sierra, a process that did not reach the much lower Coast Ranges.

The Coast Ranges consist of a very complex series of more or less independent ranges and valleys, few of which are continuous for more than 100 miles. One of the highest and longest is the Mendocino Range in the northern coast area. Also prominent is the Diablo Range, which lies just west of the San Joaquin Valley and extends from Mount Diablo (elevation 3,849 feet) for 130 miles southeast to the low Kettleman Hills on the margin of the San Joaquin Valley. The Marin Peninsula, north of the Golden Gate, is dominated by prominent Mount Tamalpais (elevation 2,604 feet). San Francisco Bay is a shallow structural and topographic depression between the Berkeley Hills (western extension of the Diablo Range) on the east and on the west the Santa Cruz Mountains, which extend southward on the San Francisco Peninsula to Monterey Bay.

San Francisco Bay separates the northern Coast Ranges from the southern Coast Ranges. West of the Diablo Range in the southern Coast Ranges is a continuous long line of depressions that make up San Francisco Bay, Santa Clara Valley, and San Benito Valley. The longest continuous valley in the southern Coast Ranges is the Salinas Valley, which

**1-1** *Northeast across Peachtree Valley in the southern Gabilan Range, southeastern Monterey County. The San Andreas fault zone here follows along Mustang Ridge on the skyline. The rounded hill, steep rocky ridge, flat-floored irrigated valley, and vegetation composed of grass and scattered live oaks are all typical of the southern Coast Ranges. (Mary Hill photo.)*

trends northwesterly as a straight depression between the Gabilan Range on the east and the Santa Lucia Range on the west.

As might be expected, streams of the Coast Ranges are smaller than those of the Sierra Nevada and tend to flow northwest or southeast, rather than west. Largest in terms of runoff is the Eel River, which drains the Mendocino Range north-northwest to Humboldt Bay at Eureka. The Russian River also originates in the Mendocino Highlands but runs southeasterly, then sharply westward to enter the ocean a few miles north of Point Reyes. In the southern Coast Ranges, the Salinas River follows the Salinas Valley northwesterly to Monterey Bay.

Offshore, the western edge of the continental margin may be taken at the 10,000-foot depth contour (see the 1:2,500,000-scale colored geologic map of California available from the U.S. Geological Survey or the California Division of Mines and Geology), which is close to the foot of the steep continental slope. The sea floor offshore from the Coast Ranges has a narrow shelf averaging about a 3-degree slope to sea and extending to a depth of about 600 feet. Beyond the continental margin is the deep-sea plain.

Two very prominent west-trending escarpments, probably of fault origin, extend offshore: the Mendocino Escarpment due west of Cape Mendocino in the northern Coast Ranges and the Murray Escarpment, which continues the westward trend of the Santa Barbara coastline. Two great fans—the Delgada Fan in the north, and Monterey Fan in the south—extend from the continental slope out onto the deep-sea plain. A series of submarine canyons, quite similar to deep canyons onshore, extends from near-shore points out across the continental shelf—some of the more notable are north to south, Delgada Canyon, Bodega Canyon, Monterey Canyon, Sur Canyon, and Arguello Canyon.

## SIERRA NEVADA

East of the Great Valley, the Sierra Nevada, the highest and most continuous mountain range in California, extends in a north-northwesterly direction for more than 400 miles. Geologically, the Sierra is a great block of granitic rocks and remnants of older rocks, which has been tilted westward. The highest peak, Mount Whitney, is near the southern end of the range; but, along the crest, there are many miles of peaks that rise higher than 12,000 feet. The western slope of the Sierra is gentle, averaging about 2 degrees; but the eastern slope, marked by extensive faults, is one of the steepest and most abrupt in the West. The southern end of the Sierra turns westward into the west-trending Tehachapi Mountains, which lie at the southern end of the San Joaquin Valley and extend eventually into the Coast Ranges. In the north, geologic structures and rock formations of the Sierra Nevada probably continue under the Cascade Mountains into the Klamath Mountains; topographically, the Sierra merges into the Cascade Range and the Modoc Plateau. From the northern end of the Sierra and the Klamath Mountains, granitic rocks and structures typical of the Sierra Nevada arc northeastward across eastern Oregon and Idaho, then northward into British Columbia.

Because of its great height and length, the Sierra Nevada forms a tremendous natural barrier, standing higher among its surroundings than most other mountain ranges in the country. Thanks to its asymmetric westward-tilted form, the high, jagged peaks of the crest are 25 to 80 miles from the western base of the range but only a few miles from the valleys on its eastern flank. The northern and southern ends of the range are lowest, with peaks about 6,000 to 7,000 feet in

**1-2** *View from eastern end of Camiaca Lake south toward Shepherd's Crest in the high central Sierra Nevada west of Mono Lake. Pyramidal peaks, serrated ridges, and small cirques in which summer snow lies are evidence of glaciation in the recent past. A typical U-shaped valley lies left of center. Metamorphosed Jurassic-Triassic rhyolite in the foreground has been rounded, striated, and polished by moving ice. Rock of the crest is Cathedral Peak granite. (Charles W. Chesterman photo.)*

elevation; in the east-central area around Lake Tahoe, Mount Tallac, Pyramid Peak, and others are in the 9,000- to 10,000-foot range. The highest peaks—Mount Whitney, Mount Williamson (14,384 feet), and Mount Langley (14,042 feet)—are in the rugged central-southern part of the range south of Yosemite National Park and west of the sharp, deep, and narrow Owens Valley.

Because of its position as a long, high, northwesterly trending, westward-tilted block, the Sierra Nevada forms a barrier to the winter storms that bring most of California's precipitation from the west. The westerly winds bearing moisture from the Pacific Ocean are forced upward across the western slopes of the Sierra; and, as they rise, condensation of moisture produces abundant precipitation, which increases rapidly upslope to a maximum of perhaps 60 to 80 inches per year at about 7,000 feet elevation. Above 7,000 feet, precipitation lessens somewhat as the air becomes more dry. In winter, the rain that is characteristic at the lower elevations falls as snow in the higher parts of the range. The eastern slopes are in a rain shadow—cut off from the western source of moisture; the floor of Owens Valley, for example, at an elevation of about 4,000 feet, is true desert,

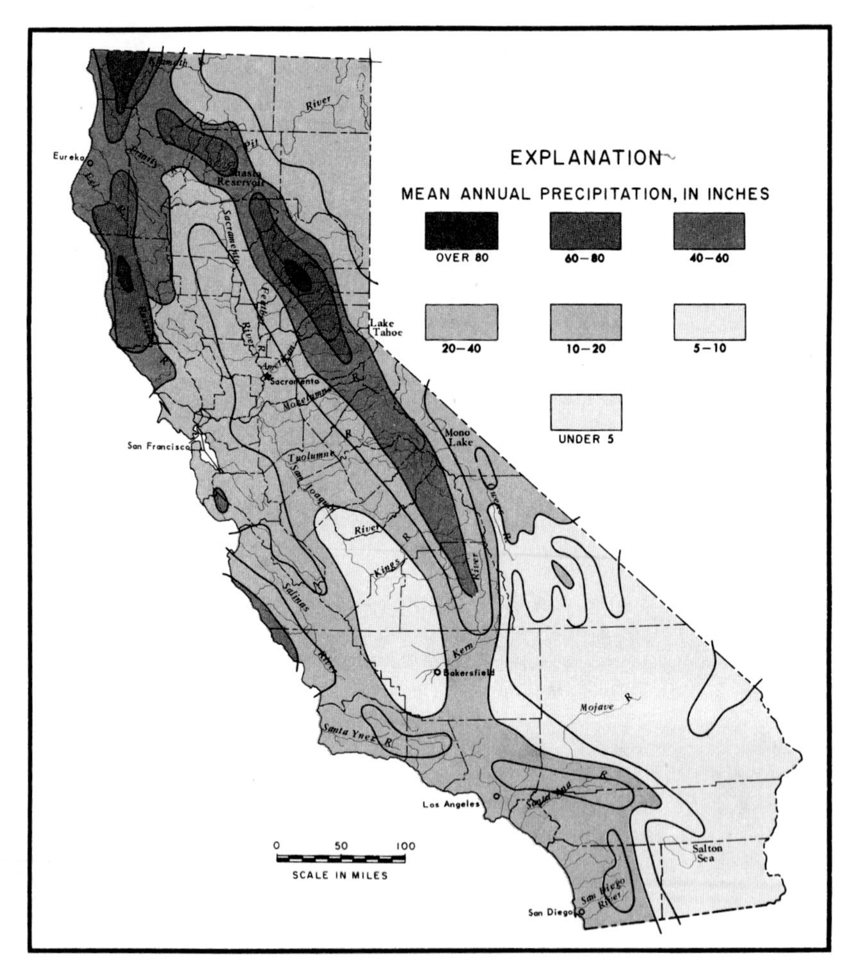


**FIGURE 1-3** *Mean annual precipitation over the state.*

with precipitation of only 3 to 4 inches per year. Because air descending the eastern slope of the Sierra is warmed, moisture is absorbed rather than precipitated. With this sort of relief and precipitation pattern, it is not strange that all the largest streams that drain the Sierra run down the western slope. The amount of annual precipitation decreases toward the south in the Sierra and also in the Great Valley; in fact, the southern San Joaquin Valley is so dry that the Kern River has no outlet to the sea but ends on the valley floor west of Bakersfield.

The principal stream on the east side of the Sierra is the Owens River, which runs southeasterly through Owens Valley and ends inland in the desert basin of Owens Lake.

## GREAT VALLEY

In predominantly mountainous California, the Great Valley, 400 miles long and averaging about 50 miles wide, is obviously a major natural region. This structural trough, almost flat-floored and lying from a few feet to a few hundred feet above sea level, is bounded on the west by the Coast Ranges and on the east by the Sierra Nevada. The valley rises very gradually eastward toward the gently sloping western foothills of the Sierra Nevada. Surface of the granitic bedrock of the Sierra is tilted gently westward and extends far out under the valley. The rise to the Coast Ranges on the west is much more abrupt and the complex bedrock formations on the eastern flank of the Coast Ranges are tilted steeply eastward under the valley. Gravel, sand, silt, and clay—derived mainly by erosion of the Sierra and, to a lesser extent, of the Coast Ranges—fill the Great Valley. The valley fill is thousands of feet deep on the western side and thins gradually to nothing on the eastern side. Maximum thickness of sediments in the Great Valley is more than 10 miles! The present flat floor of the Great Valley has been built up by sediments deposited by streams and shallow lakes during the last million years or so.

The northern part of the Great Valley—called the Sacramento Valley—is drained by the Sacramento River, fed mostly by major streams like the Pit, Feather, and American Rivers from the mountains on the north and east (Figure 1-4). Most of the southern end of the valley—called the San Joaquin Valley—is drained by the San Joaquin River. After rising in the southern Sierra Nevada, the San Joaquin River is augmented mainly by large tributaries, like the Kaweah, Kings, Tuolumne, and Mokelumne Rivers, which also flow down the western slopes of the Sierra. The Kings and Kaweah Rivers flow partly into Tulare Lake, a closed basin most of the time. One large river—the Kern, at the extreme south end of the valley—flows onto the lowest and driest part of the valley, where it sinks and evaporates in an intermittent shallow body of water called Buena Vista Lake. In wet years, at flood times, the Buena Vista and Tulare Lake basins are filled to overflowing to form one vast lake with waters of the Kern, Kaweah, and Kings Rivers all flowing northward to San Francisco Bay via the San Joaquin River. The Sacramento and San Joaquin Rivers join in a great floodplain and delta a few miles east of San Francisco Bay, which, in turn, has its outlet through the Golden Gate to the Pacific Ocean.

Nearly all the water in the rivers of the Great Valley comes from the adjacent mountains, for in the semiarid valley itself, each year only 5 to 20 inches of rain fall, mostly in the winter months.

## KLAMATH MOUNTAINS

The Klamath Mountains province covers about 12,000 square miles of northwestern California and Oregon between the northern Coast Ranges on the west and the Cascade Mountains on the east. In the Klamath Mountains province the general trend of ranges, rock formations, and structures, unlike those of provinces discussed so far, is north, rather than northwesterly. On geologic and relief maps, the broad arc-shaped form, convex to the west, is very apparent.

The Klamaths include a number of individual ranges, particularly the South Fork Mountains (which form the most westerly arc), the Trinity Alps, the Scott Mountains, the Salmon Mountains, and the Siskiyou Mountains. The Siskiyous form the most northerly arc, their trend swinging from north to northeast and east across the California-Oregon border.

Average elevation of the crest areas is between 5,000 and 7,000 feet, and Thompson Peak in the Trinity Alps rises to an elevation of 8,936 feet. The marked appearance of equality in summit levels reflects the region's late geologic history during which rather subdued hills were uplifted and then deeply dissected by rapidly flowing streams. Although lower than the Sierra Nevada crest, the higher parts of the Klamath Mountains—notably in the Trinity Alps—have also been glaciated because of their position in a more northern latitude.

In its rocks and geologic history, the Klamath Mountains province is a northwesterly continuation of the Sierra Nevada, but the direct connection is covered by alluvium of the northern Sacramento Valley and by the geologically much younger volcanic rocks of the Cascade Range and the Modoc Plateau.

**FIGURE 1-4** *Mean annual discharge of large streams.*

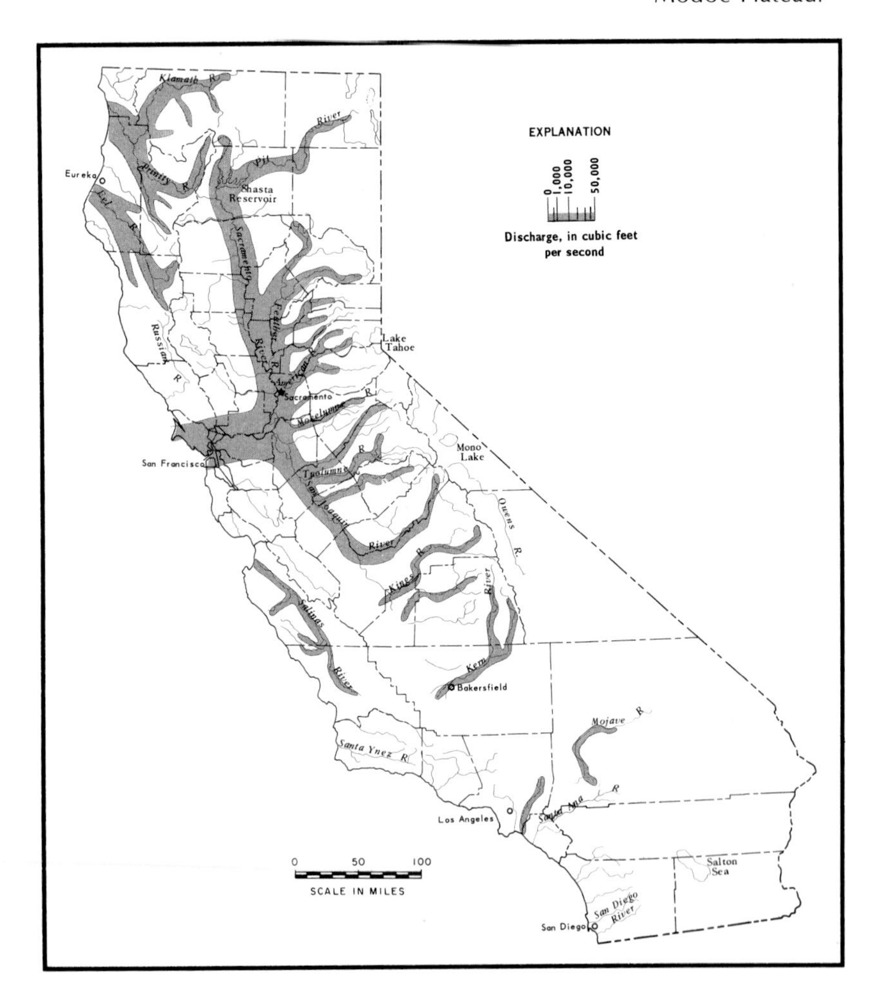

Precipitation in the Pacific Northwest is generally high, varying annually in the Klamath Mountains from 40 to more than 80 inches. As elsewhere in California, most of the rainfall occurs during the winter season, with considerable snowfall in the higher mountains. The large runoff is carried mainly by two rivers—the Klamath and the Trinity. Although both rivers follow more or less random courses across the dominant trends of the Klamath ranges, the Klamath River generally turns toward the southwest and the Trinity River eventually flows to the northwest to join the Klamath before it reaches the ocean.

**FIGURE 1-5** *Potential water power sites.*

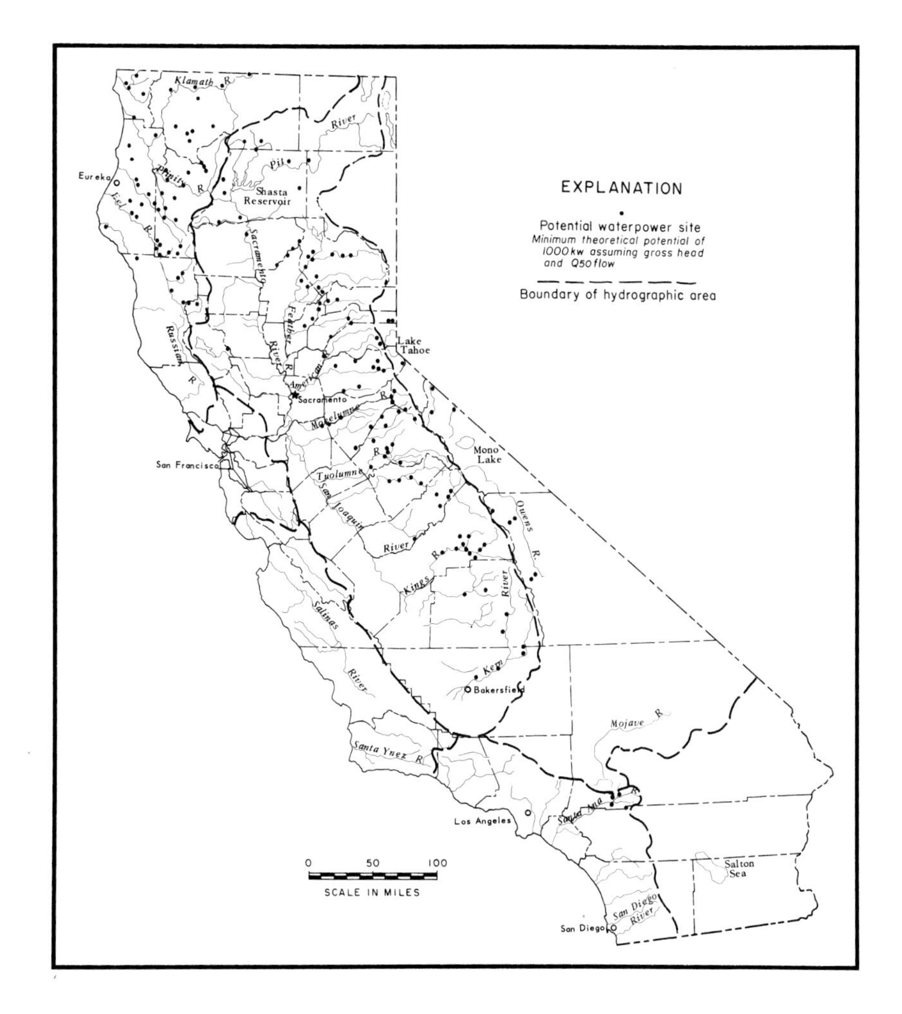


## CASCADE RANGE AND MODOC PLATEAU

About 13,000 square miles of the northeastern corner of California are covered by a thick accumulation of geologically young volcanic rocks. The province includes a great variety of volcanic rocks in many forms—lava flows, ash, volcanic fragments, mudflows, and volcanic domes and mountains. The larger part of this area is essentially an irregular, stream-dissected plateau—the Modoc Plateau—averaging 4,500 feet in elevation. It is only one small corner of the great Columbia Plateau, which covers a vast area of eastern Oregon, Washington, and southern Idaho.

The Cascade Range, trending almost due north, forms the western margin of this young volcanic province. Southernmost in this spectacular chain of volcanoes is Lassen Peak (elevation 10,453 feet), although Sutter Buttes, a young volcanic neck projecting out of the flat floor of the Sacramento Valley, may be related to the Cascades. Northward, in order, in the volcanic chain are Mount Shasta (14,162 feet) and Mount McLoughlin (9,497 feet); in Oregon; Crater Lake, the Three Sisters (10,351 feet), Mount Jefferson (10,495 feet), and Mount Hood—east of Portland—(11,253 feet); and in Washington; Mount Adams (12,307 feet), Mount Rainier (14,406 feet), and Mount Baker (10,750 feet).

The Cascade volcanoes are all composite cones—that is, they were built up by alternate lava flows and violent explosions. Lassen Peak erupted explosively and intermittently between 1914 and 1917, making it the only volcano active during this century in the continental United States (excluding Alaska). There is indirect evidence for believing that Mount Shasta erupted in the eighteenth century. All these volcanoes are

**1-6** *Castle Crags, a few miles west of Mount Shasta, Shasta County, are readily seen from Interstate Highway 5. These beautiful spires are composed of granitic rock more than 170 million years old. They mark the eastern margin of the Klamath Mountains natural province; the Cascade Range province lies just across the highway to the east. (Mary Hill photo.)*

**1-7** *Lava fields from Schonchin Butte, Lava Beds National Monument, Siskiyou County. This view west toward the Cascade Range in the far distance is typical of the landscapes of the Modoc Plateau. (Mary Hill photo, California Division of Mines and Geology.)*

beautiful and spectacular because of their conical outlines, great heights, and often snow-capped peaks. Mount Shasta is so high that it has small alpine glaciers near its crest.

The Warner Range, at the eastern margin of the Modoc Plateau in California, is an uplifted mountainous part of the plateau composed of similar volcanic materials. It is bounded on the east by north-trending faults of the adjacent Great Basin province into which the Modoc Plateau merges geologically.

**1-8** *Looking westward toward Mount Shasta from top of Little Mount Hoffman in the Medicine Lake Highland area of Siskiyou County. Little Glass Mountain, in the middle foreground, is made up of two recent obsidian flows with pumice surfaces. The white dome, in the middle distance between the flows and Mount Shasta, is Pumice Stone Mountain, a pumice-covered basaltic cinder cone. Mount Shasta, a composite volcano of the Cascade Range, erupted only a few hundred years ago. (Mary Hill photo, California Division of Mines and Geology.)*

## TRANSVERSE RANGES

One major natural region and geologic province—the Transverse Ranges—stretches from west to east directly across the dominant northwesterly trend of such provinces of the state as the Sierra Nevada, the Coast Ranges, and the Peninsular Ranges. The Transverse Ranges extend some 300 miles from the western end of the Santa Ynez Mountains along the Santa Barbara coast, first due east, then turning slightly south of east to within perhaps 60 miles of the Colorado River. The province is long and narrow—from 15 to 60 miles wide—and is characterized by some of the highest and most rugged topography in southern California.

Numbers of complex mountain ranges and intervening valleys make up the Transverse Ranges.

**1-9** *Looking north in the Simi Hills on the southern flank of the Santa Susana Mountains, Los Angeles County. Derricks of the Aliso Canyon oil field can be dimly seen on the skyline right of center. Most of the outcrops visible are folded, stratified, marine sandstones of Miocene age. This landscape is typical of the western Transverse Ranges. (Charles W. Jennings photo.)*

Some of these ranges, from west to east, are the Santa Ynez, Topatopa, Santa Monica, San Gabriel, San Bernardino, Little San Bernardino, Eagle, and Orocopia Mountains. The Santa Monica Mountains continue westward into the Channel Islands of Anacapa, Santa Cruz, and Santa Rosa. Individual ranges and basins vary in length from a few to 125 miles. Intervening basins, often hilly within themselves, parallel the ranges. The Ventura Basin lies between the Santa Ynez–Topatopa Mountains on the north and the Channel Islands–Santa Monica Mountains on the south. On the east, the Ventura Basin branches into the Soledad Basin north of the San Gabriel Mountains and into the San Fernando Valley south of the San Gabriels; the San Fernando Valley merges with the San Gabriel Valley.

The Transverse Ranges were formed by the intense folding and faulting of extremely complex rock groups very late in geologic time. Many of the ranges are individual fault blocks, also intricately folded. Relief is extreme because deep erosion has left high, usually narrow, ridges and peaks and deeply cut valleys. Crests of the western ranges are mostly 3,000 to 6,000 feet high, but some, like Frazier Mountain (elevation 8,026 feet), are considerably higher. The San Gabriel Mountains, lens-shaped and extending 60 miles east to west and 20 miles north to south, are particularly high and rugged; crests range between 4,000 and 9,000 feet, with San Antonio Peak ("Old Baldy") rising to 10,080 feet above sea level. East of the San Gabriels, the San Bernardino Mountains form an equally high and continuous range about 55 miles long. This range contains San Gorgonio Mountain (elevation 11,485 feet), the highest mountain peak in southern California. Some of the crest areas of the San Gabriel and San Bernardino Mountains have broad, gentle summits, showing that erosion by running water has not yet had time to completely dissect them.

Rain is brought to the southern California area that includes the Transverse Ranges by Pacific storms from the west and southwest. Normal rainfall for the region is low and varies from 10 to 40 inches per year, most of it coming in the three winter months of December, January, and February. The highest continuous ranges, like the San Gabriel and San Bernardino Mountains, form extreme barriers; rainfall well up on their southern slopes is a maximum, but their low northern flanks merge into true desert. Thus, the principal streams are short and intermittent in their flow, like the Santa Ynez, Santa Clara, Los Angeles, and San Gabriel Rivers, all of which reach the sea. The northern slopes of the San Bernardino Mountains are drained by the Mojave River, which flows northward into the Mojave Desert where it sinks and evaporates.

## GREAT BASIN AND MOJAVE DESERT

The Great Basin province is a large area that extends through much of far eastern California, across Nevada, and into Utah. The province is characterized by north- and northwest-trending mountain ranges (many in Nevada trend north or northeast), which alternate with intervening basins and valleys; climates are arid, but modified greatly by local topographic relief; and drainage is interior —streams run into inland valleys and end in saline lakes and playas without reaching the sea. Geologically, it is characteristic of this basins-and-ranges region that the ranges are uplifted blocks (horsts) bounded by faults while the valleys are sediment-filled downdropped fault blocks (grabens), although geologic history of the province is not that simple in detail.

**1-10** *Looking southeast from the Sierra Nevada across Mono Lake at the Mono Craters. The largest, light-colored, island is Paoha. The valley in the Sierra in the right-center foreground is Lundy Canyon, a typical glacial valley; the ridge on its right is a lateral moraine. (Charles W. Chesterman photo.)*

In northern California, the fault-bounded Warner Range and Surprise Valley are part of the Great Basin. Further south, the western boundary of the Great Basin, specifically, is the series of major faults that bound the eastern face of the Sierra Nevada (see the 1:2,500,000-scale colored geologic map of California available from the U.S. Geological Survey or the California Division of Mines and Geology), like the Honey Lake and Sierra Nevada faults.

Many geologists treat the Mojave Desert as a separate natural province by designating, as the boundary between the Great Basin and the Mojave Desert, the northeast-trending Garlock fault and a line representing its projection to the California border. However, the characterizing features of the Great Basin province hold, with little interruption, especially in the eastern part, through the Mojave Desert; the Garlock fault changes direction sharply at its eastern end to join the north-south pattern of the Death Valley fault zone. The western Mojave Desert makes a sharp westward-pointing triangular wedge, bounded on the northwest by the Garlock fault and on the southwest by the San Andreas fault. The southern end of the Great Basin–Mojave Desert province is cut off by the Transverse Ranges and their east-west fault systems.

One of the most spectacular portions of the Great Basin in California lies along the eastern base of the Sierra Nevada. Directly at the base of the Sierra is the long, narrow basin of Owens Valley, which is only 2 to 8 miles wide but 100 miles long. Its floor is at an elevation of about 4,000 feet, while the nearby crest of the Sierra is at 13,000 to 14,000 feet in elevation. To the east, the White and Inyo Mountains tower—with the White Mountain peak at an elevation of 14,242 feet. Eastward, in order, are Saline and Panamint Valleys; the Panamint Mountains; Death Valley; and the Black, Amargosa, and Funeral Mountains. Death Valley is a completely enclosed, downdropped fault basin about 130 miles north to south and 6 to 14 miles wide. To the west, the Panamint Mountains rise to 11,045 feet at Telescope Peak, which overlooks the bottom of Death Valley, 282 feet below sea level.

Much of the western triangular part of the Mojave Desert is a plain, which slopes gently eastward from the town of Mojave (elevation 2,750 feet) to Barstow (elevation 2,105 feet). The plain is dotted with numerous hills and small mountainous groups, such as Soledad Mountain (4,183 feet), Red Mountain (5,270 feet), and the Shadow Mountains (4,039 feet). Between the isolated hills are plains of alluvium, bare rock, and dry lakes or playas. East of Barstow, the same general pattern is seen, but the hills and mountains are higher, longer, and more linear; and some are oriented mainly in the northwesterly trend typical of the Great Basin. Two of the more prominent ranges are the Calico Mountains (elevation 4,520 feet) and the Ord Mountains (elevation 6,250 feet). Far east, toward the Colorado River, are the Turtle, Maria, and Chuckwalla Mountains, and finally, the Chocolate Range, east of the Salton Sea.

Rainfall is low and vegetation scarce in this entire desert province. All drainage is interior; and saline lakes—like Soda, Bristol, Cadiz, and Danby Lakes—are common. All streams are intermittent but may carry large volumes of water during occasional heavy rainstorms. Minimum rainfall average is the 1.4 inches per year recorded for Death Valley, but the desert as a whole averages less than 5 inches per year. The Amargosa River, which makes a U-turn and enters Death Valley at the southern end, and the Mojave River, which rises in the San Bernardino Mountains and runs far out into the Mojave through Barstow, are the two largest streams of the area. The Owens River, which carried a large volume of meltwater from

**1-11** *Death Valley from a high point near Dante's View in the Funeral Mountains. The valley floor, below the girl, is 282 feet below sea level. Rocks in the foreground are Precambrian. (Mary Hill photo.)*

the Sierra Nevada before it was diverted for use by Los Angeles City, runs southward through Owens Valley into the saline Owens Lake. The Colorado River, which forms the southeastern boundary of the Mojave Desert, receives a very minor part of its flow from the California side.

**1-12** *A typical desert landscape, near Mitchell Caverns State Park, San Bernardino County. The extremely steep, rocky Providence Mountains in the background are almost devoid of vegetation, except at high elevations. The rocky floor in the foreground, at elevation 4,000 to 5,000 feet, is nearly covered with desert grasses, low shrubs, and many varieties of cactus, including barrel cactus, the prickly cholla, and treelike yuccas with their swordlike leaves. (Mary Hill photo.)*

## SALTON TROUGH

The Salton Trough is a northwestern extension of the basin in which the Gulf of California lies. The bottom of the Salton Trough is presently occupied by the Salton Sea, whose shoreline fluctuates around 200 feet below sea level. The area below sea level extends south from the town of Indio and is about 85 miles long and 30 miles wide. The trough is bounded on the northeast and the east by mountains of the Transverse Ranges and the Mojave Desert, and on the southwest by the Peninsular Ranges.

The Salton Trough is extreme desert, with very hot summers and warm winters. The only significant streams are the New and Alamo Rivers, which carry overflow from the Colorado River; in 1905, these were the channels that partly filled the trough to form the present-day Salton Sea, which

**1-13** *A high aerial oblique photo, looking northeast across the Salton Trough. High Peninsular Ranges are in the lower left; the Colorado River can just be seen in the upper right. In the upper half of the photo are the ranges and basins of the southern Mojave Desert. Checkerboard patterns in the Coachella Valley and Imperial Valley parts of the trough are irrigated agricultural plots where much of the state's early vegetables and fruits are grown. Note that relative depths of water and its circulation can be seen. (Gemini V photo, U.S. Geological Survey.)*

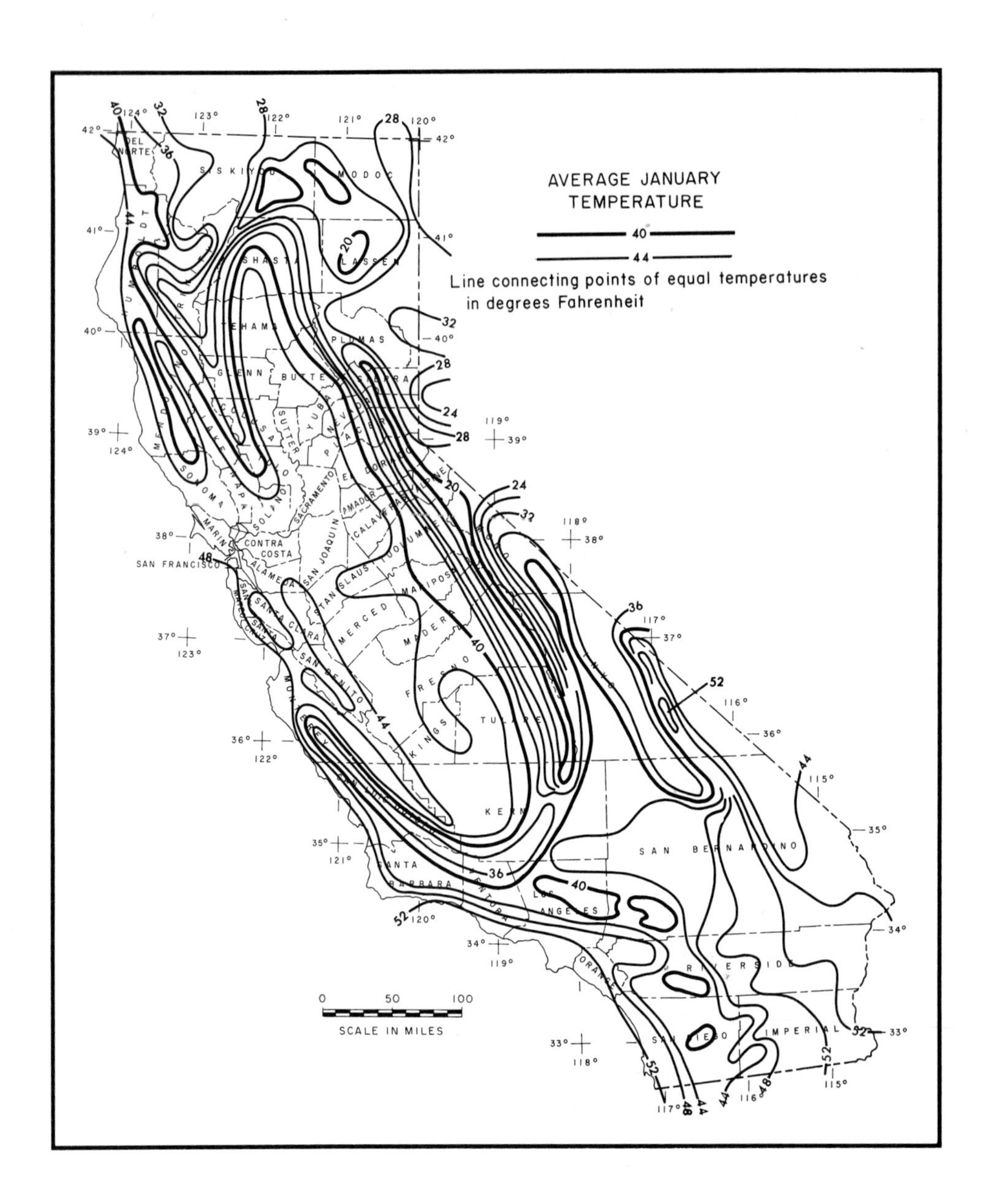


**FIGURE 1-14** *Average January temperatures.*

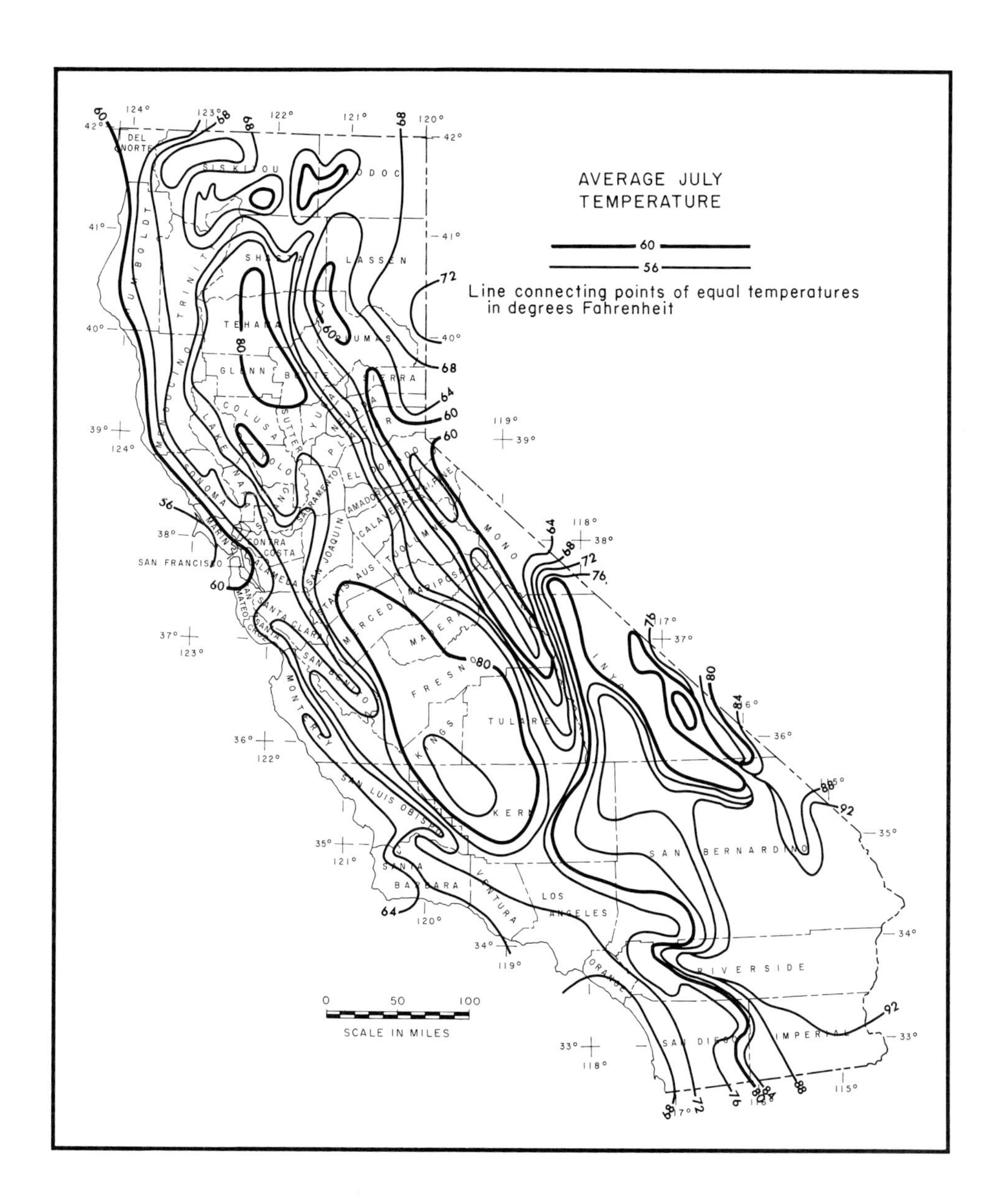


**FIGURE 1-15** *Average July temperatures.*

now receives overflow irrigation water from Imperial Valley. Thus, although the area is rainfall deficient and the high rate of evaporation could be expected to dry up the sea totally, the excess of inflow of irrigation water has stabilized the lake level.

**1-16** *View northeast across Peninsular Ranges toward the Elsinore fault zone in Banner Canyon. This active fault zone extends northward to join the Whittier fault zone in the Los Angeles Basin and continues southward into Mexico. Granitic and metamorphic rocks across the canyon form a high, resistant mass; Julian Schist forms the low hills on the near side (southwest of the rift valley). (F. W. Weber photo.)*

## PENINSULAR RANGES

The Peninsular Ranges, which constitute a prominent natural province in southwestern California, are characterized by steep, elongated ranges and valleys that trend northwestward. On the north, the Peninsular Ranges are cut off sharply by the crosscutting Transverse Ranges. The Peninsular Ranges extend through about 140 miles in California and continue southward to form the backbone of Baja California, Mexico, for another 750 miles; width of the ranges is 50 to 140 miles.

The northern end of the province includes the Los Angeles Basin and offshore an 80-mile-wide series of northwesterly trending basins and ranges—the basins forming channels below sea level and the ranges forming the islands of San Nicolas, Santa Barbara, Santa Catalina, and San

Clemente. Principal ranges on the California side, from north to south, are the Santa Ana, San Jacinto, Santa Rosa, Agua Tibia, and Laguna Mountains. San Jacinto Peak (elevation 10,805 feet) is the highest in the Peninsular Ranges; highest peak in the Santa Rosa Mountains is 8,705 feet.

The eastern face of the San Jacinto-Santa Rosa ranges, which drops off extremely steeply to the floor of the Coachella Valley, is reminiscent of the eastern face of the Sierra Nevada. As with the Sierra, the topography of the western slopes of the Peninsular Ranges is much less rugged than that of eastern slopes, descending relatively gradually to the broad marine terraces that lie back of the shore line along the coast. Not only is the asymmetric profile of the Peninsular Ranges analogous to that of the Sierra Nevada, but the similarity extends strikingly to the rock formations and geologic history. Thus, the Peninsular Ranges, Sierra Nevada, and Klamath Mountains form three quite similar but geographically distinct natural geologic and topographic provinces.

The Peninsular Ranges province has a semiarid, warm climate with average winter rainfall, between 10 and 30 inches per year, markedly influenced by topography. Summers are pleasantly warm, in contrast to the heat in the Salton Trough to the east. Snow covers the highest peaks in winter.

As in the Sierra Nevada, and for similar reasons, all streams flow westward, although none carries much water and most are intermittent. Among the larger streams are the Los Angeles, San Gabriel, and Santa Ana Rivers, which rise in the Transverse Ranges and run across the Los Angeles Basin, and the Santa Margarita, San Luis Rey, and San Diego Rivers, which drain the Peninsular Ranges directly to the Pacific Ocean.

*A cluster of quartz crystals from the Mother Lode, western Sierra Nevada.*

# 2 california's rocks and minerals

■ The ultimate objective of the science of geology is to reconstruct the history of the earth and its life. Our principal purpose in this book is to scan the highlights of California's history through the vastness of geologic time from the "beginning" to the landscapes of today. In so doing, we will follow the evidence and trace the steps of geologists as they reconstruct earth history in this part of the world. The most recent events of geologic history have left their obvious imprints in the landforms—mountains, valleys, and other topographic forms—and in the vegetation and animal life that all together form the present landscapes. But all these things are short-lived and transitory—even the greatest mountain ranges last only a few million years before yielding to the processes of weathering and erosion that eventually destroy them. All but the most recent history, then, must be deciphered from the imprints of ancient events that can be recognized in the rocks of the earth's crust and from fossils—the remains of past life—that are preserved in some rocks.

## ROCKS

Rocks are the solid materials that make up the fundamental units of the earth's crust with which the geologist deals. While generally solid, they range from soft (chalk) to very hard (unweathered granite).

For purposes of grouping rocks in the field and for geologic mapping and interpretation, large associations of rock of similar type, age, and origin are referred to as rock formations. Formations are usually given combined geographic and rock-type names, such as "Monterey Shale," or if several rock types are involved, a name like "Franciscan Formation." A formation can be distinguished from others in the field and may be shown on a geologic map by color and pattern. The words "Shale" and "Formation" are part of the proper name approved by the U.S. Geological Survey and initial letters are capitalized.

Many hundreds of rock types have been described and given names. Many of the distinctions, however, are not major, and the common rocks are relatively few in number. To make sense of a profusion of names, classification is necessary. The most meaningful and fundamental classification—certainly if the objective is to reconstruct geologic history—is one based on genesis. Indeed, this is precisely the classification almost universally used. On the basis of origin, rocks are divided into three classes: igneous, sedimentary, and metamorphic. *Igneous rocks* were formed from molten material (for example basalt, the fine-grained, dark-colored rock that makes up many lava flows, and granite, the coarse-grained rock that, though now at the surface, was formed many thousands of feet below the surface). *Sedimentary rocks* were formed from the consolidation of sediments like mud, sand, and gravel (for example, shale, sandstone, and conglomerate). *Metamorphic rocks* were formed from any previously existing rock type under conditions of pressure, temperature, and chemical changes that result in the development of new minerals and textures (for example schist and gneiss). The word "metamorphic" comes from the Greek roots *meta,* meaning "change," and *morph,* meaning "form."

## MINERALS, COMPOUNDS, AND ELEMENTS

Examination of most rocks by eye, or with a hand lens, shows that they are composed of distinct particles, or minerals. A mineral is a naturally occurring inorganic substance of generally characteristic and constant chemical composition and crystal structure. Minerals are recognizable by distinguishing characteristics such as color, luster, hardness, specific gravity, color of their powder (streak), patterns of fracture and cleavage, and crystal form. A rock may be composed of one or a number of minerals; but most rocks are composed of just a few minerals.

While minerals may often be readily distinguished by obvious features (such as appearance and weight), chemical composition and crystal structure are much more fundamental characteristics. Most minerals are chemical compounds, composed of two or more elements that are chemically bound together; a few are composed of a single element. Elements are the simplest parts into which minerals can be divided by ordinary physical and chemical means.

The fundamental building blocks of earth materials are atoms, of which all substances are composed. The atom is the unit of reference for any element and is the smallest particle that maintains all the physical and chemical properties of the chemical element. Building units within the atom include protons, electrons, and neutrons. Atoms are almost infinitesimally small; diameter of a silicon atom, for example, is less than 1 *angstrom* (4 billionths of an inch). The "big" oxygen atom has a diameter of slightly less than 3 angstroms. The crystal structure and crystal form of a mineral result from the orderly internal arrangement of its atoms; different kinds of atoms and different arrangements of atoms result in minerals with different chemical and physical characteristics. More than a hundred elements have been identified, but only a few occur in common rock-forming minerals (Table 2-1).

**TABLE 2-1 Average composition of the earth's crust, by weight**

| Element | % |
|---|---|
| Oxygen (O) | 46.6% |
| Silicon (Si) | 27.7 |
| Aluminum (Al) | 8.1 |
| Iron (Fe) | 5.0 |
| Calcium (Ca) | 3.6 |
| Sodium (Na) | 2.8 |
| Potassium (K) | 2.6 |
| Magnesium (Mg) | 2.1 |
| All other elements | 1.5 |
| | 100.0% |

## THE ROCK-FORMING MINERALS

**FIGURE 2-1** *Model of one of the simplest of all atoms—helium. Protons are solid black, neutrons are dotted circles, and the orbiting electrons (bearing negative charges) are indicated by the elliptical paths.*

Oxygen and silicon are overwhelmingly the most abundant elements in the earth's crust. Together, they constitute by weight about three-quarters of the crust's rocks and minerals that have been recognized and named. These two elements have strong chemical attraction for each other. Four oxygen atoms combine with one silicon atom in a pattern such that the oxygen atoms are at equal distances from the silicon atoms. This is the silicon-oxygen tetrahedron (*tetra* means "four"), written chemically as $SiO_4$. It is the fundamental unit of construction of the silicate minerals, by far the most common minerals in the earth's crust. Adjacent silica tetrahedrons share oxygen atoms between them in a variety of patterns and so build up mineral crystals. Some of the common, rock-forming silicate minerals are quartz, feldspar, mica, pyroxene, and amphibole (hornblende).

*Quartz* is one of the most common of all minerals. Its chemical formula is $SiO_2$ (silicon dioxide), which is commonly referred to as silica. In quartz, each tetrahedron shares four oxygen atoms in such a way as to form an extremely strongly bonded framework. Quartz has no planes of weakness along which splitting, or cleavage, can occur; it breaks irregularly, or with a curved fracture surface like glass. Quartz is a hard mineral, usually colorless or white, which often shows its characteristic crystal form—a six-sided prism topped by a six-sided pyramid.

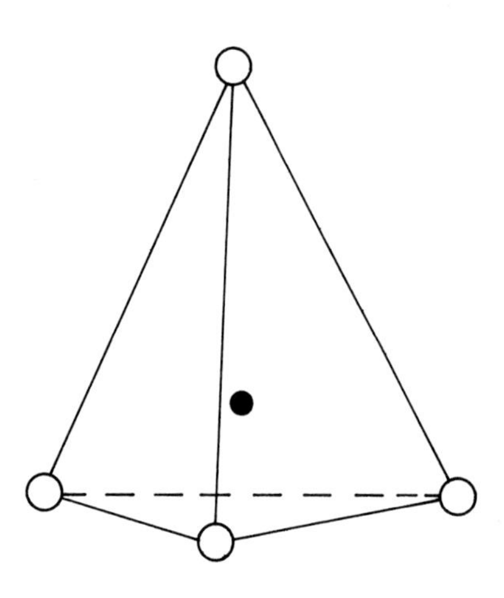

**FIGURE 2-2** *A single silica tetrahedron with oxygen atoms at four corners and equidistant from the single silicon atom (solid black circle). Tightly bonded groups of these tetrahedrons form the silicate minerals.*

*Feldspars* are the most common of all minerals in the earth's crust, and many rocks are composed of little else than quartz and feldspar. The name "feldspar" is applied to a family of minerals that are made up of shared silica tetrahedrons in which aluminum takes the place of some of the silicon atoms. Differences among the various feldspars are determined by varying proportions of potassium, sodium, and calcium. In feldspars, the complex arrangement of atoms is such that there is a prominent crystal plane of weakness (cleavage) along which the minerals tend to split. This cleavage property is one of the best means of identifying the feldspars in rocks. The feldspars are usually white or have a grayish or pinkish cast and are hard but not as hard as quartz.

*Micas* are also a family of minerals consisting of complex silicates. Their outstanding physical characteristic is one perfect cleavage; the micas can be split into very thin plates. The colorless variety is known as muscovite; biotite is black or brown, due to its iron content.

*Pyroxene* and *amphibole* (hornblende) are very common families of minerals, and are somewhat alike in chemical composition; that is, they are complex silicates of various compositions involving most of the list of common elements in the earth's crust. Colors are usually black, brown, or dark green, but others are common. The pyroxenes and amphiboles differ in crystal structure, which determines the form of the mineral particles in rocks. Both form long prismatic crystals or needles, but the pyroxenes show two good cleavage faces forming a right angle, while the amphibole faces meet at an angle of about 120 or 60 degrees.

Some common minerals are not silicates. The most important of these are calcite (calcium carbonate), which is the principal mineral constituent of limestone; dolomite (calcium magnesium carbonate), which may make up large masses of sedimentary rock, with or without calcite; the black magnetic iron oxide magnetite; and yellow, brown, red, and black oxides of iron such as limonite (yellow) and hematite (red or black).

## STATE MINERAL AND STATE ROCK

In 1848, James Marshall discovered native gold (the element by itself, not in combination with other elements) at Sutter's Mill on the American River,

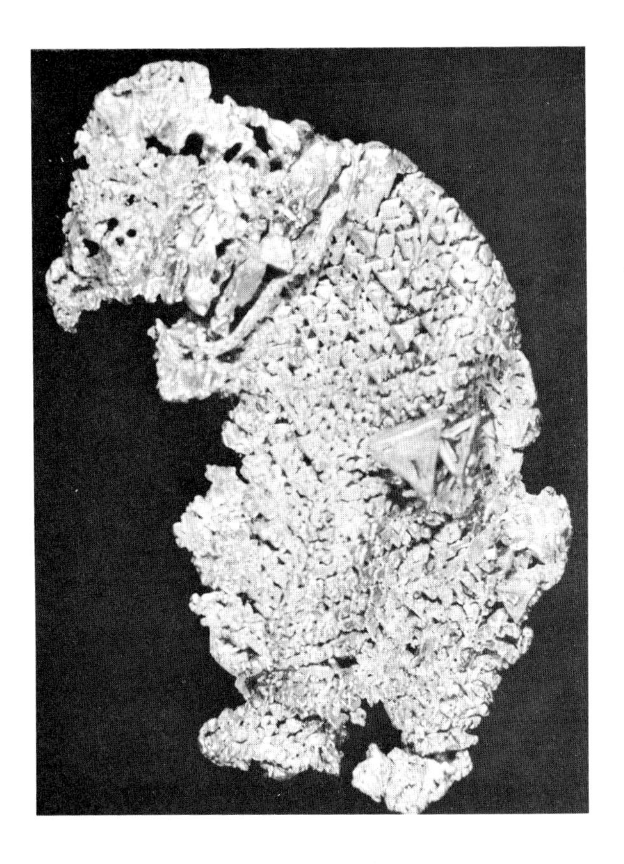

**2-3** *The famous Golden Bear nugget—crystalline gold in natural form—now in the mineral exhibit of the California Division of Mines and Geology, Ferry Building, San Franciso. (California Division of Mines and Geology photo; approximately 3 inches long.)*

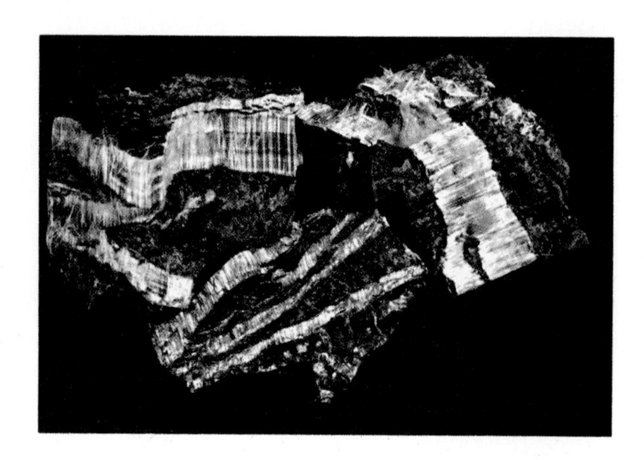

**2-4** *Serpentine—the state rock—with veinlets of chrysotile asbestos of commercial quality. (California Division of Mines and Geology photo.)*

touching off one of the greatest population stampedes in history. The value of gold mined annually in California jumped to $81 million by 1853 (about $140 million at the price of gold in the 1960s). Gold is still mined in California, but the value of the annual production is less than $1 million. In 1965, the state legislature passed a bill designating gold as the official state mineral of California. In the same bill, the green, greasy-appearing rock serpentine, common in the Coast Ranges and in the foothills of the Sierra Nevada, was designated the state rock.

No other state approaches California in the number and diversity of minerals, commercial and noncommercial, found within its borders. There are more than 600 known minerals in California, 45 of which have not been found elsewhere. Eighty mineral products, including petroleum, are mined yearly at a value of about $3.5 billion. All minerals that have had commercial value in California and their geologic settings are discussed in California Division of Mines and Geology Bulletin 191, *Mineral Resources of California*. All minerals found in California are fully described by the Division in Bulletin 189, *Minerals of California*.

## IGNEOUS ROCKS

Geologists and astronomers are generally agreed that the earth was once essentially a hot, molten mass, which cooled gradually to form a crust and finally became solid, except for a zone around the inner core (see Figure 6-1). The earth, then, is a great body of igneous rock. Any other rock types we see on the surface must have been derived from igneous rock. Table 2-2 shows principal types of igneous rocks.

To see igneous rock in the making, we could watch molten lava erupting from the active volcano Kilauea on the island of Hawaii or from volcanoes in the Philippines. In 1943, Mexican farmers watched the first explosive eruptions of an entirely new volcano, Paricutin, spewing out fragments of volcanic material in the form of dust, "ash," and larger fragments and blocks. Within a few years, Paricutin's activity had changed to extrusion of molten lava. Literally hundreds of active volcanoes on the earth today demonstrate the making of volcanic rock, often called extrusive because the molten rock has been extruded to cool at the surface of the earth. Californians have not been so fortunate as to observe volcanoes in action since the most recent (but perhaps not the last) eruptions of Lassen Peak in 1917. Active volcanoes today form a "rim of fire" around the margin of the Pacific Ocean.

The next best thing to seeing a volcano in eruption is to study the form and materials of recently active volcanoes, of which California has an abundance. In the Cascade Range, Lassen Peak and Mount Shasta are composite volcanic cones, made up of dark gray to black flow rocks (such as basalt and andesite) and rocks more or less solidified from the fragmental products of alternating violent eruptions (such as volcanic tuff, which is cemented from dust; ash; and the frothy, glassy pumice). Some volcanoes, like the Cinder Cone near Lassen Peak, are steep-sided cones consisting of fragments only. Quite different from

**TABLE 2-2 Principal types of igneous rocks**

High silica → Low silica
Low density → High density
Light colored → Dark colored

| | | | | | | | |
|---|---|---|---|---|---|---|---|
| Principal minerals | | K-feldspar<br>Less Na-Ca-feldspar<br>Quartz | Nearly equal K-feldspar and Na-Ca-feldspar<br>Quartz | Na-Ca-feldspar<br>Less K-feldspar<br>Quartz | Na-Ca-feldspar<br>Little K-feldspar<br>No Quartz | Na-Ca-feldspar<br>No Quartz | Pyroxene<br>Hornblende<br>Biotite |
| Volcanic (extrusive and shallow intrusive rocks) | Textures are fine-grained, porphyritic | Rhyolite | Quartz Latite | Dacite | Andesite | Basalt<br>Diabase<br>(Diabasic texture) | |
| | Glassy | Volcanic glasses—obsidian, pumice, perlite | | | | | |
| | Fragmental textures | Tuff, volcanic breccia, agglomerate | | | | | |
| Intrusive (large bodies, deep in crust) | Textures are coarse-grained, even-grained, granitic | Granite | Quartz Monzonite | Quartz Diorite | Diorite<br>Anorthosite (Feldspar only) | Gabbro | Peridotite<br>Serpentine (serpentine minerals only) |

**2-5** *Domes and mudflows at Lassen Peak, Shasta County. Peak at upper right of center. Lassen is an elevated, filled crater, or plug dome, surrounded by banks of talus. Numerous parasitic volcanic vents can be seen in the area. Chaos Crags, about 2 miles from the peak toward the lower right, is a jumble of cylindrical bodies of viscous volcanic rock. In 1915, Lassen erupted some lava and a great series of mudflows took place down its northeastern flank (toward the roads in lower left) forming the "devastated" area. Note the curved flow banding near the roads. (Vertical aerial photo, U.S. Geological Survey Prof. Paper 590, 1968; scale approximately 1 inch equals 6,945 feet; south is at top.)*

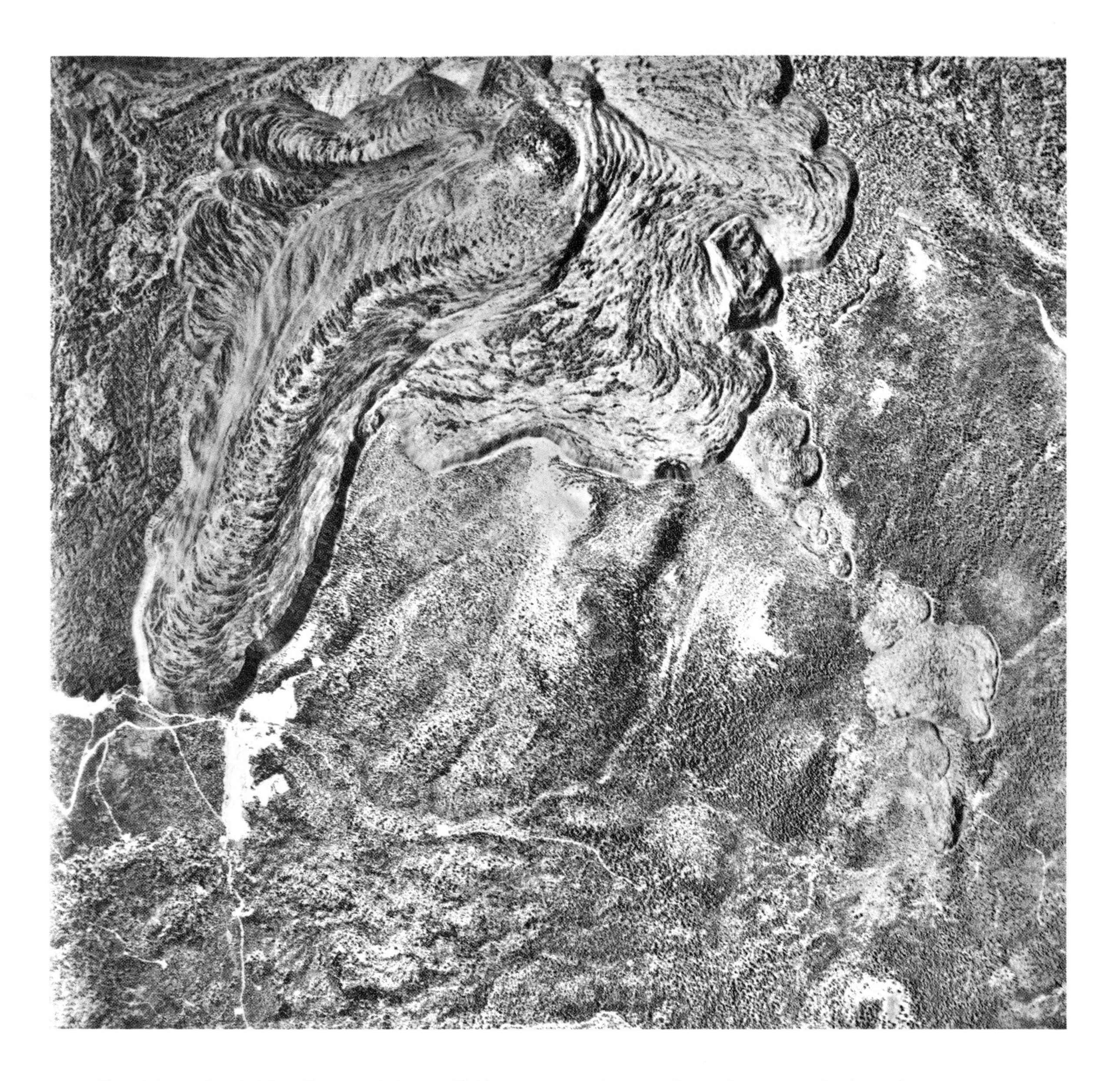

**2-6** *Glass Mountain obsidian flows and domes, Siskiyou County. Glass Mountain consists of two obsidian flows. The younger runs northeast (toward lower left of photo) from a summit dome. Flow structure and the steep margins of the flows are spectacularly shown. White patches on the older flow are pumice that has gathered in depressions on the surface. Note the line of small obsidian domes that trends northwesterly, perhaps along a fault or fracture zone. The flows are less than 1,000 years old. (Vertical aerial photo, U.S. Geological Survey, Prof. Paper 590, 1968; scale approximately 1 inch equals 2,330 feet;* south is at top.)

these are the recently extinct Mono Craters, east of the Sierra Nevada, which are volcanic domes made up of high-silica, light-colored rocks like rhyolite and the volcanic glasses—obsidian, perlite, and pumice.

Volcanic rock seems to come to the surface along zones of fractures and weakness in the earth's crust from depths of 20 to 40 miles, where pockets or chambers of magma (molten rock) are formed. It often happens, however, that magma moves upward within the crust but fails to reach the surface before cooling, thus forming intrusive rock, so called because the body of magma intruded other rock of the crust already in place. An example is granite.

**2-7** *Quarry face at the Coleman pumice operation, Clear Lake Highlands, Lake County. Divergent flow banding has been developed in Quaternary obsidian. The vesicular structure of the obsidian shows that this volcanic glass was chilled at or near the surface. (Charles W. Chesterman photo.)*

**2-8** *Close-up of pumice at the Crownite deposit in the Coso Range, Inyo County. The size of the particles, lack of clear stratification, and vesicular structure of individual particles show up well in this photo of land-laid, air-deposited pumice. The material is used in lightweight aggregate and for making pumice blocks. (Charles W. Chesterman photo.)*

**2-9** *Face of the Taylor pumicite quarry, Sierra Nevada foothills in Madera County. Pumicite is fine-grained, water-deposited pumice or volcanic ash (particles of glass). The perfection of thin bedding (lamination) in this pumicite of the Pleistocene Friant Formation suggests deposition in a lake. It is quarried for use as lightweight aggregate and is also sold as an insecticide carrier. (Charles W. Chesterman photo.)*

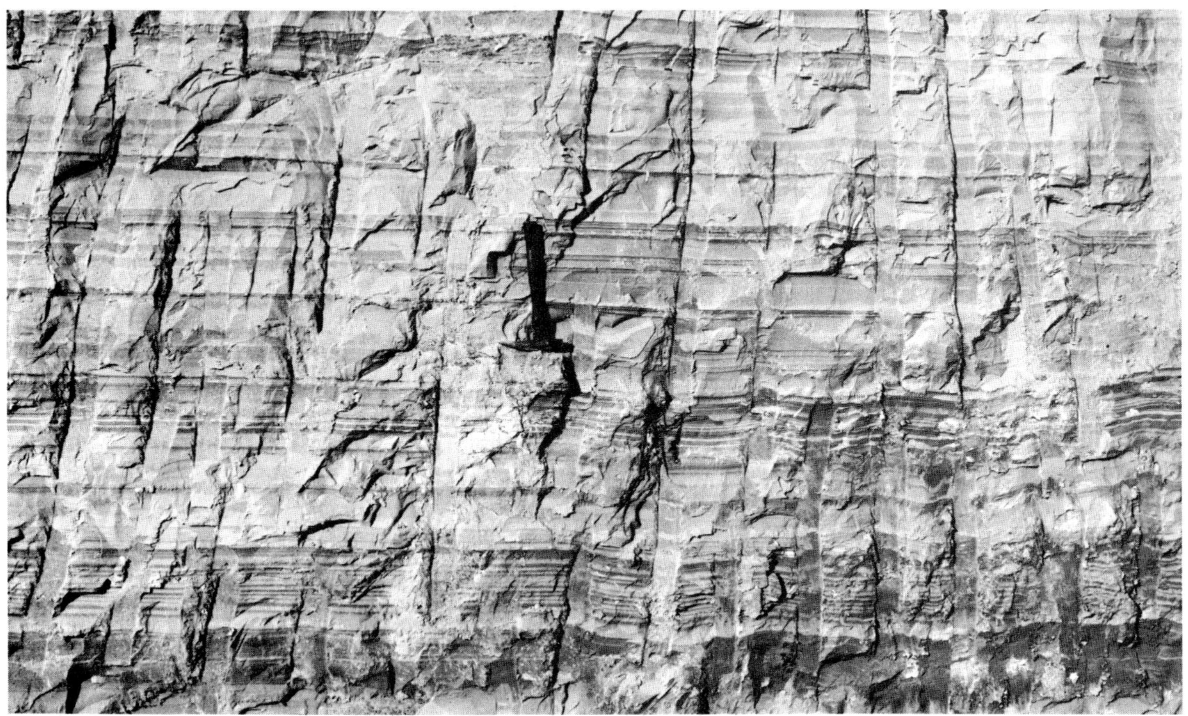

What are igneous rocks chemically, and what minerals are found in them? Chemically, they are composed of the same elements that are common in the makeup of the crust of the earth; that is, they are high in silicon, oxygen, and aluminum, with lesser amounts of iron, calcium, sodium, potassium, and magnesium. For example, basalt, the most common of all volcanic rocks, is about 50 percent silica ($SiO_2$). Granite, the most common of all intrusive rocks, is about 70 percent silica. The higher the silica, the more likely that the rock will also be high in sodium and/or potassium.

Low-silica rocks tend to be higher in iron, magnesium, and calcium oxides. Because iron imparts a dark color and is heavy, the lower-silica rocks are likely to be dark in color and to have a relatively high specific gravity.

The minerals found in igneous rocks are the common rock-forming minerals we have mentioned. The feldspars are the most common: potassium and sodium feldspars in the high-silica rocks and calcium feldspars in the low-silica rocks. Quartz is present in all high-silica rocks such as rhyolite and granite. The amphiboles and micas

**2-10** *A remarkable orbicular gabbro, near Rough-and-Ready, Calaveras County, western Sierra Nevada. Size of the orbs, or spherules, is indicated by the pick. Gabbro is common among the more basic, coarse-grained intrusive rocks of Mesozoic age in the Sierran foothills. The orbicular structure is probably the result of mineral separation in magma; the white spherules are the high-calcium feldspar anorthite. (Charles W. Chesterman photo.)*

**2-11** *Close-up of Cathedral Peak Granite, near Spiller Canyon in the High Sierra Nevada, Tuolumne County. The rock has a porphyritic texture, with large randomly-oriented phenocrysts of potash feldspar in a medium-grained groundmass (pencil shows scale). When deeply weathered, the phenocrysts tend to separate out in well-formed crystals up to several inches long. They can be collected quite easily on Sonora Pass Highway 108 at elevation about 8,000 feet, east of the Dardanelles. (Charles W. Chesterman photo.)*

are most common toward the high-silica end, while the pyroxenes are found most often in the lower-silica rocks. Some igneous rocks are ultramafic; the best examples are peridotite and serpentine. "Mafic" is a coined word made up of *ma*gnesium and *f*erric (or *f*errous) for iron. Peridotite is a heavy black rock extremely high in magnesium and iron oxides; serpentine is a dark green, greasy-appearing rock composed almost entirely of hydrous magnesium silicates.

The chemical and mineral compositions and textures of the common igneous rocks are shown in Table 2-2. *Texture* refers to the size, shape, and arrangement of minerals in the rocks and will be discussed in Chapter 5, in which the features of rocks most useful in interpreting geologic history are examined more closely.

**2-12** *A cobble conglomerate of nonmarine origin in the Oligocene Titus Canyon Formation, Death Valley. (Charles B. Hunt photo.)*

## SEDIMENTARY ROCKS

We do not know how many thousands of years have elapsed since some intelligent human being first realized that certain layered rocks were very much like the loose sands of the beaches, gravels in stream beds, and muds of lakes and lagoons. Running water, moving air, and ice in motion on the lands all pick up and transport particles of broken and weathered rock. As long as the velocity of a fluid current is great enough, it can transport rock fragments. In general, the greater the velocity of a stream (or of wind), the larger the particles that can be carried. As a stream slows down, it deposits its load; first, the biggest boulders and cobbles, then the sands and muds, and finally in quiet water the finest of clays.

All the waters of the earth also carry in solution an invisible load of such chemicals as calcium and magnesium carbonates, silica, and sodium chloride (common salt). These materials in solution are separated from water, not by deposition as current velocities lessen, but most often by evaporation of water, by the action of organisms that use the materials to build their shells, by changes in temperature that change solubility, and by chemical reaction with other substances. Under favorable circumstances, time, pressure, loss of water, and chemical action change sediments into sedimentary rocks.

Water and air, which surround the earth, transport rock particles everywhere, depositing them particularly in low spots, or basins. Thus, the sedimentary rocks form a thin veneer over three-quarters of the globe. At maximum, sedimentary rock formations are only a little more than 10 miles thick—an inconsequential layer when compared to the igneous-rock mass of the earth itself. Although the sedimentary rocks constitute only a very small portion of the earth, that portion is all at or near the surface; and for creatures (including man) that live on earth, sedimentary rocks are the most common and most often seen of the three great rock classes.

**2-13** *Coarse, poorly sorted conglomerate of the Oligocene Bealville Formation exposed along Highway 58 in the southwestern Sierra Nevada, a few miles east of Bakersfield. The irregular sizes of the fragments, the angular to slightly rounded shapes, and the predominance of granitic rocks and gneisses show that the material is fanglomerate, formed in thick fans by rapid erosion and deposition along the Sierran front. (Charles W. Chesterman photo.)*

These rocks, like igneous rocks, are most logically classified on the basis of genesis and texture (grain size and form). Sediments are rarely pure in composition or uniform in grain size. It follows, then, that sedimentary rocks are variable and gradational from one type to another in all their characteristics. The common sediments and the sedimentary rocks formed from them are listed in Table 2-3.

## METAMORPHIC ROCKS

All types of rocks may be subjected to high pressures, high temperature, and chemical changes. If a rock has been transformed so that its mineral composition, texture, and structure are quite different from the original, it is said to have been metamorphosed. The principal characteristics of metamorphic rocks are (1) the presence of newly formed minerals, (2) platy and foliated structures, and (3) recrystallized interlocking granular textures. If these structures are fine, the rock is referred to as schistose. If the structures are coarse, the rock is referred to as gneissic.

Metamorphism usually occurs in connection with the intense folding, faulting, high pressures, temperature changes, and chemical changes that are directly associated with mountain building. Therefore, metamorphic rocks are most common in mountain ranges and in the eroded roots of former mountain ranges.

### kinds of metamorphic rocks

*Slate* is a fine-grained metamorphic rock whose most prominent and characteristic feature is that of splitting into thin layers. These layers are not necessarily parallel to any layering that may have been present originally but tend to be developed at right angles to the directions of the pressures that formed them. Slate can be developed from any fine-grained rock, perhaps especially commonly from shale and tuff.

*Phyllite* is intermediate between slate and schist in its degree of metamorphism and development

**TABLE 2-3 Sediments and sedimentary rocks**

| CLASTIC SEDIMENTARY ROCKS* (made up of broken particles) | | | |
|---|---|---|---|
| *Fragment or particle* | *Diameter* | *Loose material* | *Rock* |
| Boulder | Over 10 in. | Gravel | Conglomerate |
| Cobble | 2.5 to 10 in. | Gravel | Conglomerate |
| Pebble | 0.1 to 2.5 in. | Gravel | Conglomerate |
| Sand | Just visible to 0.1 in. | Sand | Sandstone |
| Silt | Microscopic (0.002 in. to barely visible) | Silt or mud | Siltstone and mudstone |
| Clay | Less than 0.002 in. | Clay | Shale |

| ORGANIC AND CHEMICAL SEDIMENTARY ROCKS | |
|---|---|
| *Sediment* | *Rock* |
| Calcareous (mostly calcium carbonate) | Limestone |
| Dolomitic (calcium and magnesium carbonate) | Dolomite |
| Diatoms (single-celled silica-secreting plants) | Diatomite |
| Silica | Chert |
| Carbonaceous plant remains | Peat and coal |
| Salt and calcium sulfate | Salt; gypsum |

*The clastic sedimentary rocks are rarely composed of one grain size; it is most common to find the rock made up of larger grains in a matrix of smaller ones.

**2-14** *Metaconglomerate (metamorphosed conglomerate) —Jurassic or older—exposed in the Matterhorn Peak area of the High Sierra Nevada, west of Mono Lake. The original stratification seems to be inclined steeply toward the lower left, but planes of gneissic layering, produced by the pressures involved in metamorphism, seem to be tilted gently toward the lower right of the photo. (Charles W. Chesterman photo.)*

**2-15** *Close-up of chiastolite schist in the Pampa Schist in the southern Sierra Nevada. Chiastolite is a variety of andalusite (aluminum silicate) that contains black inclusions. The chiastolite rods, up to an inch or so long, have crystallized in a matrix of mica, feldspar, and quartz. The high alumina content of the schist suggests that it was formed by metamorphism of a clay shale. (Charles W. Chesterman photo, California Division of Mines and Geology.)*

of new minerals. Platy surfaces are often highly lustrous because of new micaceous minerals.

*Schist* is similar in origin to slate, but has been more intensely metamorphosed. Schist is foliated and often crinkly; and flaky, newly developed minerals like biotite, muscovite, chlorite, and amphibole are clearly visible. Schist can form from slate by higher-grade metamorphism. Often, new minerals that are not flaky develop also—garnet, quartz, and feldspar, for example.

*Gneiss* is coarsely granular and banded. Its bands are often distinguished from one another by color. It develops commonly from the granitic rocks, coarse-grained sedimentary rocks like conglomerate, and thin-bedded sedimentary rocks such as alternating layers of shale and sandstone.

**TABLE 2-4 Metamorphic rocks**

| FOLIATED | MASSIVE—NONFOLIATED |
|---|---|
| *Fine-grained; schistose* | *Fine-to coarse-grained* |
| Slate | Marble |
| Phylite | Quartzite |
| Schist | Anthracite Coal |
| *Coarse-grained; gneissic* | Graphite (often foliated) |
| Gneiss | *Fine-Grained* |
| | Hornfels |

Some rocks, particularly those composed of essentially one mineral (such as limestone, dolomite, and quartz sandstone), do not become schistose or gneissic on metamorphism but simply recrystallize. Limestone and dolomite recrystallize to form the coarse granular marble, sandstone forms quartzite, and coal forms the hard anthracite variety of coal or, if metamorphosis is extreme, graphite (pure crystalline carbon). Hornfels is a hard, massive, fine-grained rock—not foliated—formed from the metamorphism of a fine-grained igneous or sedimentary rock.

Descriptive names for metamorphic rocks are formed by prefacing the general metamorphic rock names with the names of constituent minerals or the rock type from which the metamorphic rock was derived; for example, mica schist, chlorite schist, garnet-mica schist, granite gneiss, diorite gneiss, and dolomitic marble.

We find, quite reasonably, that the processes of metamorphism tend to cause the development of new minerals, textures, and structures *that are better adjusted to the new conditions of temperature, pressure, and chemical environment;* in other words, the metamorphic rock has adjusted itself to a condition of stability in its new environment.

*Erosion by running water ceaselessly changes the face of the land. Vernal Fall, Yosemite Valley.*

# 3 the face of the land changes

■ It has been said that "Nature abhors a vacuum." Nature also appears to abhor differences in elevation, or relief, on the surface of the earth. No sooner is a land surface elevated above its surroundings than geologic processes begin to tear it down; no sooner is a basin developed than it begins to be filled. The tearing down process or wearing away of the land surface is erosion. The loosening of the rock material preparatory to removal is weathering, and the all-powerful force that moves all weathered rock fragments ever down to lower and lower levels is gravity. The products of erosion are deposited as sediments in the low spots, depressions, or basins. The ultimate and greatest basin of all is, of course, the ocean basin. Collectively, all these activities are often called the leveling processes, for the net result, after sufficiently long periods of uninterrupted erosion, would be the eventual reduction of all land surfaces to featureless plains.

Why, then, do not these surficial processes of weathering and erosion reduce the mountains to near plains and fill all basins to their brims? Certainly, geologic time has been long enough to reduce a whole series of mountain ranges like the Sierra Nevada to sea level. However, there are, within the earth's crust and mantle, internal forces that result in great uplift and the repeated building of new and younger mountain ranges. At the crest of the Sierra Nevada, as high as 14,000 feet, there are gently rolling granite surfaces believed by geologists to be approximately those that were formed when the ancient Sierra was eroded to a low hilly land surface, 10 million to 15 million years ago. Uplift of at least 10,000 feet in the past few million years has allowed running water and moving ice to erode canyons thousands of feet deep into the old erosion surface. Over much of the Sierran crest, however, erosion has progressed so far that the old, gently sloping, uplifted surfaces are gone. Uplift and changing climate, in the last 3 million years, have added the eroding power of moving ice, and the face of the Sierra has profoundly changed. In the last few thousand years, alpine glaciers have dwindled and running water again erodes the valleys of the High Sierra Nevada.

## WEATHERING

In the 1870s, when California's great naturalist and author, John Muir, camped alone in Yosemite Valley, he heard loud reports of granite blocks broken away from the precipitous walls by frost action (water entering fractures and joints by day, and freezing and expanding by night). Great piles of angular granite blocks form fans and cones along the bases of the cliffs in Yosemite. These accumulations are called talus; they are the product of the mechanical type of weathering in which rock is broken, in place, into smaller fragments without change, except in size and shape of the pieces. Growing tree roots and the activities of animals also contribute to mechanical weathering. One of the important factors here is that all breaking of rock *exposes more surface area,* which permits weathering to proceed at an accelerated pace.

To use the granite of the Sierra Nevada again as an example, there are numerous places along Interstate Highway 80 and elsewhere on the western slope where quite a different type of weathering can be seen. The rock seems to be decayed or decomposed and fresh granite in the deep road cuts seems to grade upward through the decomposed material into a black soil full of the remains of decaying vegetation. This is an example of chemical weathering, in which a major process in the disintegration of the solid rock has been chemical change. Chemical changes bringing about such decay are usually quite complex. Air with its content of oxygen penetrates fractures, openings, and pore spaces well below the surface of the earth. The oxygen readily combines with mineral substances (iron, for instance, especially when wet) to form soft oxidized products. Iron oxides impart much of the yellow, brown, and red coloration of rocks and soils. Travelers along State Highway 49, in the Mother Lode country of the western Sierra Nevada foothills, often remark about the red soils, which have been formed by the weathering of metamorphosed volcanic rocks.

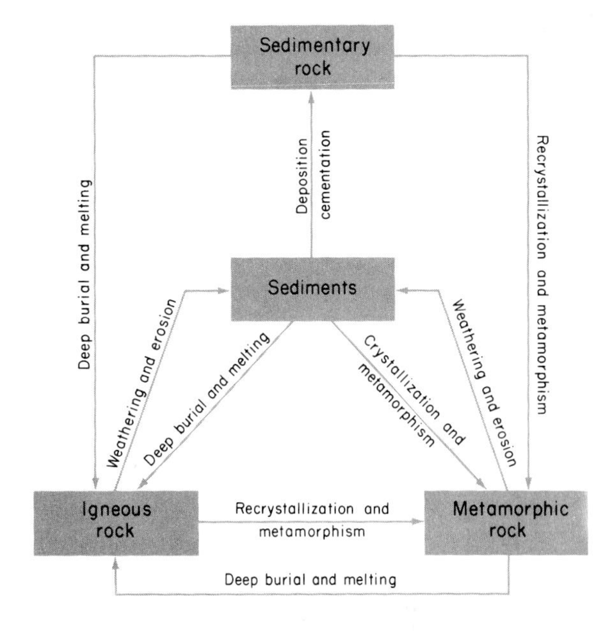


**FIGURE 3-1** *Rock cycle. This suggests many of the interrelationships among the three major rock groups and sediments. For example: an arrow shows that igneous rock may be weathered and eroded to form sediments; another, that sedimentary rock may be recrystallized and metamorphosed to form metamorphic rock. Could additional arrows be drawn?—sedimentary rock to sediments, for instance.*

**3-2** *View east up Yosemite Valley; El Capitan on left, and Half Dome in middle distance with Tenaya Canyon to its left. Bridalveil Fall, on right, emerges from a hanging valley. Sentinel Rock is the ragged rock mass to right of Half Dome. John Muir, camped at the base of Sentinel Rock, was awakened at 2:30 on the morning of March 26, 1872, by tremendous rock falls caused by the Owens Valley earthquake. To give proper perspective, the sheer face of El Capitan is about 3,000 feet high. (Mary Hill photo.)*

**3-3** *Jointing in granitic rock, eastern Sierra Nevada foothills. Weathering and erosion along joint systems at right angles. (Bateman and Wahrhaftig, California Division of Mines and Geology Bull. 190.)*

**3-4** *Talus slope of huge blocks of obsidian, Glass Mountain, Siskiyou County. Some of these obsidian fragments are such perfectly clear glass that they have been polished as mirrors for telescopes. (Charles W. Chesterman photo.)*

**3-5** *Jointing in granitic rock, eastern Sierra Nevada foothills. Slotlike small valley eroded along a joint system. (Bateman and Wahrhaftig, California Division of Mines and Geology Bull. 190.)*

Water is extremely active, both because it is a solvent of minerals and because it combines with them chemically. Particularly potent in chemical weathering is the combination of water with carbon dioxide of the air to form carbonic acid. Organic acids, formed by plant decay, are also strong agents of weathering.

Not all the common minerals in rocks and not all rock types respond in the same way to weathering processes. Quartz, for instance, is highly resistant to chemical weathering. The feldspars quite readily undergo chemical weathering to form clay. The pyroxenes and amphiboles are often oxidized to form red- and brown-colored soils and weathered rock. Thus, differential weathering takes place to form a vast variety of interesting rock forms.

## EROSION

Once weathering has loosened rock fragments, the force of gravity takes over to move the rock particles down slope. Gravity is the *reason* that weathered rock particles move down slope, but the *agents* and the *media* by which and in which the fragments are transported are water, moving air (wind), and moving ice (glaciers).

Sometimes wind has a high velocity but the density of air is low, and therefore eroding power is limited. Wind does move vast quantities of dust; and because it carries rock particles, it scours, abrades, and polishes other rocks. Pebbles scoured, polished, and faceted by wind have been found on the floor of Death Valley and in the extremely arid Mojave Desert. Wind also moves sand-sized fragments and deposits dunes back of beaches and in inland deserts or in any other place where there is a source of sand and scant vegetation to hold sand in place. A limited part of the floor of Death Valley, near Stovepipe Wells, shows beautiful desert dunes. Back of Pismo Beach, on the Santa Barbara coast, and at Carmel, near Monterey, are some of the best examples of coastal dunes.

Moving ice, in glaciers, has tremendous carrying power and transports great volumes of weathered rock, depositing its load where the ice melts. However, ice covers only 10 percent of the land surface of the earth, and all but a small percentage of this is in Greenland and Antarctica.

Overwhelmingly, water is the most powerful agent of erosion. Water has everything working for it: its universal occurrence over the earth, its relatively high density, its high dissolving power (especially when charged with carbon dioxide), and its concentration in streams and valleys. Where slopes are steep, stream velocity is high

**3-6** *Fantastic erosional form in Joshua Tree National Monument, Riverside County. The rock is White Tank quartz monzonite, a granitic rock about 150 million years old (Middle Jurassic). Such forms as this are produced by erosion along joint systems in the rock. Rain, floods, sun, and freezing water gradually round the corners of plane-surfaced joints. Wind is only a minor, modifying agent of erosion. (Sarah Ann Davis photo, California Division of Mines and Geology,* Mineral Information Service, *April, 1964.)*

and erosion goes on rapidly; where slopes become gentler, velocity lessens and streams tend to deposit sediments. Where stream gradients are steep, most of the energy of erosion is expended in downcutting, but where gradients are gentle, streams erode sideward by undercutting their banks and may even build up the bottoms of their channels by deposition.

Because volume and velocity control the amount and extent of stream erosion and deposition, those streams that have great seasonal variation show great differences in their habits from time to time. The Los Angeles River, in semiarid southern California, for example, is usually completely dry in summer, but during heavy winter rainstorms it becomes a raging

**3-8** *Glacially striated, grooved, and polished rock in the high Sierra Nevada. (Charles W. Chesterman photo.)*

**3-7** *Low terrace, Artists Drive, Death Valley—probably the shore line of a Holocene lake. The rough surface below the terrace is due to heaving action as rock salt crystallizes from solution; this is part of the "Devil's Golf Course." The desert floor above the terrace is smoother and is made up of rock fragments and gypsum. (Charles B. Hunt photo.)*

torrent as big as the Colorado River. Most streams vary considerably in their seasonal flow, depending on when the rains come or when the snow melts.

Running water on the land aided by mass wasting—the downslope movements of soil and rock—probably performs 90 percent or more of the erosion that takes place. But along the seacoasts of the world, wave erosion is an important factor. Much of California's coast is steep and rugged, backed by sea cliffs and terraces. In many places, as along the coast of San Diego County, the sea cliffs are receding by several inches a year as a result of a process of undercutting by wave action and sloughing or sliding from the cliff face. In April, 1968, an earthquake in the Borrego Mountains, 80 miles inland, triggered landslides in these sea cliffs. Waves, mostly the result of winds at sea, can actively erode only a narrow band between a few feet below low tide and the highest levels of storm waves; this concentrates their effective energy. Wave erosion is so rapid along parts of the coast—Mendocino County for example—that segments of the coastal terraces have been isolated from the land as sea stacks. A recent

**3-9** *"Beehive" or "chimney rocks" near Surprise Siding, Modoc County. These conical peaks have been developed by the erosion of horizontally stratified, firmly compacted, fine-grained rhyolite tuff of the late Miocene Upper Cedarville Formation. (Charles W. Chesterman photo.)*

study and analysis of retreating sea cliffs along the Santa Barbara coast has shown that, under present wave and sea level conditions, the average rate of recession is about 50 feet per century!

## EVOLVING LANDFORMS

The landscape is the sum total of all the features at the surface of the earth—landforms (such as hills, valleys, mountains, vegetation, even animal life) and cultural forms (like cities, canals, and other man-made features). In a more restricted sense, a landscape is that part of the earth's surface that can be seen in a single view.

Two great opposing forces are constantly acting to change the surface of the land: the surficial processes of erosion and deposition, which tend to level the land, and the internal processes of folding, faulting (breaking and displacement of the earth's crust), volcanic activity, uplift, and subsidence, which form mountain ranges and "structural" valleys. In California, which is in a particularly active orogenic, or mountain-building, belt, erosion is extremely active because of the elevated land surfaces. Although a mountain peak, for example, may not appear to alter during a man's lifetime, the form of the land is constantly changing. Constant change, but change at varying rates in time and place, is one of the basic geologic truths. Equally fundamental—a fact we have already hinted at and will discuss in Chapter 4—is the vast amount of time during which geologic processes may act to form and to modify the land surface. The present rugged crest of the Sierra Nevada has resulted from uplift and subsequent erosion, primarily by running water, with that erosion augmented and modified by mountain-valley glaciers over the past few million years. The high rugged character of the Sierra indicates that uplift and mountain building in that area have kept pace with erosion.

There is every reason to believe that the processes and the geologic forces that are operating today have acted throughout the 3 billion years and more of the solid earth's history. We have found old erosion surfaces (fossil landscapes) buried within rock formations many hundreds of millions of years old. Stream gravels and ocean wave and beach deposits of similar vast age can be identified.

The other great mountain ranges of California are structural features; like the Sierra Nevada, they have been formed by folding, faulting, and uplift. Yet the canyons and the ridges between them, in all their great detail and complexity, were formed principally by running water. Many valleys near the crest of the range and the famous Yosemite Valley farther west have been greatly modified in the past couple of millions of years by glaciers moving down the stream valleys already formed by running water. Mountain streams tend to cut downward very sharply in their channels and to form valleys that are V-shaped in cross section. Solid, moving ice smooths this characteristic cross section to a U form.

## LANDFORMS SHAPED BY RUNNING WATER

Most valleys have been formed by running water and, nearly always, have been at least modified by water erosion, but the great valleys and depressions are initially structural; that is, they have been formed by movements originating within the crust and interior of the earth. The Great Valley of California is a huge trough formed by downwarping, and the Salton Trough evolved from the structural complexities of the San Andreas fault zone and the East Pacific Rise. Many others, like Death Valley and Owens Valley, are primarily fault troughs—down-dropped blocks between faults. Still others, like San Francisco Bay, Salinas Valley, and the Los Angeles Basin have been formed by complex faulting and downfolding.

Why do some streams—for example, the lower Sacramento River, the lower Russian River (in the Coast Ranges north of San Francisco), the Los Angeles River and the Santa Ana River—run directly *across* ranges of hills? In some cases, such streams were there *before* the hills and were able to maintain their courses to the sea as slow uplifting to form the hills took place.

**3-10** *View west across Death Valley to the Panamint Mountains and Telescope Peak (elevation 11,045 feet). Note the beautiful coalescing alluvial fans. (Charles B. Hunt photo.)*

Plains, plateaus, and terraces are nearly flat land surfaces that are also common landforms. Plains are the low-lying, near-level surfaces, like much of the Great Valley of California. Many of these have been built up by deposition of sediments to form river floodplains, but in some, bedrock is exposed and it can be seen that they were formed by erosion—perhaps by streams or by wave action along a former seacoast. Plateaus are plains that have been elevated or perhaps were left behind as their surroundings were eroded. Terraces are relatively narrow near-level surfaces, most often found above stream levels or at the tops of sea cliffs along the coast. They also usually owe their flatness to deposition or to erosion, coupled with relative uplift.

Alluvial fans and deltas are always features of deposition. A delta is formed in water when a stream's velocity is abruptly reduced as it enters a quiet body of water—a sea, lake, or slower-moving river. They are cone-shaped in cross section, and in plan look like the Greek letter delta (Δ), with one point upstream. Fans typically form at the foot of a mountain valley where a stream's velocity is abruptly slowed as it extends onto the flat floor of a valley or basin.

**3-11** *Looking down Hungry Valley toward Interstate Highway 5, southeast of Gorman, Los Angeles County. The stream first built up its broader, flat valley by deposition of sediments; then a geologic change occurred to increase the stream's power of erosion and cause it to entrench itself to form the new, sharp, tree-lined gulley. (John S. Shelton photo.)*

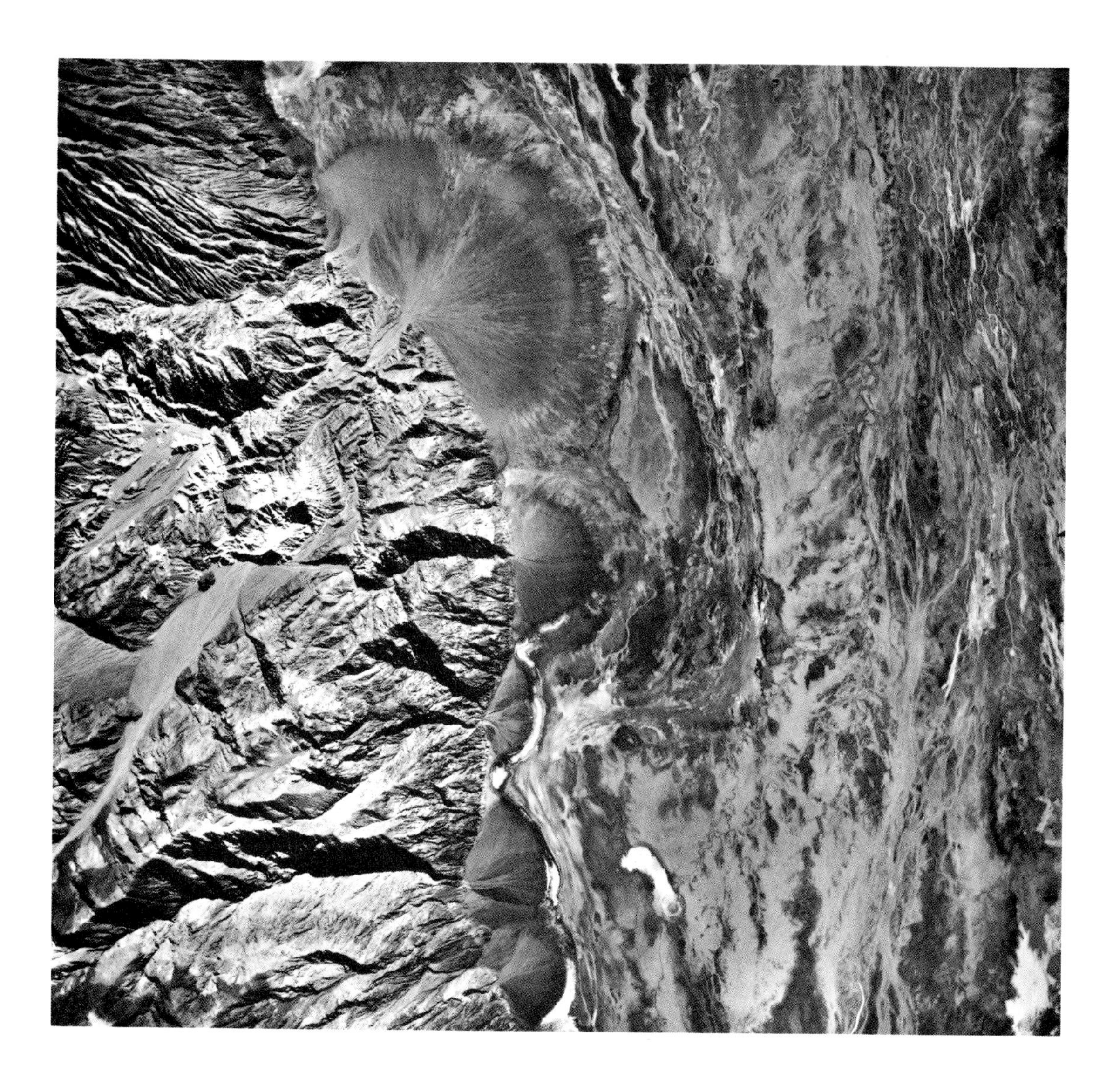

**3-12** *Death Valley and the Black Mountains. The extremely steep, intricately eroded slopes of the Black Mountains on the left (east) drop precipitously to the flat floor of Death Valley. The Black Mountains are made up of blocks of Tertiary sedimentary rocks faulted against Precambrian metamorphic rocks. The extraordinarily symmetrical, largest alluvial fan has been formed at the mouth of Copper Canyon. The floor of Death Valley, here about 200 feet below sea level, is intricately patterned by salt pans, dry saline beds (white), and the meandering, ever-changing, tiny loops of the Amargosa River. The curved, sharp, knifelike contact between recent fans and the steep, rocky front of the Black Mountains is a segment of the Death Valley fault, probably still active. (Vertical aerial photo, U.S. Geological Survey, Prof. Paper 590, 1968; scale approximately 1 inch equals 5,880 feet; south is at top.)*

**3-13** *Convict Lake, eastern Sierra Nevada, Mono County. Highway 395 crosses lower center from left to lower right. The lake, about 1 mile long, lies in a deep, glacially eroded basin at the foot of the extremely steep slopes of Mount Morrison (not in photo). The lake basin and the higher slopes in the photo consist of extremely folded and faulted, slightly metamorphosed marine sedimentary rocks of Ordovician age. The mottled surface across the highway is Quaternary rhyolite. One of the major features shown is the succession of glacial moraines along the front of the mountains. An older moraine consists of a massive embankment extending toward the left center side of the photo on the same trend as the long axis of Convict Lake. The old moraine is crossed almost at right angles by two large, later, lateral moraines; these form prominent ridges on the two sides of Convict Creek, which drains the lake. Both lateral moraines swing toward their right downstream; between the two are low ridges that represent terminal moraines left as the ice receded. (Vertical aerial photo, U.S. Geological Survey, Prof. Paper 590, 1968; scale approximately 1 inch equals 6,190 feet;* south is at top.)

**3-14** *Virginia Peak from high ridge south of Camiaca Lake, Matterhorn Peak area of High Sierra west of Mono Lake. Serrated ridge and pyramidal peaks due to glaciation. Talus accumulations, formed after the principal alpine glaciers melted, lie along the base of the ridge. Low ridge in foreground consists of glacially rounded and polished granite. (Charles W. Chesterman photo.)*

## LANDFORMS SHAPED BY GLACIERS

California was never covered by a continental ice sheet, like those over Greenland and Antarctica today. Nevertheless, during the Ice Age, glaciers were formed in the higher mountain valleys. Their effects are seen most prominently in the Sierra Nevada above 4,000 feet elevation, where the saw-toothed ridges, U-shaped valleys, pyramid-shaped peaks, and rock-basin lakes (tarns) were carved by glacial erosion. There are also the visible effects of deposition by glaciers—the long, narrow ridges (moraines) made up of rock fragments that parallel some of the U-shaped valleys and the end, or terminal, moraines that form ridges of glacial debris left *across* valleys as the ice melted and retreated. Fine examples occur at the eastern base of the Sierra Nevada, notably south of Lake Tahoe and near Mono Lake.

## LANDFORMS SHAPED BY WIND

Like water, wind carries rock fragments, erodes the earth's surface, and deposits sediments—but on a much lesser scale. Strong winds that carry sharp rock fragments polish and groove rock formations in the desert; and winds pick up dust, sand, and small pebbles to carry them over wide areas. These effects are most noticeable where the climate is arid, vegetation is scanty, and there is a large source of sand, as in a desert basin or along a beach.

Any action in which wind may pick up and transport sand is, of course, erosion, but the most spectacular and striking effects of wind are deposits in the form of sand dunes. Shapes, sizes, and varieties of dunes depend on such variables as wind direction and velocity and the source and amount of sand available. The extensive Algodones Dunes, some as much as 300 feet high, are found in the desert on the eastern side of the Salton Trough in southeastern California (Illustration 3-15). Stovepipe Wells is a famous dune area on the floor of Death Valley in Inyo County, east of the Sierra Nevada. Although sand dunes are common in these desert areas, they form only a small fraction of the desert surface.

## LANDSLIDES

In 1956, at Portuguese Bend in the Palos Verdes Hills in Los Angeles County, some of the beautiful hillside homes began to move down slope, straining and tearing apart as they went. Movements in the eastern 300 to 400 acres of an old landslide, which had been recognized by geologists mapping the area many years before, have continued. By 1968, damage from landslides in that area had amounted to about $10 million.

A few miles north of the Palos Verdes Hills, Pacific Palisades lies along a sea cliff backed by the Santa Monica terrace. A slope failure occurred there late in the 1960s, and, in a few seconds' time, a huge mass of sandstone and shale of the Modelo Formation moved across State Highway 1 and down onto the beach.

To the residents, property owners, and users of the local highways, these occurrences were unusual and, for some, catastrophic. Geologically, however, they are routine—typical of the thousands of recent landslides throughout California. Every imaginable rock type, structure, landform, and climate has been marked by landsliding of one type or another in California. The shale in the Modelo Formation in the Transverse Ranges (located in Los Angeles County, in southern California), the shale of the famous Franciscan Formation of the Coast Ranges, and the Orinda Formation mudstone and sandstone (east of San Francisco Bay) are among California rock formations that are particularly susceptible to sliding.

All manner and types of "landslides," from rock falls to thin mudflows, are found throughout the state. Particularly the high mountains (such as the Sierra Nevada, the Klamaths, and the San Gabriels) are areas of extensive and frequent rock falls, as evidenced by the characteristic talus cones and aprons (coalesced alluvial fans) at the bases of the cliffs. Many of the Sierran lakes, like Kern Lake in the Kern River Canyon and the beautiful little Mirror Lake in Yosemite Valley, were formed by landslide dams across streams. Some loose rock masses in the High Sierra occupy the valleys and move like glaciers down their slopes. Mudflows are common in the deserts, the Great Basin, and on the eastern slopes of the Coast Ranges. A gradually moving landslide mass with a bulk of about 18 million cubic yards has been threatening to move across part of the little mountain community of Wrightwood, a winter resort area on the northern slope of the San Gabriel Mountains. Mudflows are very common in association with volcanic eruptions because of the presence of quantities of fine debris and abundant water. Eruptions of Lassen Peak in the period 1914 to 1917, in Lassen County in northern California, were followed by mudflows.

Type of rock, steepness of slopes, amount of water, and a variety of geologic conditions are factors in landsliding. Rock falls are relatively dry, while mudflows may be mostly water; but every form of mass movement in between can be found.

Working back through the geologic ages, landslide debris has been recognized in California in rocks as old as the Precambrian. Clearly, then, landsliding, or mass wasting (to use a more comprehensive term), is a normal geologic process of erosion that shapes and has shaped the California landscape. In the Transverse and Coast Ranges, mass wasting is perhaps the *dominant* mode of erosion that shapes the form of the land. The rate at which such soil and rock movements

**3-15** *Algodones Sand Dunes, along the San Andreas fault zone, east of Imperial Valley, Salton Trough. The sand hills are crossed by Interstate Highway 8. The patterns of ripple marks can be clearly seen. Which direction are the prevailing winds, relative to the viewer? (Southern California Visitors Council photo.)*

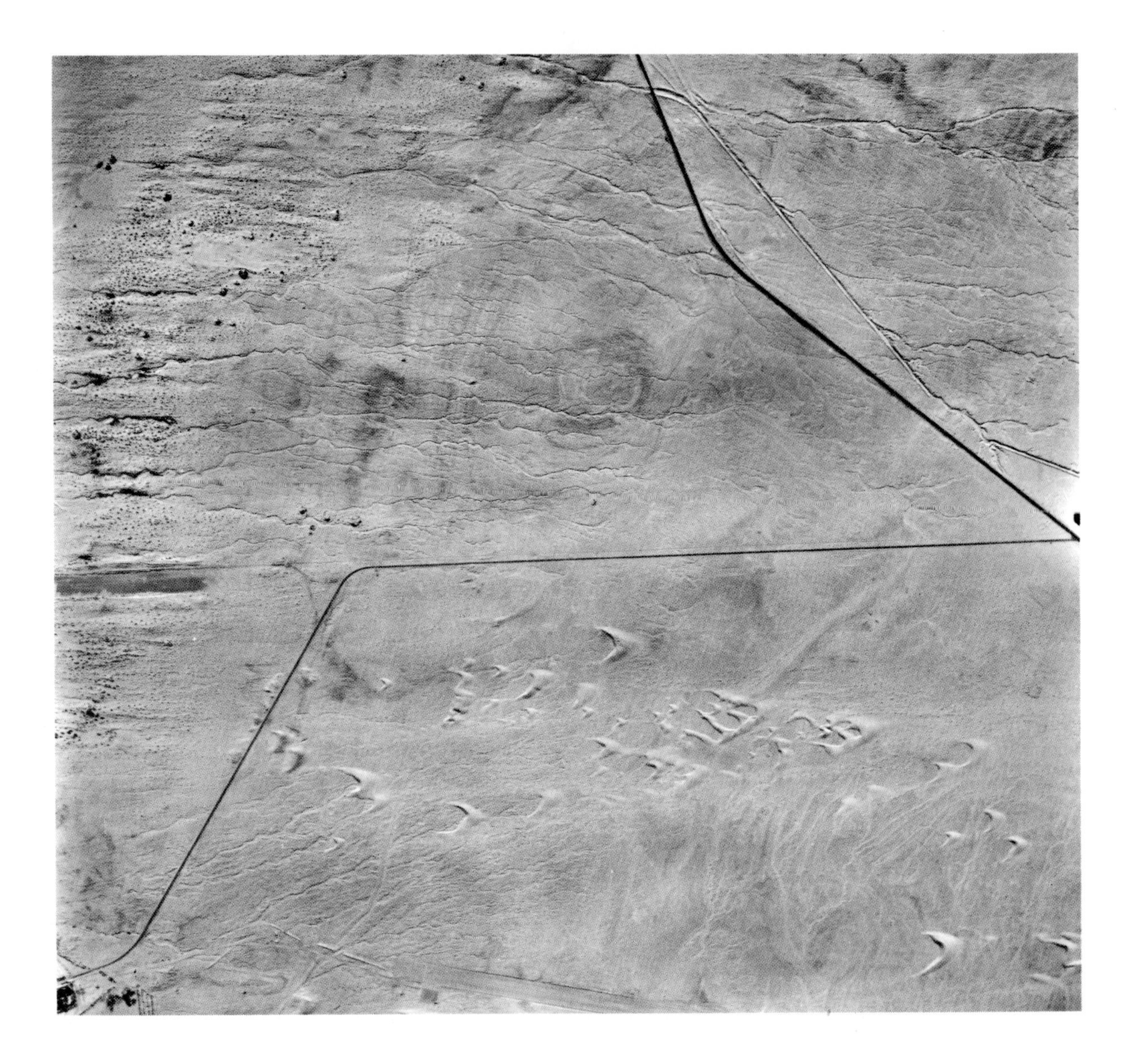

**3-16** *Sand dunes west of Salton Sea. The sea is just off the photo to the left (east). State Highway 86 runs across the upper right quadrant. The other paved highway leads to the sea in lower left corner. The barchans (crescent-shaped dunes)—open to the east—show that the prevailing winds are from the west. They moved with the wind from 325 to 925 feet in seven years, from 1956 to 1963. The dunes rest on a smooth, gently sloping plain that descends to the Salton Sea. They range in size, horn to horn, from about 150 to 800 feet and in height from 10 to 30 feet. The area of the photo is underlain by older lake beds; some of the curving shorelines appear in upper center of the photo. (Vertical aerial photo, U.S. Geological Survey, Prof. Paper 590, 1968; scale approximately 1 inch equals 2,330 feet;* south is at top.)

take place is highly variable, from slow soil creep over many years to mass movements of more than 100 miles per hour.

The problem of mass wasting is a matter of great concern in the developing urban areas of the state. For several years, federal and state geologists have intensively studied the problems created by landsliding, and engineering geologists have been learning to cope with them. These studies of the origin of landslides and of their geologic settings have saved millions of dollars for the people of the state.

Landslides occur whenever loose rock and soil become unstable on a sloping surface, but California's frequent earthquakes contribute by triggering slides. Evidence of particularly numerous landslides is found in and near the zones of active faults. All earthquakes—even mild ones—in the historic record have been accompanied by numerous landslides. Landslide topography along the major fault zones is evidence that slides have been caused by earthquakes throughout geologic history.

**3-17** *Landslide at Point Firmin, Palos Verdes Hills, Los Angeles County. Photo, taken in 1959, shows total movement after thirty years. Borings have shown that movement has taken place down the low dip of stratified sandstone and shale toward the sea. Individual slide blocks have remained practically level, and movements of less than 1 inch a day seem to be associated with heavy rainfall seasons. (John S. Shelton photo.)*

*As time passes, life evolves. This unusual mass of the bivalve* Buchia *in sandstone from the northern Coast Ranges demonstrates the rock's Late Mesozoic age.*

# 4 geologic time and its measurement

■ In today's society, time is extremely important. We come and go by time, we do our jobs and, indeed, order our whole days by the clock, which is one of many mechanical ways of keeping track of time. The period that we call a year is the time it takes for the earth to make one revolution in its elliptical orbit about the sun; the period called a day is the time it takes for the earth to make one rotation on its axis. Very precise time measurements are made by recording the vibrations of atoms.

## GEOLOGIC TIME

In the early seventeenth century, Archbishop James Ussher, who lived in Ireland from 1581 to 1656, made the pronouncement that the earth was created in 4004 B.C.! He arrived at this startling figure by adding up the genealogies of the Old Testament families. In the nineteenth century, the great British physicist Lord Kelvin computed the age of the earth at 40 million years by calculating the time it would take for a molten earth to cool. While his figure is closer than that of Archbishop Ussher, his reasoning was based on faulty premises. He lacked knowledge of the radioactive decay of certain elements in the earth, a process that furnishes an enormous and continuing source of heat. Instead of cooling, the earth today may actually be getting warmer.

Our review of the surficial geologic processes has already given us some clues concerning geologic time. Mountains cannot be formed in a day, and the erosion of a mountain range to a near plain or the cutting of great canyons like Yosemite Valley and Kern Canyon (3,000 to 4,000 feet deep in solid granite) must require very long periods of time. At least 150 years ago, some scientists recognized that geologic time is extremely long. Later, rates of erosion and rates of accumulation of sediments, as measured in numbers of places, always showed that many millions of years were required to accomplish the geologic changes that we can see have taken place. By adding up total thicknesses of all the sedimentary rocks and multiplying by a judgment of the average time to deposit a foot of the sediments that became rock, geologists calculated that earliest deposition for the stratified rocks probably took place on the order of 600 million years ago.

Ideally, the goal of geologists is the reconstruction of the history of the earth. Achievement of this goal requires that we should not only record major events in earth history but also date such events.

## RELATIVE TIME

The relative ages of geologic events can often be determined by observation, even though the true age in years of a geologic happening may elude us.

Fundamental to relative dating of sedimentary rocks, for example, is the simple realization that the younger layers are deposited on top of the older! In a succession, or column, of stratified rocks, the youngest is at the top unless the sequence has been overturned. Even though stratified rocks have often been folded, faulted, or

**4-1** *Not all geologic processes are slow! This slide of a segment of Chelton Drive in the Berkeley Hills took place in a few minutes. Contributing factors were heavy winter rainfall; contorted, near-vertical, weak clay-shale beds; and a naturally steep slope (out of picture to left), oversteepened by removing material from the toe of the slide.* (Montclarion, *Montclair District, Oakland, photo.*)

**4-2** *Ubehebe and Little Hebe volcanic craters, north end of Death Valley. The surrounding rock consists of fragments of rhyolitic rock of explosive origin. This is evidently a series of explosion craters of very recent origin. Note the relative ages of the explosion pits. (Mary Hill photo.)*

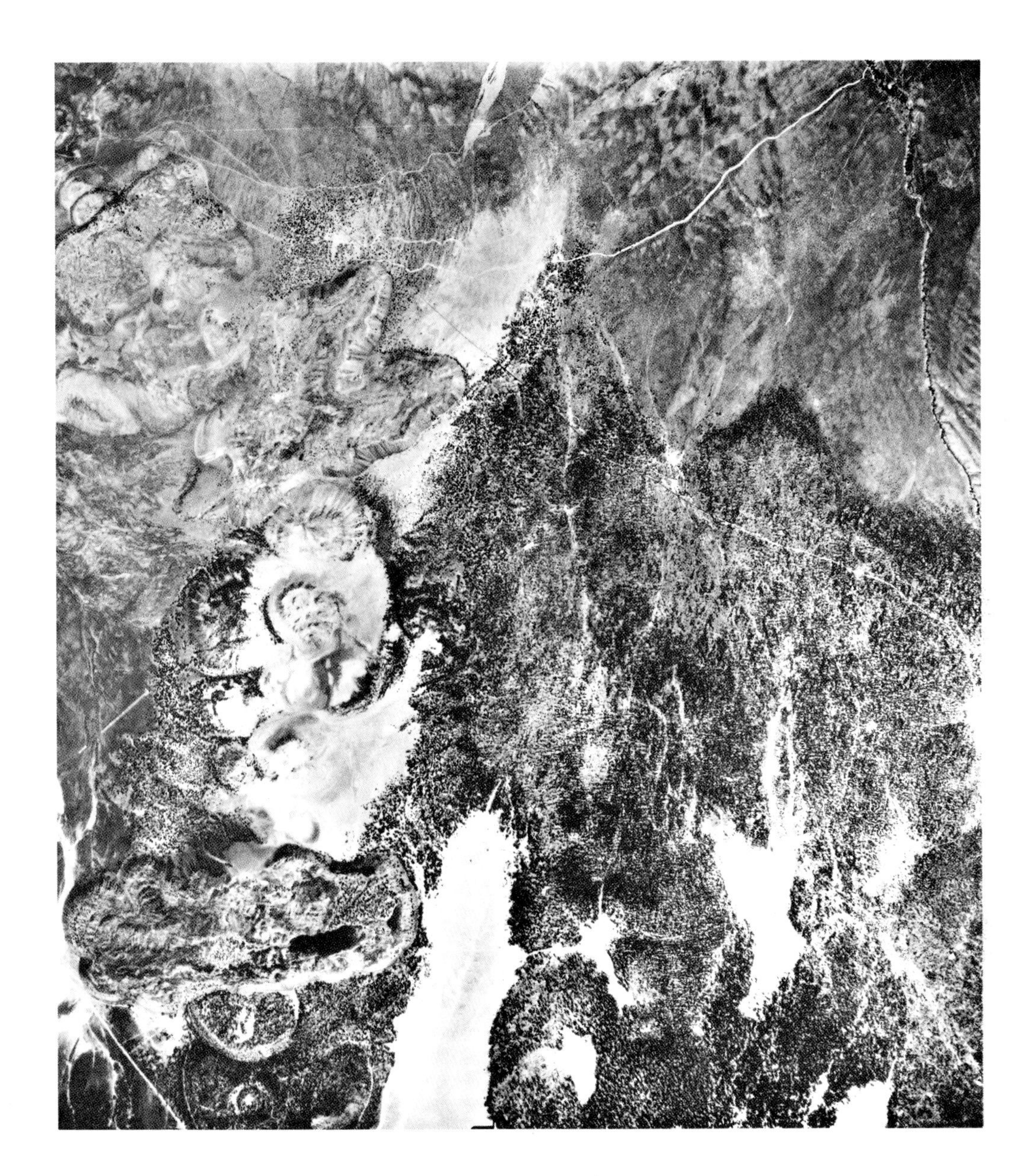

**4-3** *Mono Craters, Mono County. The craters consist of a chain of rhyolitic obsidian domes and stubby, steep-sided flows, which extend across most of the left half of the photo. Each of the more perfect domes consists of a circular crater, which was formed by explosion of pumice fragments, into which a plug of banded obsidian was intruded. Sometimes a circular moat has been left around the plug. Note how the successive ages of flows and craters can be worked out by their overlap. The areas around and away from the domes consist of pumice flats sparsely covered with small pine trees. (Vertical aerial photo, U.S. Geological Survey, Prof. Paper 590, 1968; scale approximately 1 inch equals 5,650 feet;* north is at top.)

turned completely upside down, relative age can still be determined if the geologist can correctly interpret the original position of the rock formation.

Igneous rocks can also be fitted into the age succession. Suppose a dike (sheetlike body of igneous rock) is found cutting across sedimentary strata (or any other rock type, for that matter); the dike is obviously younger than the rock it has intruded. This is true for any intrusion, whether it has the form of a dike or not. The vacationer, driving his car over California's 9,941-foot Tioga Pass on State Highway 120 across the Sierra east of Yosemite, can see irregular masses of light-colored granite cutting across the structure of dark metamorphosed rock, which is obviously older than the granite.

If some features found preserved in the rocks could be recognized as having been formed annually, then parts of geologic history could be measured in years. Such annual measurements are made possible by using two features: tree rings and glacial varves. Neither glacial varves nor tree rings can give time data for more than a few thousand years, however.

In trees, seasonal changes in supply of moisture and other factors cause the development of pairs of growth rings, each set representing one year. Some of California's redwoods in the Sierra Nevada foothills and northern Coast Ranges are more than 3,000 years old; older yet, by a thousand years or so, are the bristlecone pines that grow high in the Inyo Mountains.

Lakes associated with melting glaciers receive sediment made up of the fine "flour" that is ground from rock by moving ice. In summer, the

**4-4** *Quartz diorite at Bodega Head, intruded by white pegmatite dike. The dike has been offset by three small faults. The coarse granitic texture of the quartz diorite has been emphasized by differential weathering of the mineral grains. (Mary Hill photo, California Division of Mines and Geology.)*

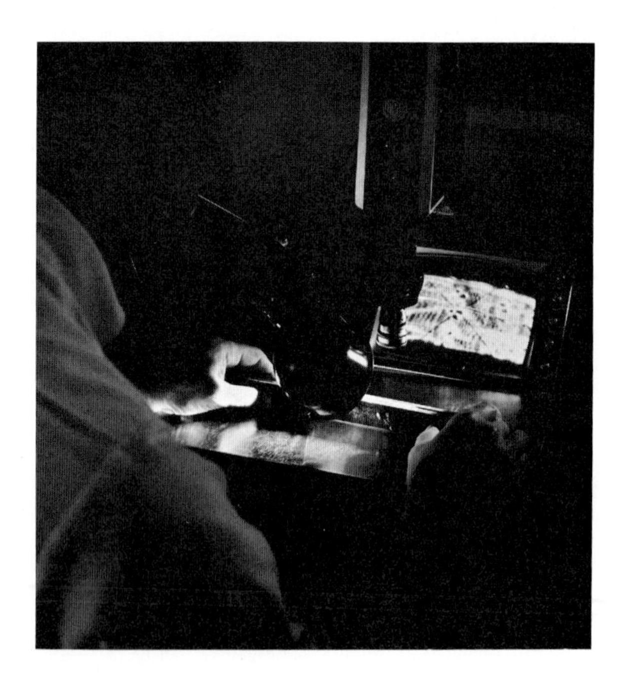

**4-5** *At tree-ring laboratory of U.S. Geological Survey, Washington, D.C., core of old tree is magnified and projected on television. Ring and growth patterns provide a "window" to past events. (U.S. Geological Survey photo.)*

coarser particles may settle in lake water and form a layer of sediment; in winter, a thinner, finer layer of sediment may slowly settle in water beneath the frozen lake surface. Counting the pairs of varves, as such layers are called, gives a measure of the life of an ancient lake in years.

## RADIOMETRIC DATING

Quite by accident, Henri Becquerel, a French physicist, discovered in 1896 that certain chemical elements, such as uranium, radium, and thorium, spontaneously emit radiation that comes from the breakdown of their nuclei. This property is radioactivity. Radiometric dating is the use of radioactivity to determine the ages of rocks in years.

Isotopes of atoms of an element differ slightly in mass (atomic weight) because of different numbers of neutrons within their nuclei. Some isotopes are stable; some are unstable and break down by natural radioactive decay. Rates of decay, which have been determined for many radioactive isotopes of the elements that occur in rock minerals, are usually given in terms of the half-life of an isotope—the time it takes in years for half

**4-6** *One of the oldest living things—a bristlecone pine 25 feet high—in the White Mountains, Inyo County. Tree rings show some to be as old as 4,000 years. (U.S. Forest Service photo, California Division of Mines and Geology,* Mineral Information Service, *October, 1958.)*

4-7 *Close-up of tree standing in soil layer buried by pumice fragments. Fluting in the tree represents scars of forest fire that burned the tree deeply before the pumice layer was deposited. Radiocarbon dating shows that the tree was burned, probably by the heat of volcanic eruption, about 1,100 years ago. This is in the Fouch pumice pit where pumice is being quarried for lightweight aggregate. (Charles W. Chesterman photo.)*

the atomic nuclei in an atom to break down. This time is a constant, not affected by chemical or physical changes. The absolute dating of rocks is based on this principle. We may now, for example, measure in a mineral the precise quantities of uranium 238 (the parent element) and of lead 206 (the daughter element) that have been derived from the uranium. We can then calculate the number of years *since the mineral crystallized.*

Uranium 238 is not abundant in minerals, but it is widespread, and its half-life is long. Table 4-1 lists the isotopes most commonly used for dating.

A great many minerals have been used for dating the rocks in which they occur. Some of those most often used are the micas, the feldspars, hornblende, pyroxene, zircon (zirconium silicate), and uranium and thorium minerals. Volcanic glass can also be dated.

Uranium isotopes are most commonly used for the older rocks, but potassium is extremely useful because of the many common minerals in which it occurs. The potassium-argon method has now been refined so that it yields dates as recent as 50,000 years ago. In materials younger than that, so little argon has been formed that analyses are not reliable.

An important development in radiometric dating for very young rocks was the discovery of the carbon 14 method in 1947. Carbon 14 is formed in the upper atmosphere as the result of

**TABLE 4-1 Isotopes used in radiometric dating**

| PARENT ELEMENT | DAUGHTER ELEMENT | HALF-LIFE (IN YEARS) |
|---|---|---|
| Uranium 238 | Lead 206 | 4,510,000,000 |
| Uranium 235 | Lead 207 | 710,000,000 |
| Thorium 232 | Lead 208 | 13,900,000,000 |
| Rubidium 87 | Strontium 87 | 47,000,000,000 |
| Potassium 40 | Argon | 1,350,000,000 |
| Carbon 14 | Nitrogen 14 | 5,730 |

the process by which cosmic rays from space bombard nitrogen 14, forcing it to accept an additional neutron into its nucleus, while emitting a proton. Carbon 14 is incorporated in all living things, and the rate of addition and disintegration stays in balance during the organism's life. After the death of an organism, no more carbon 14 is added, and that already present in the organism emits electrons and reverts to nitrogen 14. One-half of the original amount of carbon 14 present in an organism reverts to nitrogen 14 in approximately 5,700 years. This relatively short half-life permits dating of wood, bone, hair, shells, peat, and anything else that contains carbon. The carbon 14 method has been a great help to archaeologists as well as to geologists. Archaeologists have dated mummies, parchments, human bones, charcoal from fires, and Ptolemy's coffin! Dates obtained using the carbon 14 method have been checked with human cultures, tree rings, and glacial varves and are quite accurate for dates up to about 20,000 years ago. For dates between 30,000 and 50,000 years ago, the accuracy is less reliable.

**4-8** Fusulina—*a genus of foraminifera that was very abundant and widespread in late Paleozoic seas. Very useful in dating strata of late Paleozoic age. The largest of these is perhaps* $\frac{1}{8}$ *inch long.* (*Smithsonian Institution photo.*)

## FOSSILS

A long and abundant record of life (unfortunately, still with many gaps) through the past 600 million years has been preserved as fossils in the sedimentary rocks. We now know that life has evolved throughout geologic time, generally from the simpler to the more complex forms. Thus, rocks of different ages are different in the forms of life they retain as fossils.

Perhaps the first to recognize and systematically to use the differences of fossils in different kinds and ages of stratified rocks was the English surveyor William Smith (1769-1839). While digging canals, he noticed that certain fossils always occurred in certain rock layers. Soon he began to connect and relate one group of rock layers to another by means of the similarity of fossils they contained. This process of connecting and relating rock units to show that they are approximately equal in age is called correlation. Correlation of rock formations by fossils is one of the most important means of systematizing events in geologic history (at least, for the past 600 million years of history!). Baron Georges Léopold de Cuvier (1769–1832), a contemporary of Smith, was one of the first to study fossils systematically and to classify them according to their hard parts.

Life forms have flourished and disappeared from earth through geologic time, and fossilized remains of many forms provide significant clues in determination of geologic ages of rocks which contain them. Trilobites, which were marine organisms, lived, for example, only during Paleozoic time (600 million to 225 million years ago), the dinosaurs lived only in the Mesozoic Era (225 million to 70 million years ago), and the earliest fossils of horses are found in rocks approximately 60 million years old. Discovery of a fossilized trilobite external skeleton or horse's tooth can give some definite evidence for age of the rock in which it is found.

## GEOLOGIC TIME DIVISIONS

No discussion of events of earth history that are recorded in the rocks, fossils, and landscapes can be logically developed without a timetable of some sort. The need for such a timetable was realized about two centuries ago. Earliest efforts to classify rock formations by time were based on division of geologic time into Primary (earliest), Secondary, and Tertiary periods; later, the Quaternary period was designated for the youngest unconsolidated sands and gravels.

Succeeding generations of geologists have made many changes and have extended grouping of rock units by time to all continents of the earth. The modern time scale has been built upon those principles that are fundamental in geologic thought: (1) the recognition of constant change throughout the ages, along with the understanding that geologic processes operating today are the same as those that have always operated; (2) the principle of relative ages of layered rocks—the youngest are on top; and (3) the use of fossils for correlation and general age determination.

On this modern time scale, the four eras are the major units of geologic time, divided one from another by widespread and great changes in the earth's surface and life forms. The eras are divided into periods, and the periods of the most recent era are subdivided into epochs.

The geologic time scale is pieced together from geologic history gathered from all parts of the earth. Time divisions are always written down the page from youngest to oldest, just as if they were units in a section of stratified rocks. The geologic time scale is thus a composite geologic column for the earth as a whole. A time scale emphasizing California history is given as Table 4-2.

Geologic columns are often made diagrammatically to show the sequence of rock formations, rock types, and their ages in any one locality, such as southern Death Valley, the western San Gabriel Mountains, or the Berkeley Hills. For example, Figure 4-9 is a diagram of a geologic column, or columnar section, in the central part of the Mojave Desert. In this diagram, the first column shows the age of the rock formation; the second column shows the name of the rock formation; the third column ("lithology") shows the kind of rock by a standardized pattern; the "thickness" column shows the range of thickness of the formation in different parts of the area; and the next column is a combination letter symbol which shows the period (Q for Quaternary, T for Tertiary), and initial letter of the formation name (p for Pickhandle Formation).

A geologic column can also be constructed as a composite column for any large region, such as a natural province. In such constructions, all the rock formations shown are never found in any one spot; the geologic column in any given area is never complete.

**TABLE 4-2 Geologic time scale**

| ERAS | PERIODS, EPOCHS | TIME, IN MILLION YEARS | SOME GREAT EVENTS IN CALIFORNIA | LIFE ON THE EARTH |
|---|---|---|---|---|
| Cenozoic | Quaternary<br>Recent or Holocene | 0.01 | Continued faulting and mountain building | |
| | Pleistocene | 3 | Principal building of Coast and Transverse Ranges | Great land mammals; oldest man |
| | Tertiary<br>Pliocene | 11 | | |
| | Miocene | 25 | Local movements in Coast and Transverse Ranges; first movements on San Andreas fault | First apes |
| | Oligocene | 40 | | |
| | Eocene | 60 | Widespread coastal seas | First placental mammals |
| | Paleocene | 65 | | |
| Mesozoic | Cretaceous | 136 | Building of the Sierra Nevada, Klamath, and Peninsular Ranges | Extinction of dinosaurs |
| | Jurassic | 190 | | Age of dinosaurs |
| | Triassic | 225 | Shallow seas | First dinosaurs |
| Paleozoic | Permian | 280 | Volcanism and mountain building (extent unknown) | Rise of reptiles |
| | Pennsylvanian | 325 | | First reptiles |
| | Mississippian | 345 | Probably shallow seas over much of California, Cambrian to Permian | First land vertebrates |
| | Devonian | 395 | | Fishes abundant |
| | Silurian | 430 | | |
| | Ordovician | 500 | | |
| | Cambrian | 570 | | Trilobites dominant<br>First abundant fossils |
| Precambrian | Late | 1800 | Uplift<br>Mountain building in southern California | Organic tubes in marine limestone |
| | Early | 3800 | Oldest rocks and mountains on earth | Oldest fossils (algae?)<br>First life (?) |
| Crust of the earth solidified about 4,000 million years ago | | 4500 | Origin of the earth | |

| AGE | | FORMATION | THICKNESS | SYMBOL | DESCRIPTION |
|---|---|---|---|---|---|
| QUATERNARY | RECENT | SURIFICAL SEDIMENTS | 0-100' | Qs<br>Qa<br>Qc | Wind-blown sand<br>Gravel and sand<br>Clay and silt |
| | PLEISTOCENE | OLDER ALLUVIUM | 0-300' | Qoa | Gravel and sand |
| | | BLACK MTN. BASALT | 0-183' | Qb | Vesicular basalt lava flow |
| | | | | | UNCONFORMITY |
| | PLIOCENE? | LANE MTN. ANDESITE | 0-480' | Ta | Gray porphyritic andesite |
| | | | | | UNCONFORMITY |
| TERTIARY | MIOCENE | BARSTOW | 0-4500' | Tbfg | Fanglomerate of granitic detritus |
| | | | | Tbfv | Fanglomerate of volcanic detritus |
| | | | | Tbs | Stream-laid arkosic sandstone |
| | | | | Tbt | Thin beds of white rhyolitic tuff |
| | | | | | Thin beds of impure limestone |
| | | | | Tbl | Lacustrine clay and clay shale |
| | | | | Tbgb | Lense of granitic breccia |
| | | | | Tbc | Basal conglomerate (Owl Conglomerate Member)* |
| | | | | | UNCONFORMITY |
| | | PICKHANDLE | 0-2800' | Tpb | Granitic breccia and some intercalated white rhyolitic breccia. |
| | | | | Tpt | White to light colored tuff and tuff breccia. |
| | | | | Tpc | Conglomerate and fanglomerate of granitic and some volcanic detritus |
| | | | | Tb | Basalt lava flows |
| | | OPAL MOUNTAIN VOLCANICS* | OPAL MTN. VOLCS.* | Tri | Rhyolitic intrusive plugs |
| | | | | Trb | Rhyolitic flow breccia |
| | | | | Trp | Gray rhyolitic perlite |
| | | | | Tab | Porphyritic andesite breccia |
| | | JACKHAMMER* | 0-150' | Tj | Tuff, red arkosic |
| | | | | | UNCONFORMITY |
| TERTIARY OR CRETACEOUS | | QUARTZ LATITE | INTRUSIVES | ql | Porphyritic to felsitic quartz latite dikes. |
| | | BASALTIC PORPHORY | | bp | Black basaltic porphyry dikes |
| JURASSIC-CRETACEOUS &/OR OLDER | | PEGMATITE & APLITE | | p<br>a | Coarse grained pegmatite dikes<br>Fine grained aplite dikes |
| | | LEUCOCRATIC QUARTZ MONZONITE | | lqm | White fine to medium quartz monzonite |
| | | QUARTZ MONZONITE | | qm | Gray-white medium grained biotite quartz monzonite |
| | | GRANODIORITE | | grd | Gray-white medium grained biotite granodiorite |
| | | QUARTZ DIORITE | | qd | Gray medium grained quartz diorite |
| | | HORNBLENDE DIORITE | | hd | Dark gray to black medium to coarse grained hornblende diorite and gabbro |
| PRECAMBRIAN? | | WATERMAN GNEISSIC COMPLEX | 6000'± | gn<br>m | Gray quartz diorite gneiss and gneissoid quartz diorite<br>Lenses of white marble |

New stratigraphic names *

**FIGURE 4-9** *Example of a geologic column. Central Mojave Desert, near Barstow. (T. W. Dibblee, Jr., California Division of Mines and Geology Bull. 188, 1968.)*

# 5 interpreting geologic history from the rocks

*Faults and unconformities assist in interpreting geologic history from the rocks. Western San Gabriel Mountains, Los Angeles County (see Illustration 5-15).*

■ Every summer, thousands of visitors drive along U.S. Highway 395, which runs north-south east of the Sierra Nevada. Just south of Mono Lake and east of the highway is a string of volcanic domes and craters. Few people would fail to recognize these as volcanoes; the conical volcanic form, the craters partly filled by volcanic spines or plugs, and the thick stubby flows that extend onto the surrounding plateau all testify to their origin. Some of the flows extend over glacial debris and alluvium, so we know that they are geologically recent volcanoes, even though no one has seen them in eruption.

Quite different in origin and geology is the majestic composite volcano, Mount Shasta, which dominates the scene east of Interstate 5 as it crosses the northern Sacramento Valley and southern Cascade Range. Yet Mount Shasta, too, is readily recognized as a volcano.

In deciding on the origin of the Mono Craters and Mount Shasta in this manner, we are applying our knowledge of landforms. However, a geologist would not be satisfied about their origins until he had examined the rocks and had determined that they are volcanic.

Landforms like these last but a short time geologically; they are soon reduced by erosion and lose all semblance of their volcanic form. Evidence of ancient volcanism and of all types of igneous activity *is preserved only in the rocks.*

Most rocks exhibit evidence that suggests the conditions under which they were formed. Rocks, like people, reflect their heredity and environment. Not only are landforms changed by erosion or covered by later deposits, but mountain-building forces may cause folding, faulting, uplift, or depression to leave only the rocks themselves for study. Inherent features of the rocks that remain are (1) mineral and chemical composition, (2) texture (the size and arrangement of the mineral grains), and (3) structure and surface features.

In the drawing on this page the geologist sees at least three rock formations: (1) a lower, tilted, stratified unit, (2) overlying horizontal strata, and (3) unstratified exposures at top and upper right. Field examination and a study of geologic relationships in the area permit interpretation of the history outlined on page 89.

## IGNEOUS ROCK STRUCTURES

By definition, igneous rocks have been formed by the cooling of molten magma, either above or within the earth's crust. One of the structures most often preserved is flow. Unmistakable flow structure is best preserved when molten lava erupts quietly to flow along the surface of the earth as it cools and is therefore useful in recognizing that a rock was formed as a lava flow. However, movements of molten magma below the earth's surface may be "frozen" to preserve flow structures. Flow structure, in itself, is not conclusive evidence of an external lava flow; flow may be present in a granite intrusion.

### mono craters

The Mono Craters in Mono County, so striking to the traveler on Highway 395 east of the Sierra Nevada, are a series of conspicuous volcanic domes that consist mostly of obsidian—volcanic glass. They were formed as viscous obsidian plugs rising above the floors of shallow explosion pits. The domes project above a plain formed of loose pumice ejecta ranging from shot-sized particles to angular blocks several feet thick. Chemical composition of this material averages about 70 percent silica, so the rock is a rhyolite.

Why is there so much explosive material in this area, and why is it mostly rhyolitic glass? All magmas at depth contain quantities of dissolved gas, most of which is water vapor. In fact, the gas content is important in supplying some of the lift that brings magma to the surface. Because high-silica (70 percent or more $SiO_2$) magmas are viscous and become solid at relatively high temperatures, the contained gases tend to be held and released explosively. Rapid "freezing" allows no time for individual crystals to form, so the lava chills as a glass (a solid substance without crystal structure). A basaltic lava, with a composition of perhaps 50 percent silica, freezes at a lower

**5-1** *Mono Lake and Mono Craters, Mono County. Looking northeast at white Paoha Island from the Sierra Nevada. Paoha consists of volcanic explosive rocks, flows, and older lake sediments, including diatomaceous earth. Black Negit Island (left of Paoha) consists of dark red volcanic fragmental rocks in slightly eroded craters. (Charles W. Chesterman photo.)*

**5-2** *From the Carson Spur on State Highway 88—the Carson Pass Highway across the Sierra Nevada. Horizontally stratified Late Tertiary agglomerate and pyroclastic rocks here form vertical cliffs. These volcanic rocks lie on the eroded surface of Cretaceous granite, exposed in the canyon of the Consumnes River below. (Mary Hill photo.)*

temperature, is less explosive, and only rarely forms glasses.

In the Mono Craters, the volcanic rock structures characteristic of rhyolite are exemplified to the extreme: fragmental structure (large blocks, crags, and jumbled masses of collapsed spines, as well as dust-, and ash-, and pea-sized glassy particles—the last called lapilli), flow structures preserved by instant chilling while lava was in motion, and vesicular structures formed when escaping gases left holes (vesicles) and tubes. Similar rocks in short, stubby flows of this sort are also strikingly developed in the Glass Mountain area of Siskiyou County in northernmost California.

**5-3** *Red Cinder Mountain, Inyo County. A beautifully symmetrical cinder cone, formed by explosion of red vesicular fragments (cinders). The completely uneroded surface of the volcanic cone shows its recent origin. The dust cloud and piles of cinders near the base of the volcano mark the site of a quarrying operation for materials used in concrete building blocks and stucco. Dark-colored rock in the foreground is a basalt flow. (Charles W. Chesterman photo.)*

## lower-silica rocks and their features

Low-silica rocks like andesite (often about 60 to 65 percent silica) and basalt (50 percent silica) are also often vesicular and preserve excellent flow structures. Flows of basalt are often highly fluid when first erupted and may extend widely from the ruptured flanks of volcanoes or directly from fissures in the earth's surface. Pliocene lava in the flows of Table Mountain in Tuolumne County may have moved as much as 70 miles down the gentle western slope of the Sierra Nevada.

Long after the volcanic landform is gone, the characteristic structures are left in the rocks to tell the story. At the Pinnacles, in the southern Coast Ranges about 90 miles southeast of San Jose, rhyolite and andesite of Miocene age preserve many of these telltale structures that disclose their origin, although no volcanoes remain. In the Berkeley Hills, the Grizzly Peak volcanics (of

Pliocene age) include vesicular basalt with the vesicles filled with chalcedony, opal, and other minerals of interest to collectors.

At Amboy Crater in the Mojave Desert, some thin flows have extended over desert alluvium. Blisters, or huge bubbles, several feet or yards across have developed on the flows, splitting as they cooled. Some of the fragments at Amboy Crater are volcanic bombs—head-sized, melon-shaped, smooth fragments twisted at the ends, which represent masses of lava cooled as they fell through the air. Grotesque-appearing spatter cones, a few feet high, are common structures. Amboy Crater is readily seen and reached from U.S. 66 about 70 miles east of Barstow.

In Lava Beds National Monument in Siskiyou County and in Hat Creek Valley in Shasta County, beautifully preserved structures in basalt include flow structures, spatter cones, and extensive caves and tubes. Lava tubes are most commonly formed when the surface of a basalt highly charged with gas crystallizes and the still-molten lava within continues to move on. Lava of this sort forms smooth, ropy surfaces called pahoehoe after the Hawaiian lava of this kind. Less commonly, basalt flows develop a clinkery surface called aa.

### columnar jointing

Varied structures form as lava cools. One of the most striking, developed to a spectacular extent at the Devils Postpile National Monument in the gorge of the upper San Joaquin River in the high Sierra, is columnar jointing. Joints are cracks that

**5-4** *Entrance to a partially collapsed lava tube (or cave) in Lava Beds National Monument, Siskiyou County. This is one of nearly 300 tubes in the area. Such tubes form when the surface of a river of basaltic lava chills and the more slowly cooling interior continues to flow. The rock is Quaternary basalt. (Charles W. Chesterman photo.)*

**5-5** *Columnar jointing in a Quaternary lava flow, Devils Postpile, in narrow canyon of the upper San Joaquin River, Sierra Nevada. (California Division of Mines and Geology,* Mineral Information Service, *June, 1965.)*

form patterns. Here, four- to six-sided columns developed in a uniformly cooling andesite lava flow about a million years ago. The lava, about 600 feet thick, contracted about centers as it cooled, forming joints that make a polygonal pattern. The columns are from a few inches to several feet across and some are on the order of 100 feet long. The long axis of each column is perpendicular to the cooling surface.

### structures in granite

Magma, cooling below the surface of the earth, tends to develop more massive structures without the variety of forms found in surface volcanism. Flow structures are preserved in granite, for example, but the banding is less conspicuous and may easily be confused with banding developed by metamorphic processes. Perhaps the most spectacular structures found in the massive granitic rocks are joints—fractures, or incipient fractures, that form more or less regular patterns.

Nowhere are joints better seen than in Yosemite National Park in the Sierra Nevada. The great dome surfaces (like Half Dome, North Dome, and Basket Dome) were formed by curved jointing; that is, the granite peeled off in layers like those of an onion. This type of jointing may be due to relief of load as thousands of feet of overlying rock have been removed. Several other joint systems exposed in the valley consist of plane surfaces in at least three directions: horizontal and two vertical at about right angles. A nearly horizontal joint system is exposed at the base of Liberty Cap. The vertical joints that cut Half Dome in half and the vertical walls of Yosemite Valley are great planar joint systems. Joint planes at still other angles, enlarged by the erosive action of running water and ice, account for other rock forms, like the Three Brothers.

## IGNEOUS ROCK TEXTURES

Texture—the size, shape, and arrangement of mineral grains in a rock—tells us much about the history of formation of an igneous rock.

**5-6** *Upper (eastern) end of Yosemite Valley, Mariposa County. Half Dome, upper left center, appears as white area that casts a heavy shadow toward Tenaya Canyon. Nevada Fall, on the Merced River, is about $1\frac{1}{2}$ inches to the right of Half Dome. Vernal Fall, lower down on the Merced River, is about 1 inch lower on the photo. Tenaya Creek joins the Merced River at the east end of Yosemite Valley; the Merced runs toward the lower right of photo. The 3,000-foot-high vertical cliff of Half Dome was formed by a prominent joint system and by erosion by the Tenaya Glacier. A glacier moving down upper Merced Canyon left the huge glacially eroded staircase marked particularly by Nevada and Vernal Falls. One of the most striking features of this photo is the extraordinary pattern of jointing in the granitic rocks, which played such an obvious, controlling role in sculpturing the area. (Vertical aerial photo, U.S. Geological Survey, Prof. Paper 590, 1968; scale approximately 1 inch equals 5,200 feet; east is at top.)*

### fragmental and glassy textures

In discussing structures, we have already seen that instantaneous chilling of lava may develop a rock of glassy texture—that is, with no mineral crystals. Rapid cooling of particles of lava thrown into the air favors the development of glassy textures, which may also be formed as a lava flow, sill, or dike is chilled against rock already in place. In general, fine-grained and glassy textures develop in extrusive volcanic rock and in the thin sheetlike sills and dikes. Coarse-grained textures develop when slow cooling takes place.

Any rock composed of volcanic fragments is said to have a fragmental texture. Examples are tuff, made up of dust-sized to pea-sized particles; volcanic breccia, made up of larger angular fragments; and agglomerate, made up of coarse, rounded fragments. Some breccias form as the crust of an active lava flow breaks up and forms again, as well as along the front of a flow. Others form, sometimes explosively, when a hot flow enters the sea.

### textures of intrusions

Great masses of magma that cool for many thousands of years below the surface of the earth form rocks like granite, which usually have coarse, even-granular, or *granitic,* textures in which the outlines of the mineral grains can be clearly seen. As they slowly cool, magmas within the crust

**5-7** *Upper Kern River area, near Bearpaw Meadows, high Sierra Nevada. Note glacial polishing of rocks in the foreground. Avalanche chute appears as white scar left of center of photo. Vertical jointing in granitic rock has produced the prominent cliffs right of center.* (Mary Hill photo.)

retain their gases and fluids for long periods of time; this is also a factor in developing large crystals. In fact, some magmas that are especially high in fluids and crystallize late in the cooling history develop giant crystals—quartz, feldspar, and mica up to several feet in length in some instances. Coarse-grained rocks of this composition are granite pegmatite.

### porphyritic texture

In a great many lava flows and in small irregular intrusions as well as in dikes and sills, the texture of the rock formed is not glassy, fragmental, or granitic. Such flows are characterized by porphyritic texture, in which well-formed mineral crystals appear clearly in a fine-grained or glassy matrix, or groundmass. Conditions of pressure, temperature, and composition of a cooling magma were such that some mineral crystals solidified first; later, relatively rapid solidification of the whole mass took place to form solid rock. The porphyritic texture is common, because it is not unusual for cooling to occur in several stages, and different minerals separate out and crystallize at different temperatures. In particular, large crystals

**5-8** *Close-up of banded obsidian, Mono Craters, Mono County. The banding was probably developed as quick freezing took place during flow. (Charles W. Chesterman photo.)*

**5-9** *Perlite in the Castle Mountains, San Bernardino County. Perlite is a glassy rhyolite in which concentrically layered spherules, or "pearls," are formed. Perlite contains water, which causes the rock to expand on heating to form a glassy froth that is useful as a lightweight aggregate. The pencil is 4 inches long. (Charles W. Chesterman photo.)*

may form at depth, the magma may then be extruded, and the remaining liquid may then solidify as a fine-grained or glassy groundmass.

## SEDIMENTARY STRUCTURES

Why should sedimentary rocks be stratified? They have all been formed, of course, by consolidation of sediments which have settled out of water or air. Particles of different size, weight, and density settle out of a fluid at different rates. Observe any stream and note the vastly different amounts of rock fragments it carries at different times, the differences in the sizes and shapes of fragments, and the differences in volume of water, and velocity of the currents.

### layering

One of the most common structures of sedimentary rocks is layering, or stratification. Individual layers, strata, or beds may range from the thinnest lamellae (thin plates) to many feet thick. Bedding is such a common characteristic that when we see well-stratified rock, we think of it as sedimentary—not igneous or metamorphic. As a general rule, this evaluation would be correct; however, the layering produced by flow in igneous rocks and by the pressures of metamorphism can simulate sedimentary stratification.

**5-10** *Pyroclastic (explosive, fragmental) volcanic rocks of Late Tertiary age on the slopes of Leavitt Peak, Sonora Pass on State Highway 108. (Mary Hill photo.)*

As the velocity decreases, a stream in flood may deposit a layer of gravel first, then sand beds, and lastly silt and clay. So, in addition to stratification, there is also sorting—that is, the rock fragments generally tend to arrange themselves according to size. Similarly, as sediments are deposited in a lake or the sea, they tend to become sorted and layered.

### graded bedding

Imagine a stream in flood season dumping a load of unsorted sediment into the quiet waters of a lake—a common occurrence. The coarsest particles settle first, so that we may find, from bottom to top, strata of gravel, sand, and mud; therefore, within a given bed, or stratum, the coarser particles have settled a little ahead of the finer. Thus, a single sand bed may show a gradation from coarser sand at the bottom to finer sand at the top. This graded bedding, like other sedimentary structures, may be and usually is well preserved in the solid rock. Rock formations involved in mountain building may literally be so intensely folded that they become overturned. This situation can be readily recognized by the field geologist if he finds any graded bedding, because he knows that the finer grains in a graded bed were originally deposited on top.

### turbidity deposition

Particularly turbid currents—currents loaded with sediment—may deposit relatively coarse materials many miles offshore in the ocean. Such currents also perform erosion of soft sediments under

**5-11** *Stratified sedimentary rocks in the southern Mother Lode, on State Highway 49 near Bagby, Mariposa County. Slightly metamorphosed sandstone and shale of the Upper Jurassic Mariposa Formation. (Mary Hill photo.)*

**FIGURE 5-12** *Sedimentary structures formed by turbidity-current deposition and erosion in seawater about 300 meters deep. Preserved in sandstone (dot pattern), shale and mudstone (dash pattern), and pebbly sandstone (dots and circles) of the upper Pliocene Pico Formation, exposed in Santa Paula Creek, Ventura County. Fossils have enabled paleontologists to judge the depth and temperature of water. (California Division of Mines and Geology Special Report 89.)* (a) *A, scour channel at base of pebbly sandstone; B, overfold of mudstone; C, layer of pebbles along bedding plane; D, shells, mostly concave up, within a sandstone bed; E, shells, concave down; F, mudstone slabs, broken by erosion and redeposited; current moved from left to right.* (b) *Lower units I and II show cross-bedding, which dies out upward into curved laminations. Unit II is topped by two domes, with their flanks and the trough between them capped by a dark clay-silt layer. Unit III consists of cross-bedded sandstone at base, grading upward into curved laminations in mudstone and isolated load pockets.*

water and then redeposit them. Some of the great variety of minor sedimentary structures formed under the varying turbidity conditions are shown diagrammatically in Figure 5-12*a*. Turbidity currents distribute great amounts of sediments over extremely wide areas on the continental shelves and even beyond the shelves to the bottom of the continental slopes.

Because sediments are deposited in basins of one sort or another, all sedimentary beds and formations *must have horizontal limits and must be lenticular in form,* albeit in very flat lenses.

### cross-bedding

We might expect most strata to be relatively parallel, level, and flat (unless deformed by folding or faulting); however, there are many exceptions. Within nearly flat beds of sand and gravel deposited by a stream, the thinner layers and laminae often lie distinctly at various angles to the major stratification (Figure 5-12*b*). This

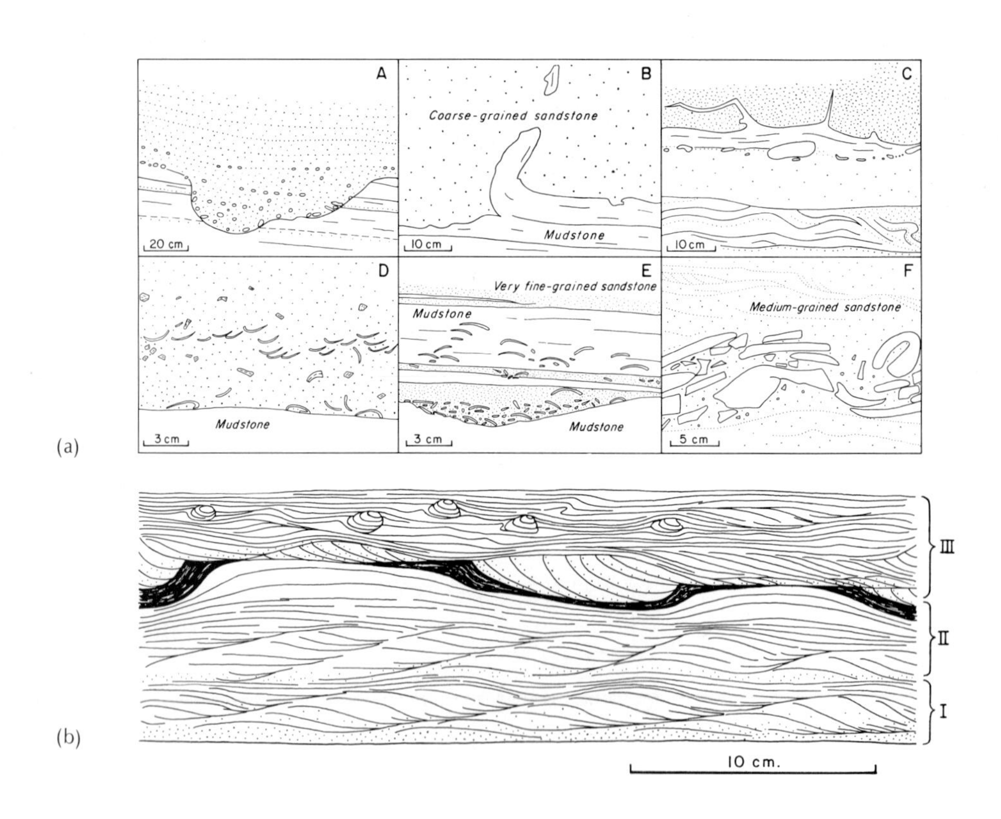

cross-bedding is the result of local, irregular processes of cut and fill in stream channels, alluvial fans, deltas, and sand dunes. Look at the sands and gravels in almost any stream bank at low water. A layer of gravel has been deposited; the next rush of water scoured a channel in it, then as the current slowed again, layers of sand were deposited in the cut channel.

Deltas are formed under water by the deposition of sediment as the currents of a stream enter a quiet body of water. Beds at the top of the delta are nearly flat, as are the beds far out at its base; but the foreset beds on the sloping delta front are deposited on slopes at varying angles. Consider sand dunes slowly marching with the prevailing wind. The sands are swept up the gentle beachward slopes of the dunes and deposited on steep leeward slopes.

Cross-bedding preserved in the rocks suggests that such rocks were formed where there was rapid and varied current action and changing slopes of deposition, probably in shallow water but perhaps under wind as it shaped dunes. The form of the cross-beds—gentle slopes toward the current, steeper slopes away from the current—enables us to reconstruct current directions in sandstones of past geologic periods.

## ripple marks, mud cracks, and imprints

Whenever currents of water or wind gently move sands to and fro, ripple marks develop. Under favorable conditions, they are preserved in solid rock.

**5-13** *Mud cracks formed in a desert playa when fine-grained sediments were dried. (Dorothy Radbruch photo, California Division of Mines and Geology,* Mineral Information Service, *May, 1967.)*

**5-14** *Desiccation cracks on the dry bed of Owens Lake, Inyo County. Irregular patterns of polygons are characteristic; one large polygon may enclose many smaller ones. Some of the largest rectangular polygons are about 1,000 feet long. (Vertical aerial photo, U.S. Geological Survey, Prof. Paper 590, 1968; scale approximately 1 inch equals 4,210 feet;* south is at top.)

In the beach sands and dunes along California's coastline, ripple marks are always alternately forming and being destroyed by erosion. They are also beautifully shown on the surfaces of dunes of the desert, like those in Death Valley, and the Algodones Dunes east of the Salton Sea. Ripple marks are also formed by gentle currents in rather deep seawater—wherever there are currents and fine sediments.

Currents that oscillate back and forth form symmetrical ripple marks. In currents of one prevailing direction, ripple marks are asymmetrical, with the gentler slope upstream. Ripple marks preserved in sedimentary rocks indicate current directions and are also useful in determining whether beds are rightside up or overturned.

Mudflats are often exposed in nature—on river floodplains, in dry river channels, in lagoons and estuaries at low tide, in dry lakes, and at low-water time along any shore line. Muds and clays dry out, contract, and crack to form mud cracks; animals walk across the mud and leave their footprints; leaves and seeds settle in the mud and either are preserved or leave impressions. All these features can be preserved in rock—usually in fine-grained rock like shale. Even raindrop imprints have been found in some shales!

## HISTORY FROM METAMORPHIC ROCKS

Metamorphic rocks are usually formed under conditions of high pressure and temperature, often accompanied by chemical changes. They are formed from all rock types, including igneous, sedimentary, and older metamorphic rocks. Metamorphism takes place over extended belts of the earth's crust during great periods of mountain building. In connection with mountain-building processes, belts of sedimentary rocks, for example, may be downbuckled to great depths where pressures are high and then folded, faulted, and elevated to thousands of feet above sea level. Great pressures may develop, not only because of the enormous load of overlying rocks, but also from tectonic forces—the pressures of folding and faulting.

Thus, the location of belts or zones of metamorphic rocks shows *where* ancient mountain ranges were built, and radiometric dating of the metamorphic minerals shows *when* the mountain building took place.

In the heart of the Coast Ranges, California's Franciscan Formation of Mesozoic age contains some unusual schists and massive metamorphic rocks. Much experimental work on the conditions under which certain minerals are formed enables us to draw some conclusions about environment of formation. The jadite-glaucophane-lawsonite-bearing rocks of the Franciscan Formation were formed under high pressures—several thousand times sea level atmospheric pressure—perhaps in narrow oceanic troughs at depths of 10 to 12 miles and at relatively low temperatures for those depths (about 200° to 300°C). These conditions would have been met if the sedimentary rocks of the Franciscan Formation were carried down to such great depths very rapidly and were so rapidly uplifted that higher temperatures did not develop in the trough.

Because metamorphic minerals are formed in adjustment with their environment, geologists are able to reach some conclusions concerning important events in geologic history.

## UNCONFORMITIES

In some places, thousands of feet of sedimentary rock lie in parallel layers without a break. Such an area is found on the northeast side of Mount Diablo, where at least 30,000 feet of Upper Jurassic and Cretaceous marine sedimentary rocks form an uninterrupted, or conformable, series. There are similar successions elsewhere in California, but nowhere is there a series of sedimentary rocks spanning any large part of geologic time without gaps. The reason is, of course, that sedimentation cannot go on forever in any region without a change in conditions. The breaks or gaps in the

**5-15** *Late Pleistocene stream gravels deposited on Cretaceous granitic rocks. This unconformity has been exposed in the past few thousand years by downcutting by Pacoima Creek in the western San Gabriel Mountains, Los Angeles County. (Charles W. Jennings photo, California Division of Mines and Geology.)*

record are due to interruptions in the sedimentation cycle, particularly due to uplift and succeeding erosion. Such a gap in the sedimentary record is called an unconformity. Unconformities can tell us something about important events in geologic history.

Let us analyze a few examples to see what interpretations can be made. Each sketch represents a geologic cross section through rock formations near the surface, such as might be exposed in a sea cliff or the steep side of a canyon.

We can deduce the following sequence of events from the sketch in Figure 5-17*a*, in order from oldest to youngest:

**1** Mud, clay, and sand were deposited below sea level and consolidated as the Monterey Formation (shale and sandstone).
**2** Uplift and tilting (folding) of the Monterey Formation exposed it to wave erosion. The surface of erosion represents the unconformity.
**3** There is evidence of wave erosion and deposition of gravel and sand (terrace deposits, *Qt*) across the folded edges of the Monterey shale.
**4** Emergence resulted in a terrace and exposed the rock formations to erosion. From the diagram, we cannot be sure that uplift exposed the terrace; perhaps sea level was lowered.

The angular unconformity ("angular" because the terrace gravels are at an angle with the older Monterey beds) represents a gap in time from the

**5-16** *Conglomerate of the early Pliocene Neroly Formation lying on the irregular erosion surface of marine Upper Cretaceous Panoche Shale, east end of Altamont Pass in the Diablo Range, San Joaquin County. The man is sampling east-dipping Panoche Shale; the unconformity is about 4 feet above him and dips to the right. The Neroly conglomerate is made up of rounded pebbles of andesite, probably transported by streams from the Sierra Nevada. (Charles W. Chesterman photo, California Division of Mines and Geology.)*

**FIGURE 5-17** (a) *Given exposure: Seaward-dipping shale (dash pattern) and fine-grained marine sandstone (dot pattern) of the middle Miocene Monterey Formation* (Tm), *Santa Barbara coast, overlain by marine conglomerate* (Qt—*circle pattern*) *of late Pleistocene age.* (b) *Given exposure: Flat-lying Pliocene shale beds containing marine shells overlain unconformably by modern stream gravels and sands. Santa Cruz coast.* (c) *Little Tujunga Canyon, Los Angeles County, western San Gabriel Mountains. Given:* Granite (Kgr) *of Cretaceous age (dated at 70 million years) faulted against land-laid sandstone and conglomerate of the lower Pleistocene Saugus Formation* (Qs) *and the upper Miocene marine Modelo Shale* (Tm). *Stream terrace gravels* (Qg) *deposited in late Pleistocene time unconformably across Saugus Formation and granite. The fault is later than the Saugus Formation since it displaces the Saugus and all older formations; the fault is older than the terrace gravels since it is overlain by the latter.*

end of the Middle Miocene Epoch to the beginning of the late Pleistocene. Something like 18 million years is missing here; we don't know precisely what went on during that time.

From the sketch in Figure 5-17*b*, we can infer the following events:

**1** Marine muds and clays of lower formation were deposited in Pliocene time.
**2** Uplift or change in sea level exposed the shale unit to stream erosion.
**3** Recent stream gravels were deposited on the eroded surface—an unconformity, but not angular. The sedimentary record, at least for Pleistocene time (perhaps 3 million years), is missing and unknown here.

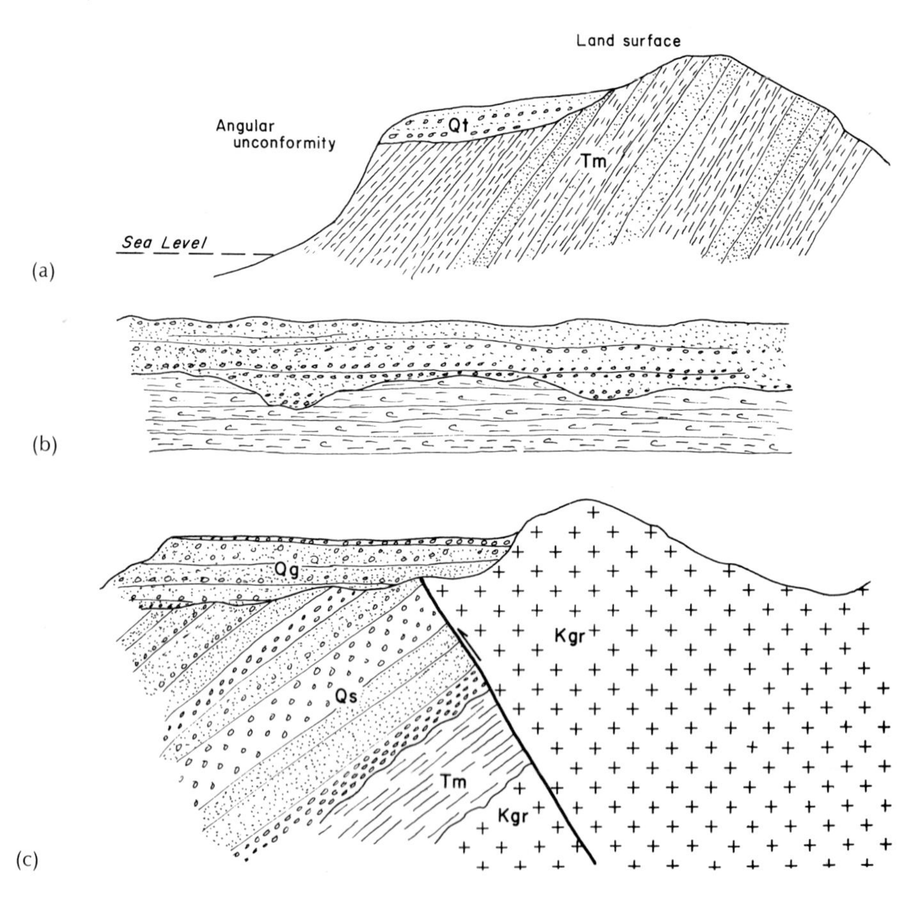

From Figure 5-17c, we can deduce the following interpretation of history:

**1** Granite was intruded in late Cretaceous time (80 to 90 million years ago), probably accompanying mountain building; erosion of the mountain range occurred.
**2** Seas advanced and fine-grained sediments (Modelo Shale) were deposited unconformably on the eroded surface of the granite. This unconformity represents more than 50 million years during which we know little of what happened. (Was the area above sea level?)
**3** There was uplift and erosion before sands and gravels, perhaps stream-laid, of the Saugus Formation were deposited.
**4** Faulting left the granite uplifted on the right and thrust over the Modelo and Saugus Formations. Tilting, or folding, of the Modelo and Saugus Formations together was followed by renewed mountain building, including faulting and folding in mid-Pleistocene time (post-Saugus and pre-terrace).
**5** Finally, the modern surface was developed by uplift and erosion.

In any one locality, there are always great gaps in the geologic history in the rocks recorded. However, somewhere, often not far away, there are rock formations that fill the gap. It is reasonable, of course, that erosion in one locality should be accompanied by deposition in another.

This is the way earth history must be reconstructed—by putting together bits and pieces of the record here and there and correlating parts of rock sequences to fit together the whole.

Graphic methods in common use to show rock formations in their proper stratigraphic and age relationships and to show structural features include principally geologic maps, geologic columns, and structure sections. Constant reference will be made in this book to geologic maps. Detailed examples of a geologic column and structure section are shown in Figure 4-9 and Figure 16-12, respectively.

□ The dim, far dawn of California's geologic history has left its meager records in Death Valley and in the heart of the rugged San Gabriel Mountains where the state's oldest rocks are exposed—rocks that were formed nearly 2 billion years ago. Thousands of feet of marine stratified rocks show that ancient seas covered the Death Valley area, and sediments were deposited then, as now, in the margins of the oceans. Remains of algae, bacteria, and wormlike creatures preserved as fossils record the life of those seas. Granitic rocks imply that mountain ranges—of which we know little—were built; sedimentary and volcanic rock sequences and erosional gaps show that the mountain ranges were reduced by erosion and were reconstructed by uplift, folding, and volcanism—again and again.

Six hundred million years ago, at the dawn of the Paleozoic Era, seas advanced from the west over Precambrian rock formations; and, from that time, rocks and fossils record in increasing detail the complex history of California's lands and life.

Rock formations of the Mesozoic Era—from 225 million to 70 million years in age—testify to the building of the structural framework of the state. This was the Age of Reptiles, with its primitive evolving land plants and its seas teeming with most of the ancestors of today's life, as well as many forms long since extinct. Toward the close of this era, enormous masses of granite magma invaded the mountain ranges that covered most of the state; and the Great Valley was depressed as a gigantic trough between the rising mountain chains of the Sierra Nevada and the Coast Ranges.

About 70 million years ago modern life and land began to emerge, and California began to assume the outlines and landscapes familiar to us today. An uneasy crust and changing landforms in geologically dynamic California today are evidenced by active volcanoes, slipping faults, earthquakes, landslides, uplift and subsidence of the land, and spectacularly high and rugged mountain ranges.

# II CALIFORNIA THROUGH THE GEOLOGIC AGES

# 6

# evolution of the crust of the earth, and california's oldest rocks

*Erosion exposes some of California's oldest rocks. Precambrian rocks in Surprise Canyon, Panamint Mountains, with a glimpse of Panamint Valley down canyon toward the west.*

## THE PRIMORDIAL EARTH: CONTINENTAL PLATFORMS AND OCEAN BASINS

During the past sixty years knowledge of the structure and characteristics of the interior of the earth has grown as more direct evidence from the passage of earthquake waves has become available. Seismologists have demonstrated that the earth is distinctly layered, with more or less sharp boundaries between layers (Table 6-1). There is an inner core with a density of about 13 grams per cubic centimeter and a radius of some 1,216 kilometers (1 kilometer equals 0.62 mile). The characteristics of earthquake waves that pass through it reveal the density and the essentially solid character of this inner core. It is surrounded by a transition zone (455 kilometers thick) and then by the dense liquid of the outer core, which is about 1,800 kilometers thick. The mantle, which surrounds the core, is about 2,880 kilometers (nearly 1,800 miles) thick and is essentially solid; its density (average $4\frac{1}{2}$ times that of water) and rigidity increase with depth from the surface of the earth. The mantle is also divisible into several layers of differing density and rigidity. The outermost, very thin layer or shell is the crust, a term held over from the days when scientists visualized a still-molten interior for the present-day earth. Average density of the crust is about 2.8 grams per cubic centimeter and of the whole earth is 5.5 grams per cubic centimeter.

Many astronomers and geologists believe that 4.5 to 5 billion years ago, our solar system was a huge cloudlike mass of gaseous and dustlike material in which the planets, including the earth, were forming by gravitational attraction around the denser centers. As the earth grew by addition of solid materials, separation of denser from less dense materials occurred, under the influence of gravitational forces, to form a layered earth with an essentially metallic core, a high-iron–magnesium silicate mantle, and a high-silica crust. We can reasonably postulate that a thin granitic-rock (high-silica) crust formed very early in the earth's history, perhaps 4 billion years ago. This thin crust must have been broken repeatedly by volcanic activity, and irregular portions of the granitic rock were engulfed and remelted countless times over wide areas. Remelting in the lower parts of the earth's crust and in the upper mantle still goes on today to form bodies of molten rock, or magma, which may solidify higher in the crust as bodies of igneous rock or reach the surface in volcanic activity. This all requires enormous quantities of heat, some of which may be heat retained from the molten earth. However, the never-ending process of radioactive decay of elements is sufficient to account for the total of the earth's heat release.

Viewing the earth as a whole, its greatest features are certainly the vast depressions of the ocean basins, which cover more than two-thirds of the surface of the globe, and the high-standing continental platforms. The continental platforms, which are made up essentially of low-density, high-silica granitic materials, began to form as the initial thin crust of the earth broke up irregularly and was rafted together to form the nuclei of primitive continents. These early continental masses were liable to migration over the molten earth. A fascinating controversy in geology today is the question of how much and to what extent such continental drift has gone on through later geologic time and is going on now.

Judging from the gases that come from molten rock and volcanic activity today, the atmosphere over the molten earth probably included such substances as water ($H_2O$), carbon dioxide ($CO_2$), sulfur (S), nitrogen ($N_2$), methane ($CH_4$), and ammonia ($NH_3$). We know little of what the proportions were. As the crust formed and the surface of the earth cooled, water could become liquid. About this time the first simple plants appeared. The process of photosynthesis, by which the sun causes green plants to form and liberate oxygen ($O_2$) from $CO_2$ and $H_2O$, began to build up an oxygen atmosphere. Geologists know that by middle Precambrian time there was abundant oxygen, for there are enormous bodies of iron-oxide rocks and other oxides in the Precambrian materials in many places over the earth.

Thus, the stage was set for the beginnings of recorded geologic history: continental platforms that were being weathered and eroded by water and atmosphere above sea level, ocean basins along whose margins the sediments formed by erosion were being deposited, and volcanic activity furnishing lavas and fragments to form layers on the land and layers within accumulating sediments. In such a setting, the first primitive forms of single-celled life developed on the lands and at the margins of the seas.

**TABLE 6-1 Structure of the earth's interior**

| DEPTH (KM) | LAYERS | | CHARACTERISTICS |
|---|---|---|---|
| 0–10 | Crust | Ocean and crust | (oceanic region) |
| 0–40 | Crust | Crust | (continental regions) |
| 40–350 | Upper mantle | | Solid but includes a "soft" low-velocity zone |
| 350–410 | Upper mantle | Transition zone | Solid; rapid increase in rigidity |
| 410–640 | Upper mantle | Intermediate zone | Solid; mild increase in rigidity |
| 640–720 | Upper mantle | Transition zone | Solid; rapid increase in rigidity |
| 720–2,700 | | Lower mantle | Solid; homogeneous region |
| 2,700–2,894 | | Transition zone | Solid; some reduction in the rate of increase of rigidity |
| 2,894–4,700 | | Outer core | Liquid |
| 4,700–5,155 | | Transition zone | ? |
| 5,155–6,371 | | Inner core | Perhaps solid |

## OUR MOBILE CRUST

Studies of the character and velocities of earthquake waves have now pretty well defined the base of the crust as being the Mohorovicic discontinuity (Moho) where rock densities and wave velocities abruptly increase with depth. Rocks in the upper mantle, below the Moho, seem to be high in iron and magnesia and relatively low in silica—the peridotites and serpentine.

There are great differences between oceanic crust and continental crust. Oceanic crust lies under an average depth of 2.5 miles of water and is only about 3 to 5 miles thick. Earthquake-wave data, a few drilled holes, and dredgings show that the top crustal material is basalt, with serpentine below. On top of the basalt in the deep oceans, are very thin layers of fine sediments—muds, oozes, and organic matter.

In great contrast, the crust of the continental platforms and their margins ranges from 12 to 35 miles thick and averages about 20 miles. The crust of the continental platforms is made up principally of granitic rocks and sedimentary rocks derived from them, all relatively high in silica and low in density. Under mountain ranges, which often have great masses of granite at their cores, the crust may be thicker by several miles. The Sierra Nevada, for example, has such a "root." It is as if deep-rooted, high-standing columns of rock of the mountain range were balanced against shallow-rooted, low-standing rock columns underneath the deep marginal valleys. This sort of isostatic (equal standing) equilibrium does, in fact, exist. The lighter crust floats on the denser mantle.

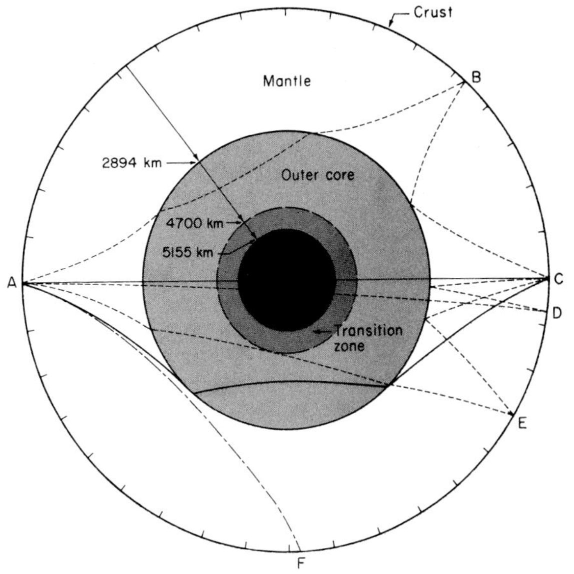

**FIGURE 6-1** *Interior of the earth, as inferred from the recorded passage of earthquake waves. Assume an earthquake takes place at A. Then about 90 degrees away around the earth the earthquake is recorded directly at seismograph station F; this P-wave has moved through the mantle only. A slightly steeper wave from A that happens to be just tangent to the outer core is refracted to station C at a distance of 180 degrees around the earth. Other lines show other ray paths as they have actually been recorded in earthquakes (see also Figure 10-11). (Diagram from Bruce A. Bolt, Professor of Seismology, University of California, Berkeley.)*

## THE THEORY OF PLATE TECTONICS

During the past decade, earth scientists have been gathering an impressive amount of data relating to the revolutionary "new global tectonics." Basically, the theory of *plate tectonics* postulates that the crust of the earth (*lithosphere*) is composed of about six huge, more-or-less rigid, but elastic, plates, about 75 to 100 kilometers thick, plus perhaps fifteen or twenty smaller plates, all of which move, shell-like, around the spherical earth over a semisolid, plastic layer, the *asthenosphere*. A slowdown of earthquake waves as they pass through the asthenosphere in the upper mantle first led seismologists a half-century ago to postulate a plastic or partly liquid "low-velocity zone" roughly at depths of between 100 and 200 kilometers. The great curved earth plates are perhaps carried along over the surface of the globe by *convection,* that is, currents in the hot plastic materials and magma that move upward, outward, away from spreading centers, and downward again as they cool. The major moving crustal plates of the earth are shown in Figure 6-2. The American plate, for example, includes North and South America plus the half of the Atlantic Ocean west of a line of spreading centers called the mid-Atlantic ridge or rise. At the spreading centers, new magma rises to the surface, spreads out, and is gradually carried in both directions away from the rise. The Pacific plate encompasses almost the entire north Pacific Ocean; west of the East Pacific rise, this plate moves northwestward. Eventually, the denser oceanic plates move under the lighter continents along *subduction zones* (Figure 6-3); or two continental plates collide, as does the Indian plate which moves northward into collision with Asia; or two plates slide past each other, as the Pacific plate does past California on its way northwestward to subduction in the Aleutian Trough.

So, the mobile crustal plates have several kinds of boundaries: (1) they spread apart and move away from each other in the midoceanic ridges (rises) where magma wells up from the upper mantle to form new crust which then moves away under the influence of convection currents at rates of a few centimeters per year; (2) the new crust formed moves outward until eventually it dives down beneath a continent in a subduction zone, as the Nazca plate dives under the west coast of South America (Figure 6-4) and the northwestward-moving Pacific plate subducts under the Aleutians; (3) junctures are formed where the plates slide past each other with little disruption of either (along the San Andreas fault, for example?); and (4) land masses are formed when violent collisions of plates, one against another, as for example, northward-cruising India

**FIGURE 6-2** *The earth's moving crustal plates. The "continental" plates—like the American—appear to move at an average rate of about 2 centimeters a year, whereas the "oceanic" plates—like the Pacific—seem to move faster, perhaps about 10 centimeters per year. Notice the arrows which show the general surface direction the plates are moving in. Remember, of course, that the plates are curved segments on a globe and therefore rotate about centers. Along the "mid-oceanic" ridges (rises) adjacent plates move away from each other; in areas like the Aleutian and Japanese Trenches the oceanic plate is sliding down under the continental plate; and along the coast of California it looks as if the American and Pacific plates are simply sliding past each other. (See also,* Oakeshott, Volcanoes and Earthquakes.)

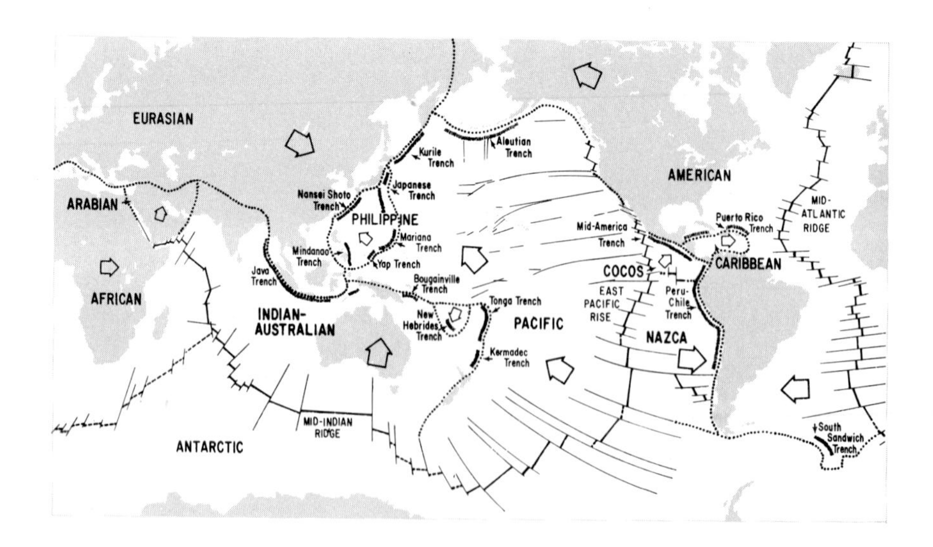

crashing into Asia to form the Himalaya Mountains.

Figure 6-5 is an idealized model (not to scale) which shows the elemental relationships of the parameters of plate tectonics.

## plate tectonics—a revolution in thinking

The first vague notion that the continents might have drifted apart appears in the writings of several nineteenth-century scientists, but strongly reached the attention of the scientific world through a classic publication by A. Wegener in 1912. Wegener noted that South America and Africa fit so closely as to preclude the elements of chance. His ideas lay dormant—largely for lack of a mechanism—for many years, but hit the scientific world again at the right time, beginning with J. Tuzo Wilson in 1962.

Many studies since Wegener's day have fitted together the continental platforms of all continents. There is attractive evidence that all continents were welded into a single great land mass—Pangaea—existing about 225 million years ago, that is, at the close of the Permian Period. No rock formations older than about 200 million years (late Triassic) have been found in the present ocean basins of the world. This suggests strongly that our ocean basins are of Mesozoic age and younger.

The concept of Pangaea is attractive because it solves many problems: matching rock formations in different continents; late Paleozoic patterns of

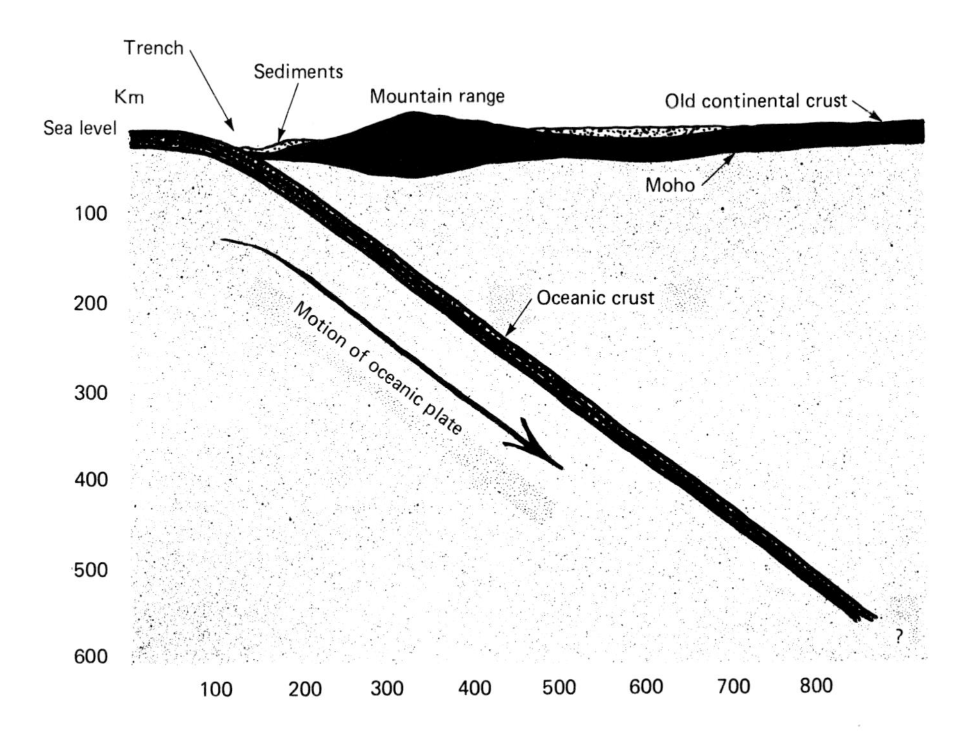


**FIGURE 6-3** *Subduction model. This is an idealized model of what plate tectonics theory says is happening at the continental margin of eastern Asia and the Aleutian Islands region. Oceanic crust is being pushed downward at various angles—perhaps 10 to 30 degrees—under the continental margin, thus developing a trench in back of an island arc. A mountain range is developing inland by the thrusting, uplift, and rise of newly formed magma. (From Oakeshott,* Volcanoes and Earthquakes.*)*

**FIGURE 6-4** *Earthquake foci, west coast of South America. Diagrammatic section through the Andes and the west coast of South America, where the oceanic plate is actively moving into and under the continental plate. Breakup of the subducting plate develops a pattern of earthquake foci of different depths, something like that shown.*

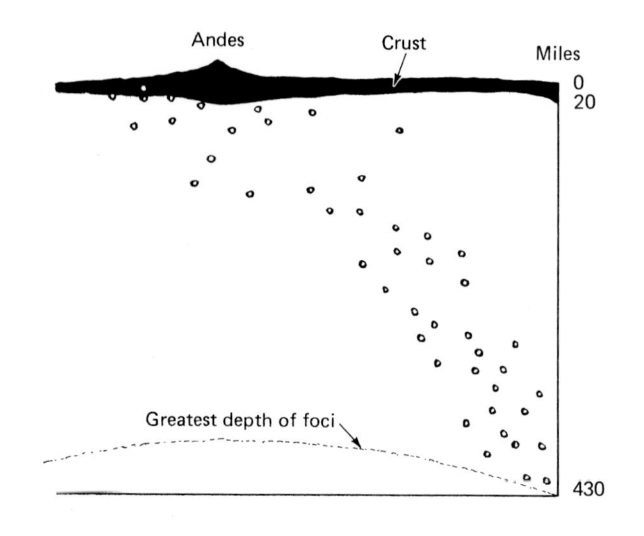


**FIGURE 6-5** *A diagrammatic model (not to scale) illustrating some of the features and motivation of sea floor spreading and plate tectonics theory. This model has been used by P. C. Bateman (*California Geology, *January, 1974) to show broad relationships of the California Coast Range–Sierra Nevada subduction zone and island arc as in early Late Jurassic time.*

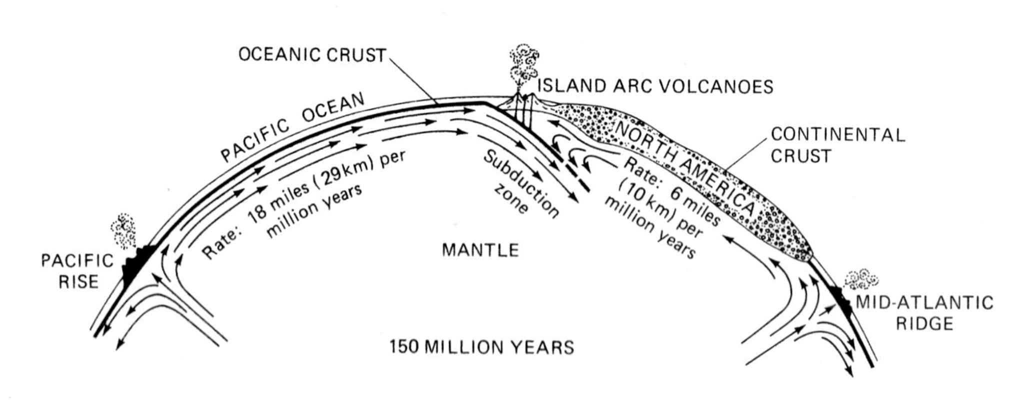

glacier movement; ancient distribution of climates, plants, and animals; continuations of folded mountain ranges; and ancient fault systems.

The concepts of crustal plates and sea floor spreading, with action at the plate junctions, account for the formation of our mountain ranges in Mesozoic and later time (see Chapter 8), volcanoes, earthquakes, great fault systems, and island arcs. The related ideas of plate-tectonics theory shed new light on some of our greatest geologic problems and give new bases for geologic research.

During the last decade, new evidence for drifting continental plates has developed from intensive exploration of the sea floor. Bands of rocks of reversed magnetic polarity have been measured on the sea floor, running parallel to the zones of sea floor spreading. Measured ages of rocks on the ocean floor do, indeed, seem to increase in belts away from the spreading centers. These and other specific and impressive new data have been obtained by the drilling vessel *Glomar Challenger* in 573 holes drilled into the sea floors. This important program, under the auspices of the National Science Foundation and involving most of the world's leading oceanographic institutions, is believed to have demonstrated the validity of the concept of sea floor spreading.

### theory or fact?

Although the accumulation of information bearing on plate tectonics is becoming overwhelming, many problems remain. To be valid, *all* local and regional geologic facts must fit. Plate tectonics has not yet been accepted by all earth scientists as "truth"; a competent minority still see great fallacy in it. Particularly difficult is the question of a driving mechanism: Does convection really take place in the upper mantle, as postulated, and is it an adequate mechanism to drive crustal plates? How does a crustal plate maintain its internal rigidity? And there are many other questions!

## CALIFORNIA'S OLDEST ROCKS

California's oldest rocks are incomprehensibly old—on the order of 1.8 billion years—yet they are by no means the oldest rocks in the world or even in North America. The age of the earth is perhaps 4.5 billion years, judging from radiometric dates obtained from meteorites and the moon; and the time since beginning of normal geologic processes at and near the surface of the earth may be about 4 billion years. We can think of the beginning of geologic processes as marking the start of the Precambrian Era of geologic time. The oldest rocks dated by radiometric means are granitic rocks in the Barberton Mountain Range in South Africa and are on the order of 4 billion years old. In southwestern Minnesota some rocks have been dated at 3.8 billion years.

Starting with California's oldest rocks, we can determine something of what this part of the earth may have been like for the past 2 billion years, even though there are great gaps in the record. But, what about that vast period from the time of the earth's origin, 4.5 billion years ago, to the time of formation of our oldest rocks? Even here, there is some evidence from which the earliest history of California can be reconstructed.

The geologic map of California shows the distribution of the major rock groups and their ages as inferred from more or less scattered outcrops. A check of the map shows that known Precambrian rocks occur only in the southern part of the state, in three principal areas: (1) Death Valley-Panamint Valley area, (2) the central and southeastern Transverse Ranges within and along the San Andreas fault system, and (3) scattered localities across the Mojave Desert.

Of course, being the oldest of all rock groups, Precambrian rocks should lie at the very bottom of the geologic column, underlying all other rock formations. They should and they do—*except* where great uplift has allowed erosion to strip away overlying rocks or possibly where high-standing Precambrian terrain was of such nature that deposition of younger sediments could not take place. Thus, it is no coincidence that the Precambrian rocks in southern California are found chiefly in the hearts of mountain ranges such as the Transverse Ranges and the mountains around Death Valley.

## recognizing precambrian rocks

How can rock formations be recognized as Precambrian in age? The most positive evidence is found where the rocks can be seen exposed in their normal position underlying Cambrian stratified rocks that have been dated by the characteristic fossils they contain. The next best means of determining a rock as Precambrian is by radiometric dating showing that the rock crystallized (or recrystallized by metamorphic processes) more than 570 million years ago.

It might be thought that these most ancient rocks should "look old"; that is, that they would be most likely to show the banding, contortions, development of new minerals, textures, and structures induced by several epochs of metamorphism and mountain building. This is a hazardous criterion because rocks that are only 100 million years old may have been very intensely metamorphosed in some areas and may also "look old." The negative criterion of lack of fossils is not much good, either, because extensive groups of young rocks may contain few or no visible remains of past life.

## death valley area

In the southern Death Valley area, in the Black Mountains east of the valley, and in the Panamint

**6-6** *Basal conglomerate of the Late Precambrian Crystal Spring Formation. It rests unconformably on gneisses and schists which have been dated radiometrically at 1.7 billion years. This unconformity is at a high angle and represents a long gap in the sedimentary record. (Bennie W. Troxel photo.)*

**6-7** *Conglomerate bed (pebbly mudstone) in Late Precambrian Kingston Peak Formation, southern Death Valley. Note its unsorted, nonstratified character. This formation looks much like glacial till, but it is interlayered with strata that most geologists have interpreted as marine deposits! (Bennie W. Troxel photo.)*

Mountains west of the valley, both lower and upper Precambrian rocks are exposed. Early to middle Precambrian time is represented by highly metamorphosed rocks including gneiss, schist, and amphibolite, as well as by granitic rocks that have intruded them. Radiometric dating from zircon in these rocks shows them to be as much as 1.8 billion years old. Some of the schists and gneisses suggest that they were originally sedimentary rocks such as shale and sandstone; amphibolite is a dark amphibole-rich rock that most likely was of volcanic origin. The contorted and highly complex character of the rocks and the large amounts of intruded granite suggest their involvement in an important epoch of mountain building.

A long period of erosion followed the mountain-building epoch, for unconformably on the earlier Precambrian rocks lies a section of unmetamorphosed sedimentary rocks more than 7,000 feet thick called the Pahrump Group. A belt of Pahrump strata continues for about 75 miles to the southeast. At least three distinct rock formations can be recognized and mapped within the Pahrump: the Crystal Spring Formation, consisting of coarse-grained sedimentary rocks such as sandstone and conglomerate, with an intruded layer (sill) of the igneous rock diabase (similar to basalt in composition); above that, limestone and dolomite called the Beck Spring Dolomite; and at the top, the Kingston Peak Formation, made up mostly of quartzite derived from sandstone and conglomerate. These rocks were sediments deposited in a northerly trending marine trough. The sediments were probably derived from higher-standing lands on the east and northeast; they thicken toward the west.

Following deposition of the Pahrump Group, tilting, uplifts, and erosion must have occurred, for the Noonday Dolomite lies unconformably on Pahrump strata. Farther north, in the Funeral Mountains, strata equivalent to the Pahrump Group are much thinner and are metamorphosed to schists and gneisses. Above the Noonday Dolomite are something like 10,000 feet of sedimentary rocks—sandstone, shale, and dolomite—which form a conformable series (no erosional breaks) transitional from the late Precambrian into the early Cambrian. The earliest Cambrian strata contain abundant remains of marine life, including trilobites typical of lower Cambrian rocks elsewhere in North America; the trilobites occur in shale of the Wood Canyon Formation. A complete transition from Precambrian to Cambrian rocks has rarely been found; usually there is a great unconformity between Precambrian and Cambrian time.

About 75 miles southeast of Death Valley in the Providence Mountains, about 1,000 feet of hard sandstone called the Prospect Mountain Quartzite lie with great angular unconformity on lower Precambrian schist, gneiss, and granite. The Prospect Mountain Quartzite may be latest Precambrian or earliest Cambrian in age.

## transverse ranges

Extensive masses of very distinctive rock types crop out in the core of the Transverse Ranges, typically in the western San Gabriel Mountains. Oldest of these complex highly metamorphosed crystalline rocks are found in an area of outcrops of coarse-grained augen (containing eye-shaped structures) gneiss on the northern side of the range, which has been radiometrically dated at about 1.7 billion years. It has been intruded by Cretaceous granite. An odd, dark-colored, granitic-textured rock called syenite, in the same area, is about 1.22 billion years in age.

Just north of the important San Gabriel fault, the Mendenhall Gneiss, about 1.45 billion years old, lies in a belt 10 miles long. This gneiss is a high-grade metamorphic rock consisting mainly of blue quartz, soda-lime feldspar (high in sodium and calcium), potash feldspars, and some dark minerals. The Mendenhall Gneiss has been intruded by a group of coarse-grained igneous rocks that include anorthosite, various types of dark gabbro gradational into anorthosite, and some black rocks made up mostly of titanium-bearing magnetite. Anorthosite is a light-gray to white rock composed of more than 90 percent plagioclase (soda-lime) feldspar. The anorthosite-gabbro group of rocks has been radiometrically dated at about 1.22 billion years. Anorthosite of the San Gabriel Mountains was formed by crystallization from a magma at about 1300° C.

The extraordinary thing about large bodies of anorthosite (more than 50 square miles in the San Gabriel Mountains) is that it is known to occur *only* in areas of Precambrian rocks, and it occurs in large areas of rocks of this age in *every* continent! This unique occurrence still awaits explanation. The only other suite of similar rocks in California is in the Orocopia Mountains, northeast of the Salton Sea, at the extreme southeastern end of the Transverse Ranges province. Precambrian rocks of the Transverse Ranges are all between major faults and so their original, "natural" relationships with younger rocks are uncertain.

**6-8** (a) *Vertical aerial photo showing fault contacts between gabbro (dark) and anorthosite (light) of the Precambrian anorthosite-gabbro complex in upper Pacoima Canyon, western San Gabriel Mountains, Los Angeles County. East-west distance shown is about 2 miles. (California Division of Mines and Geology Bull. 172, south is at top.)* (b) *Key to aerial photograph 6-8a.*

(a)

On the large geologic map one rock group is designated "Pre-Cenozoic metamorphic rocks of unknown age." This group *may* include some Precambrian rocks. An example is the Pelona Schist—metamorphosed sedimentary and volcanic rocks in the form of green schists—which crops out extensively in the Garlock and San Andreas fault zones in southern California. Since it lacks fossils and is found in belts bounded by faults, stratigraphic relationships to adjacent rock formations are obscure. Its age is unknown; however, many geologists consider the Pelona to be Mesozoic.

Professor Leon T. Silver, of the California Institute of Technology, has done much of the radiometric dating of the Precambrian rocks in southern California. He has reconstructed Precambrian history (all dates are approximate and tentative) in that province somewhat as shown in Table 6-2.

## relationship to southwestern united states

It is interesting to note that the oldest rocks in the Transverse Ranges, although quite different, are probably about the same age as those in the southern Death Valley area. Although research on Precambrian rocks and history is in an early stage, we already know that the Precambrian history of southern California can be extended into Arizona. The oldest rocks exposed in the bottom of the Grand Canyon, for example, are about 1.8 billion years old and events of 1.8 billion to 1.4 billion years ago in the geologic history of Arizona *seem* to be similar to those of southern California.

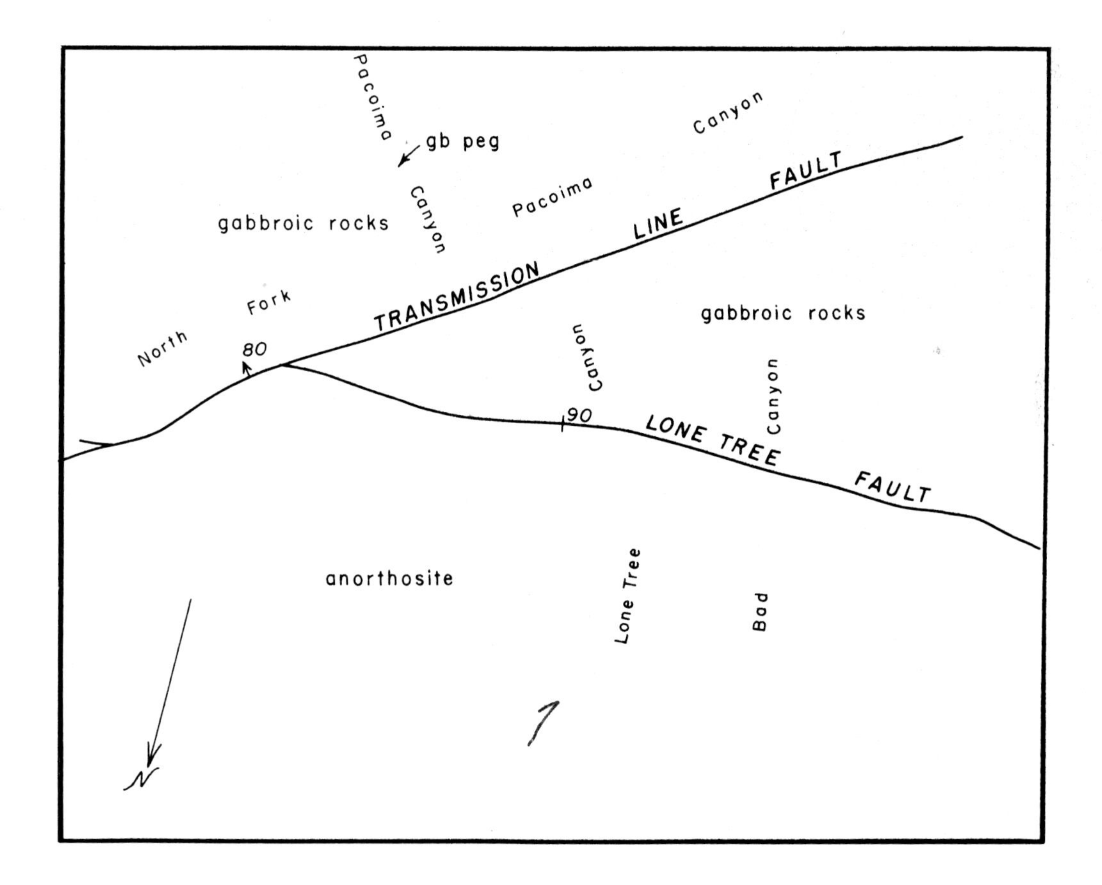


(b)

**TABLE 6-2 Precambrian history of southern California**

| | |
|---|---|
| 1.7 billion years | Accumulation of sedimentary and volcanic rocks to great thicknesses in troughs below sea level; deformation of these rocks. (The rocks now are the augen gneiss and similar types.) |
| 1.66 billion years | Intrusion of the older rocks by granitic bodies. |
| 1.45 billion years | Great and widespread deformation and metamorphism accompanying mountain building. (Rocks presently are like the Mendenhall Gneiss.) |
| 1.22 billion years | Intrusion of the Mendenhall Gneiss and other older rocks by the complex anorthosite-gabbro-syenite group; probably some mountain building. |

Taking a still broader look at the Precambrian rocks of North America and keeping in mind that (1) granite is the major stuff of the continents, and (2) granite is most often directly intruded into zones of major mountain building, study Figure 6-9, which shows the general outlines and ages of granitic rock provinces in the continent. The data are from outcrops and wells and show radiometric ages of granitic rock. Two areas stand out as the oldest parts of the continent at 2.5 billion years—Great Slave Lake area in northwestern Canada and a broad northeasterly trending area from Wyoming across Lake Superior and Hudson Bay. These probably represent ancient mountain systems constituting the early Precambrian nuclei of the continents. More or less surrounding the 2.5 billion-year-old continental platform, as if it had been added by accretion, is a much broader continental platform, which underlies the central plains and the Rocky Mountains and extends across southern California. In this younger platform, rock ages are middle Precambrian—1.0 billion to 2.5 billion years; many of these are in the range from 1.2 billion to 1.8 billion years. Beyond the middle Precambrian continent are rocks shown as 0 to 600 million years (mostly much younger than the Cambrian period), which make up the Appalachian Mountain system and the Pacific mountain systems.

If southern California was part of the continental platform of North America in middle-to-late Precambrian time, what was the rest of California like? Perhaps it was part of the Pacific Ocean Basin and had a floor of basalt and serpentine; much later deposition of sediments and volcanics, uplift, folding, and faulting added to Precambrian California to form what we see today.

## evidences of life in precambrian rocks

None of the Precambrian rocks exposed in southern California contain positive evidence of life, but there are some highly suggestive occurrences. Algal-like forms and tubes that may have been formed by gases from organic matter are known in the latest Precambrian Noonday Dolomite of the southern Death Valley region. Some have interpreted the tubes as worm borings. Some bacteria like forms have also been recognized in the Crystal Spring Formation.

Elsewhere over the earth, fossils in Precambrian rocks are extremely rare. This fact is noteworthy because marine fossils in earliest Cambrian strata are abundant and represent highly developed forms of life. In the Death Valley area, for example, the transition from no fossils to abundant fossils takes place abruptly within the Wood Canyon shale beds. Where Precambrian fossils have been found in North America (Rocky Mountains, Lake Superior), they consist of calcareous algae, bacteria, very doubtful primitive corals, and equally doubtful worm borings. The oldest fossil remains of life are probably about 3.5 billion years old.

**FIGURE 6-9** *General outlines and ages of granitic rock provinces in the continent as shown by present-day outcrops and borings.*

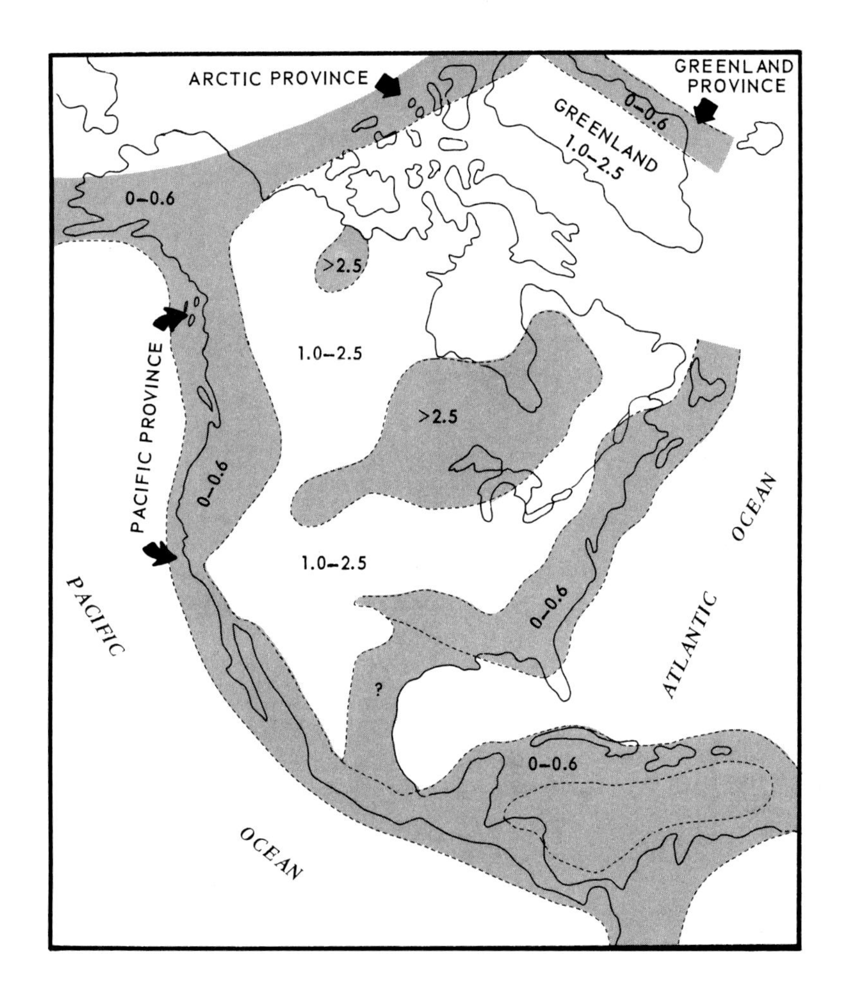

**6-10** *Late Precambrian Noonday Dolomite in Galena Canyon, southern Panamint Mountains.* Scolithus *(worm) tubes show the existence of multicelled life.* Scolithus *remains are best known from Cambrian shale in the northern Rocky Mountains. (Charles B. Hunt photo.)*

# 7 rocks and life of the paleozoic seas

*Trilobites—typical life forms of Paleozoic seas. This species of* Olenellus *from shale of the southern Death Valley region shows the rock's Early Cambrian age.*

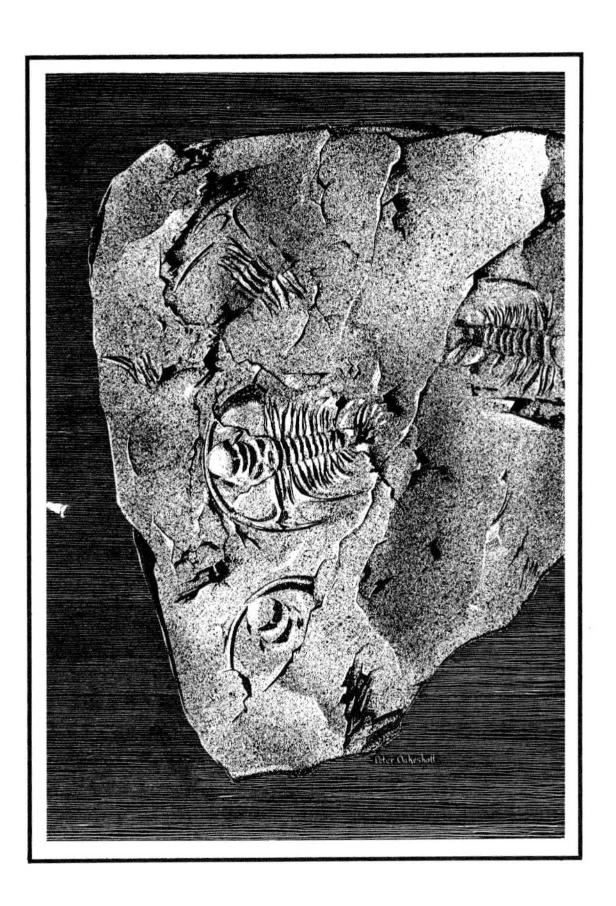


■ During the latest stages of the Precambrian Era, lands of the primitive continental platform of North America were largely above sea level and were mountainous in local areas. These high-standing Precambrian lands were the areas from which the earliest Paleozoic sediments were derived. There is evidence—but not as positive as we would like—that a Pacific Ocean existed west of the continent. Outlining that ocean and the geography of western lands in North America during late Precambrian time presents almost insurmountable problems.

The record left in Paleozoic rocks in California is overwhelmingly marine. Seas apparently encroached from the west into the Great Basin where Cambrian formations are oldest and thickest.

Where are the Paleozoic rocks found today? What do they show of the distribution of land and sea, and what can they tell of the 400 million years of the Paleozoic Era?

A logical starting point for the study of rock formations and geologic history is the 1:2,500,000 colored geologic map of California (available from the U.S. Geological Survey or the California Division of Mines and Geology). This map's explanation, or legend, shows all the complex rock types of different ages in California grouped into eleven units. One of these units is "Paleozoic sedimentary and volcanic rocks."

This map shows that there are three broad areas where Paleozoic rocks are extensively exposed: (1) the Great Basin-Mojave Desert province of southeastern California, east and southeast of the Sierra Nevada; (2) the Sierra Nevada, chiefly in a northwestern belt; and (3) the Klamath Mountains. Outside these areas, there are isolated outcrops of metamorphosed rocks, of doubtful age, in the Coast, Transverse, and Peninsular Ranges. Some of these rocks, shown on the map and marked "Pre-Cenozoic metamorphic rocks of unknown age" are thought to be Paleozoic. To learn more about the distribution of Paleozoic rocks in California, it is necessary to examine more detailed geologic maps—maps drawn to a scale of 1 inch equals 4 miles or less (for example, *Geologic Map of California,* also published by California Division of Mines and Geology).

Is it reasonable to assume that the thick and extensive Paleozoic rock formations exposed so widely were once connected as one? Quite so; for seas that existed long enough for the deposition of thousands of feet of sediments could not have been isolated—they must have had ocean connections. Precisely locating such connections hundreds of millions of years after profound geologic changes is one of the challenges geologists face!

## A BROAD VIEW OF CALIFORNIA IN PALEOZOIC TIME

At the dawn of the Paleozoic Era, water of the ocean basin began to move in over the marginal lands. Shallow seas spread rapidly during early Cambrian time into a prominent area of downsinking of the earth's crust that shortly extended over all the western cordillera, or mountain systems, from the site of the present Rocky Mountains to the Pacific margin. This great trough, or geosyncline (geo means "earth"; *syncline* means "downfold") became the site of marine deposition for the greater part of the nearly 400 million years of Paleozoic time.

The early Paleozoic seas advanced over Precambrian rock formations that had been folded, metamorphosed, and eroded; over most of the continent, a profound unconformity exists between the earliest Cambrian strata and the Precambrian rocks. In the southern Death Valley area, a transitional group of stratified rocks suggests that a marine trough had already been formed locally in late Precambrian time and that it persisted into the Cambrian period. In this trough, the oldest Cambrian rocks in North America rest conformably upon Precambrian sedimentary rocks. From this ancient trough, shallow seas spread rapidly until they occupied the whole Cordilleran geosyncline, and extended eastward to the mid-continent. These seas were shallow and warm throughout the Paleozoic Era, as shown by the abundance of fossil corals, which are warm-water organisms. Sediments deposited were predominantly carbonates, and as a result vast thicknesses of limestone and dolomite were built up.

The oldest Cambrian rocks in North America are found in the Waucoba district of the Inyo Mountains and in southern Death Valley. In the latter area, stratified rocks are conformable and continuous in their succession from Late Precambrian to Early Cambrian. Most geologists now place this important Cambrian-Precambrian boundary in the middle of the Wood Canyon Formation (sandstone, shale, and dolomite) at the base of the first (lowest) trilobite zone. Total thickness of the Cambrian strata is approximately 8,000 feet.

The Early Cambrian shore line ran roughly north-south approximately through the California-Nevada-Arizona junction; the Middle Cambrian shore line was near the eastern border of Arizona; and the Late Cambrian shore line was still farther east across central New Mexico and extended as far northeast as southern Wisconsin. The seas persisted longest and the geosyncline became deepest in the Basin Range region east of the Sierra, including the Inyo-Panamint-Death Valley area.

Rocks of the Ordovician System, in the same general area, include dolomite and the very striking light gray middle Ordovician Eureka Quartzite (hard silica-cemented sandstone). The Eureka Quartzite is very widespread and readily recognized in this part of eastern California and western Nevada. Very thick sections of Ordovician strata appear as more or less isolated remnants in the Sierra Nevada and extensively in the Klamath Mountains. Silurian and Devonian rocks are mostly

**7-1** *West side of Death Valley. From lower left: Cambrian, Ordovician, Silurian, and Devonian marine stratified sedimentary rock formations on the south side of Tucki Mountain. (John H. Maxson photo, U.S. Geological Survey, courtesy of Charles B. Hunt.)*

limestone and dolomite. The rocks of the Mississippian System also include a great deal of limestone. From late Mississippian into Pennsylvanian time the seas must have become shallower and more muddy, perhaps because of beginning uplift exposing higher lands; shale becomes the dominant rock type.

The Pennsylvanian and Permian periods brought changes to coarser sediments and increasing volcanism and unconformities, suggesting uplift of the lands and orogenic activity in the Great Basin, Sierra Nevada, and Klamath Mountain provinces. Some bodies of granitic rocks in the Sierra Nevada, Klamath Mountains, and Transverse Ranges have recently been radiometrically dated as Permian. This is strong confirming evidence of rather widespread mountain building in California toward the end of the Paleozoic Era.

West and southwest of the Klamath Mountains, Sierra Nevada, and Mojave Desert little is known of Paleozoic history. Marble and metamorphosed volcanic and sedimentary rocks of questionable late Paleozoic age in the Coast and Peninsular Ranges suggest eugeosynclinal deposition, but nothing is known of what went before.

We have just sketched the broad aspects of California's Paleozoic history. What evidence has led geologists to these conclusions? Where may it best be examined? Part Three of this book undertakes a *regional* approach to the state's geologic history—province by province—but it is worth while at this point to look, briefly, at some of those regions or areas where rock exposures contain most of the concrete evidence used by geologists to build up the history of the Paleozoic Era in California.

**7-2** *Eight thousand feet of limestone and dolomite of Cambrian, Ordovician, Silurian, and Devonian ages are exposed here at Panamint Butte on the east side of Panamint Valley, Inyo County. (Photo by Wayne E. Hall and Hal G. Stephens, U.S. Geological Survey.)*

## GREAT BASIN–MOJAVE DESERT

In the mountains marginal to Death Valley, there are marine sedimentary rocks representing all periods of the Paleozoic Era, from earliest Cambrian to the Permian Period. The lower Cambrian are clastic (made up of particles or fragments) sedimentary rocks, like shale and sandstone for the most part, followed by the later Cambrian to Mississippian formations, which are largely limestone and dolomite. The greatest thickness of carbonate rocks in North America is found in this area. Total thickness of Paleozoic stratified rocks in the Death Valley area is about 20,000 feet; in the Inyo Mountains, to the west, it is about 23,000 feet. In these extremely thick sections there are few gaps or unconformities.

Rocks of the Permian System are heterogeneous sandy, pebbly limestone; shale; sandstone; and conglomerate. There is a local angular unconformity between the Permian and underlying Pennsylvanian strata, suggesting folding and uplift at the end of Pennsylvanian time and

**7-3** *A flat thrust fault—called the Burro Trail fault—separates two Cambrian sedimentary rock formations in the Death Valley region. (Charles B. Hunt photo, U.S. Geological Survey.)*

probably continuing on into the Permian. The coarse, clastic sediments of the Permian System were derived from older exposed rock formations nearby.

In the eastern Mojave Desert, southeast of Death Valley, rock formations and relationships are similar to those in southern Death Valley but only Cambrian and late Paleozoic strata are present; the Ordovician and Silurian rocks are not represented. In the southern and western Mojave Desert all of the lower Paleozoic and Devonian strata are missing; only the Mississippian to Permian Periods are represented by marine limestone and sandstone, entirely metamorphosed to marble and quartzite.

How to explain such gaps in the sedimentary rock sequence? There are two principal possibilities: the areas may have been land above sea level, or uplift and erosion may have caused the removal of sediments once deposited. Perhaps the former is more reasonable in this case.

**7-4** *Flat thrust fault separating two tilted Cambrian sedimentary rock units. Below the fault is Zabriskie Quartzite (on left) and shale and limestone of the Carrara Formation. Above the fault is dolomite of the Bonanza King and Nopah Formations. Death Valley Canyon. (Charles B. Hunt photo, U.S. Geological Survey.)*

## COAST, TRANSVERSE, AND PENINSULAR RANGES

In the Coast Ranges, Transverse Ranges, and Peninsular Ranges, there are remnants of quartzite, schists, and marbles that contain very scant fossil remains; these remnants are probably metamorphosed equivalents of marine rock formations of late Paleozoic age. The Furnace Limestone (really marble) in the San Bernardino Mountains, 4,500 feet thick, is the best dated of these rock formations. It contains fossil brachiopods, horn corals, and foraminifera of Mississippian and perhaps Pennsylvanian age. The Sur Series marble, schist, and quartzite in the Santa Lucia Mountains in the Coast Ranges south of Monterey Bay may well be metamorphosed late Paleozoic marine rocks. Very scattered, isolated remnants of quartzite, schist, and marble are found in the vicinity of Point Reyes, west of the San Andreas fault a few miles north of San Francisco. These are probably remnants of the once-extensive Sur Series.

Radiometric dates of approximately 245 million years ago have been obtained for some of the granitic rocks in the central San Gabriel Mountains in the Transverse Ranges, a few miles northeast of Los Angeles. This finding strongly confirms Permian mountain building in the Transverse Ranges province.

## SIERRA NEVADA

Large bodies of granitic magma intruded older rocks in the Sierra Nevada in the Mesozoic Era. Before being largely removed by erosion, a roof of Paleozoic rocks lay over the granitic rocks. Later deep erosion has left many fragmented sections and small bodies as roof pendants in the granite. They consist mostly of schists, limestone, quartzite, and slate, which record the history of Paleozoic times. The thickest and best-known section consists of 32,000 feet of such rocks in the Mount Morrison area. The lower part is a very thick and complete succession of Ordovician strata; the upper part consists of fossiliferous Pennsylvanian and Permian rocks.

In the western foothill belt of the Sierra Nevada, mostly east of the Mother Lode fault zone, is the thick Calaveras Formation. The original rocks—all now metamorphosed—were shale, dark sandstone, chert, some limestone, and much submarine volcanic rock. At the north end of this western metamorphic belt, in the Taylorsville region, fossils of Silurian to Permian age have been found in similar rocks. Angular unconformities have been found between Silurian and Devonian strata and between Permian and Triassic strata, suggesting orogenic epochs at those times.

## KLAMATH MOUNTAINS

Paleozoic rocks are abundantly represented in the Klamath Mountains in three great arc-shaped zones, with the most complete section in the eastern Klamath Mountains. There are more than 25,000 feet of marine stratified rocks, which range in age from Ordovician to Permian. The rocks are metamorphosed but were originally graywacke (dark sandstone containing feldspars and rock fragments), shale, chert, conglomerate, rhyolite, basalt, and explosive volcanic rocks such as tuff and breccia. The volcanics were extruded under water to form "pillow lavas" such as those shown in Illustration 8-10; the interbedded sedimentary rocks contain abundant marine fossils.

Two Paleozoic periods of orogenic movement, widespread metamorphism, and intrusion of igneous rocks can be recognized in the Klamath Mountains. The first took place in the Devonian Period, and a second in late Paleozoic time between late Pennsylvanian and Triassic time. So, again, in the Klamath Mountains, we have evidence for emergence of the land and mountain building during the middle and near the close of the Paleozoic Era.

**7-5** *Split Mountain in the Sierra Nevada about 25 miles south of Bishop, Inyo County. The dark formation is a roof pendant—part of the older roof rock that erosion has left—intruded by light-colored granite of the Sierran batholith. (John S. Shelton photo.)*

## LIFE OF THE SEAS

Remains of life preserved in the Precambrian rocks are extremely scant and are of simple forms only. With the dawn of the Paleozoic Era there came a literal life-form explosion! Fossils in marine Paleozoic rocks are abundant and of highly diversified forms. Invertebrate marine fossils are the principal representatives of life forms.

Fossils are a major source of information in the reconstruction of geologic history. Some of the things they tell us are (1) where the rock formation belongs in the geologic column—its age; (2) climatic and other environmental conditions in which the organisms lived; (3) much about the distribution of lands and seas; (4) depth of the seas; (5) the nature of now-extinct life forms; and (6) the progress of evolution of plant and animal life through the ages.

In early Paleozoic time, although invertebrate forms continued to be most numerous, vertebrate life began to evolve. The fishes, which first appeared in Ordovician time, came into prominence in the Devonian seas. On the lands (of which there were *very* limited areas in California in Paleozoic time) amphibians developed in the late Paleozoic Era and the first reptiles evolved as a prelude to their overwhelming importance in the Age of Reptiles—the Mesozoic Era.

The invertebrate creatures whose fossils are most useful in the interpretation of geologic history through all eras are listed in Table 7-1.

**7-6** *A slab of Silurian* Bryozoa, *very typical of Paleozoic seas and often one of the organisms in "coral" reefs. Slab shown is about 7½ inches wide. (Smithsonian Institution photo.)*

## COMMON FOSSILS IN CALIFORNIA ROCKS

The trilobites—three-lobed, segmented marine arthropods—are among the most distinctive fossils found in Paleozoic rocks (Illustration 7-7). These fossils occur in the earliest Cambrian rocks in the Death Valley area and are especially abundant in younger Cambrian and Ordovician rocks. They are less common in later Paleozoic rocks, and the trilobites became extinct at the end of the era. Trilobites are good Cambrian guide fossils

**TABLE 7-1 Invertebrate fossils most useful in interpreting geologic history**

| PHYLA (MAJOR LIFE GROUPS) | CLASSES (SUBDIVISIONS OF PHYLA) | DESCRIPTION OF COMMON FORMS |
|---|---|---|
| Protozoa | | Single-celled animals: |
| | Foraminifera | Tiny chambered shells; mostly calcium carbonate |
| | Radiolaria | Tiny silica shells |
| Porifera | | Sponges; simple cell groups |
| Coelenterata | Anthozoa | Corals; calcium carbonate reef builders |
| Bryozoa | | Moss animals; all marine |
| Brachiopoda | | Marine |
| Mollusca | | Shellfish; in all environments: |
| | Pelecypoda | Two-valved shellfish, like oysters and clams |
| | Gastropoda | Coiled shells; marine, freshwater, and land snails |
| | Cephalopoda | Nautilus, squid, nautiloids, ammonoids |
| Arthropoda | | Invertebrates with jointed legs and external chitinous skeletons: |
| | Trilobita | Three-lobed, segmented; Paleozoic only |
| | Crustacea | Lobsters, crabs; freshwater and marine |
| | Insecta | Rare as fossils |
| Echinodermata | | Prominent five-sided symmetry, like starfish; all marine only; usually skeletons of crystalline calcite: |
| | Cystoidea and Blastoidea | Irregular plated, perforated by pores or slits |
| | Crinoidea | Stone lilies, with arms |
| | Echinoidea | Sand dollars and sea urchins |
| Chordata | Hemichordata | Graptolites; extinct; chitinous parts; colonial; all marine. Especially abundant in Ordovician and Silurian black shale |

everywhere on earth. Their abundance, their rapid evolution, and wide distribution make them excellent markers, or guide fossils, in rocks of Paleozoic age.

Although trilobites are extremely abundant and beautifully preserved in lower Paleozoic rocks, these rocks also contain many other kinds of fossils. We shall mention only some of the most common. Most abundant protozoa are fusulinids (see Photograph 4-8), a type of foraminifera—single-celled animals with calcareous shells—which became very abundant in the late Paleozoic, particularly in Pennsylvanian time. They are known from all marine formations of this age in California. Fusulinid shells are about the form and size of a wheat grain and make up an important part of some limestones. They are especially good marker (guide) fossils of late Paleozoic time.

Corals have been abundant and important builders of reefs from the Ordovician Period to the present day (see Illustrations 7-8 and 8-23). Their favorite environment today is warm shallow seawater; we accept their fossil presence as indication that these conditions prevailed at the time of deposition of the sediments from which

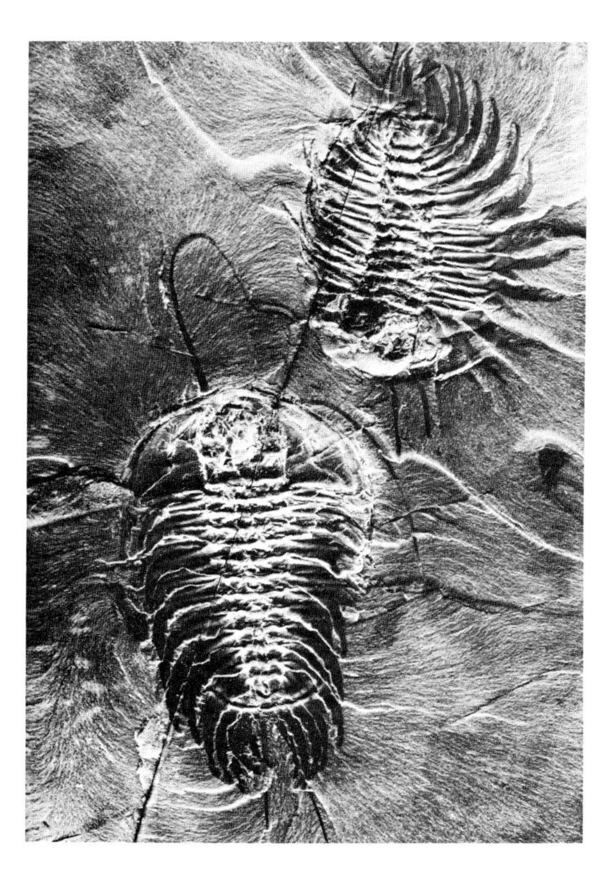

**7-7** *Trilobites, genus* Olenellus, *very characteristic of Cambrian time. About 1 inch long. (Smithsonian Institution photo.)*

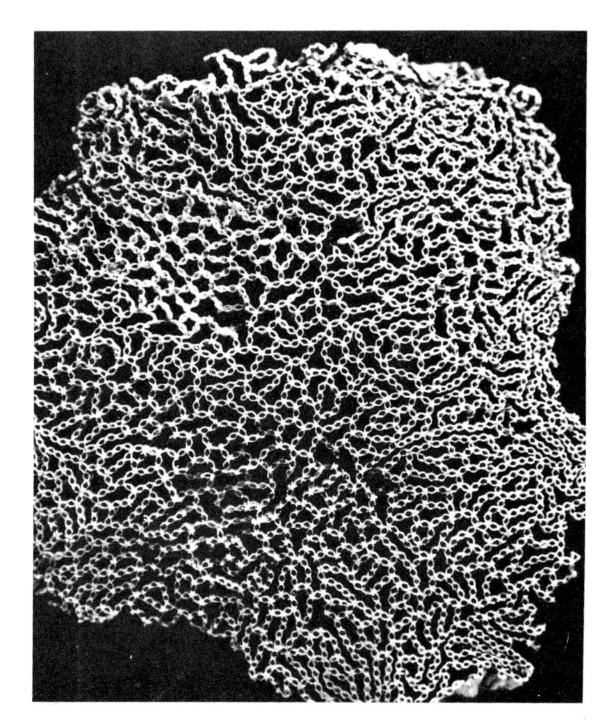

**7-8** *The chain coral* Halysites, *typically found in Silurian rocks. Section shown is about 9 inches long. (Smithsonian Institution photo.)*

rocks were formed. Sponges, bryozoans, echinoids, and crinoids are often found with corals in Paleozoic rocks. Archaeocyathans—primitive spongelike animals—are found in rocks of Cambrian age in the Death Valley area. They are extinct.

Brachiopods were very numerous, second only to the trilobites, and particularly characteristic of Paleozoic rocks. They are symmetrical bivalves, but the two valves are of unequal size. They attached themselves to the bottoms of shallow seas and so are best known as fossils in shales and limestones of the Paleozoic Era. Spirifers and productids are among the best-known fossil forms. Although something like 3,500 species of brachiopods are known from Paleozoic rocks, they declined rapidly and today are unimportant members of the marine shellfish fauna.

The major groups of molluscs—pelecypods (symmetrical bivalves like oysters and mussels), the spirally coiled gastropods, and the straight to flat-coiled cephalopods are all known from lower Paleozoic rocks. The molluscs are the most highly developed and numerous shellfish found in the marine environment today. The nautiloids reached their greatest number and variety in Silurian time.

**7-9** *A crinoid—*Aesiocrinus*—found in Pennsylvanian rocks. About 7 inches in diameter. (Smithsonian Institution photo.)*

**7-10** *Typical brachiopods of Devonian Period. The large "winged" forms are* Spirifer*—very widespread in Paleozoic seas. Largest* Spirifer *about 7 inches long. (Smithsonian Institution photo.)*

In Late Paleozoic time, the most highly advanced molluscs—the ammonoids—developed from an earlier nautiloid stock. Some of the largest shells ever formed were those of straight-shelled Ordovician nautiloids—up to 15 feet long. The ammonoids were developing rapidly at the close of the Paleozoic; their successors—ammonites—became the dominant shellfish in Mesozoic seas.

Graptolites were most prominent during the Ordovician and Silurian Periods. After first appearing in the Middle Cambrian, they became extinct in Early Mississippian time. As the highest products of evolution among the invertebrates, they had close affinity with some of the Chordata, the phylum that includes the vertebrates. Their great variety and abundance make them good guide fossils for age divisions within Ordovician and Silurian rocks; moreover, they were held together in colonies by floats of their own construction and so became very widely distributed.

The fishes, although not the most advanced vertebrates that lived during the Paleozoic Era, were in all probability the most abundant. The first primitive fishlike forms were the ostracoderms, whose bony plates are found in Ordovician rocks. The true fishes followed ostracoderms shortly and became the highest and dominant forms of life in the Devonian seas. Fishes are the oldest animals with internal bony or cartilaginous skeletons and a central nervous system—distinguishing characteristics of the vertebrates. Fossil fishes are not abundant, but under special conditions where their skeletons may be preserved in fine-grained shaly rocks they have been found in great numbers. The teeth of sharks (which have no bony skeleton, as they are not the most typical fishes) are locally very numerous as fossils in upper Devonian and younger formations.

The first amphibians developed in the Devonian by evolution of lung-bearing, lobe-finned fishes and, in turn, the first reptiles (Pennsylvanian) were developed from amphibian

**7-11** Worthenia, *a typical Late Paleozoic gastropod about 2 inches tall. (Smithsonian Institution photo.)*

**7-12** *Graptolite colonies in Paleozoic shale. Rods are about ½ inch long. (Smithsonian Institution photo.)*

stock. A group of reptiles eventually gave rise to the birds and mammals in the Mesozoic Era. Amphibians and reptiles are rare as fossils, even in land-laid Paleozoic strata, and are unknown in the California Paleozoic.

Commonest plants in the early Paleozoic were the marine algae—single-celled plants—including seaweeds. Many of the algae were (and are) calcareous and helped build limestone reefs. "Coral" reefs, built in warm, shallow seas since early Paleozoic time, usually have included algae, as well as higher invertebrate animals that make calcium carbonate shells. Although land plants first appeared in the Silurian Period and became abundant and luxuriant later in the Paleozoic Era, the fact that California's Paleozoic rocks are almost all marine means that land plants as fossils of this age are apparently nonexistent in the state.

# 8
# building california's structural framework in mesozoic time

*Part of California's Mesozoic framework exposed today. Rounded granite joint-blocks of the Alabama Hills in the foreground, dark metamorphic rocks to Lone Pine Peak (left), and Mount Whitney at the very crest of the Sierra Nevada—these rocks were all formed in Mesozoic time.*

■ Dominating the relief of the state (inside cover) and forming a massive, double structural backbone, are two northwesterly trending mountain systems: the Klamath-Sierra Nevada-Peninsular Ranges, and the Coast Range system. The numerous lesser ranges of the Great Basin and Mojave Desert also follow this trend. The Transverse Ranges cut west to east across southern California, "chopping off" the Sierra Nevada, Coast Ranges, and Peninsular Ranges.

*The overwhelming event of the Mesozoic Era was the initial building of this structural framework of California.* By the close of the Era—about 70 million years ago—all the elements of the mountain systems and the Great Valley had been constructed. Our principal objective in this chapter is to reconstruct the history of this important event.

Although mountain building took place in limited areas in Late Triassic to Mid-Jurassic time, the major mountain-building epoch occurred near the close of the Jurassic Period. Because this important episode is best known in the Sierra Nevada, it has been called the Nevadan orogeny. Mountain building was by no means ended with the late Jurassic; it was continued and renewed periodically into late Cretaceous time, with the characteristic accompaniment of massive intrusions of granitic rock, volcanic activity, folding, faulting, and uplift of portions of the earth's crust.

Despite the fact that all California's major structural features were outlined during the Mesozoic Era, they have been so profoundly altered by later uplift and renewed orogenies, as well as by deep erosion, that today's landscapes are useless in reconstructing geologic history of the Mesozoic. The imprint of history must be sought in the rocks themselves.

Where are the rock formations of Mesozoic age? If they had not been disturbed, they would be found lying on top of Paleozoic and older rocks and beneath Cenozoic formations over most of the state. The exception to this, of course, would be those limited elevated areas that were undergoing erosion throughout the era. Actually, Mesozoic rocks are widely exposed in all the mountainous parts of the state (and this is most of it!).

Five of the eleven rock units shown on the large geologic map of California are Mesozoic. The oldest of the Mesozoic units groups together those sedimentary and volcanic rocks that are older than the Nevadan orogeny; that is, they date from the Triassic Period to early parts of the Late Jurassic Period. Rock formations of Late Jurassic and Cretaceous age are divided into four additional rock units. These are (1) *granitic rocks,* which are widely distributed through the Sierra Nevada, Klamath Mountains, and the Coast Ranges (west of the San Andreas fault only) and in all natural provinces in southern California; (2) *ultramafic rocks* (peridotites and serpentine), which are most widespread in the Klamath Mountains but are also distributed along the major fault zones in the Sierran foothills, Coast Ranges, and the Klamath; (3) *eugeosynclinal sedimentary and volcanic rocks* of the Franciscan Formation, which are found in the Coast Ranges only, except that there are similar rocks on Santa Catalina Island; and (4) *shelf and slope sedimentary rocks* (miogeosynclinal), which are most extensive in a belt on the eastern flanks of the Coast Ranges, but are also widely distributed within the Coast Ranges, southern Klamath Mountains, western Transverse Ranges, and western Peninsular Ranges. These four units not only comprise, collectively, the largest area of the state, but they have entered profoundly into building the geologic framework of California as we see it today. Solutions to some of the state's deepest geologic problems are intricately involved in understanding the origin and history of just these rock units. Consideration of the characteristics of these rock groups and their structure will give some appreciation of what it took—geologically—to frame today's California.

What are the rocks of each of the five Mesozoic groups like and what inferences does each give on California history during the Mesozoic Era?

## PRE-NEVADAN ROCKS AND THE HISTORY THEY SHOW

Rock formations older than the Nevadan orogeny are sedimentary and volcanic types, nearly all of marine origin as far as is known. Their very wide distribution from end to end of the state, although principally in narrow belts along and within the mountain ranges, strongly suggests that most of California was covered by seas in Triassic and Early to Middle Jurassic time. Triassic rocks lie in many places, conformably on Permian rock formations, and reflect the continuation of somewhat similar environments from Permian into Triassic time.

The geologic map shows that in northwestern California there are two belts of "Mesozoic sedimentary and volcanic rocks older than the Nevadan orogeny": one in the west bordering the Klamath Mountains and Coast Ranges provinces, the other in the far east on the flanks of the Klamath Mountains.

The western belt consists principally of the Galice Formation—3,000 feet of metamorphosed dark mudstone and shale, graywacke, and some conglomerate, plus 7,000 feet of metamorphosed volcanic rocks at the base of the section. Fossils show that this formation is of marine origin and of early Late Jurassic age. The Galice Formation probably correlates with the Mariposa Slate of the western foothill belt of the Sierra Nevada.

In the eastern belt, northeast of Redding, there is a continuous, conformable section of marine dark shale, siltstone, limestone, sandstone, and interbedded lavas, volcanic tuff and breccia, and tuffaceous sedimentary rocks ranging in age from Late Permian to Middle Jurassic. The lower thickness of about 8,000 feet of this section has Triassic fossils; the top 6,800 feet is Jurassic. This well-stratified, fossiliferous Triassic-Jurassic section has been divided by geologists into eight formations, which can be separately mapped. The great angular unconformity between these rocks and all later formations marks the uplift, disturbances, and erosion accompanying the Nevadan orogeny.

Throughout the High Sierra, numerous roof pendants of pre-Nevadan strata occur in a northwesterly zone for more than 150 miles. Metamorphosed shale and graywacke are abundant, but very thick sections of volcanic rocks of flow and explosion origin are characteristic. Being partly metamorphosed and containing scanty fossils, many of these roof pendants are of uncertain age and have been mapped by geologists as "Jura-Trias metavolcanics." In the high Ritter Range of the eastern Sierra there is a

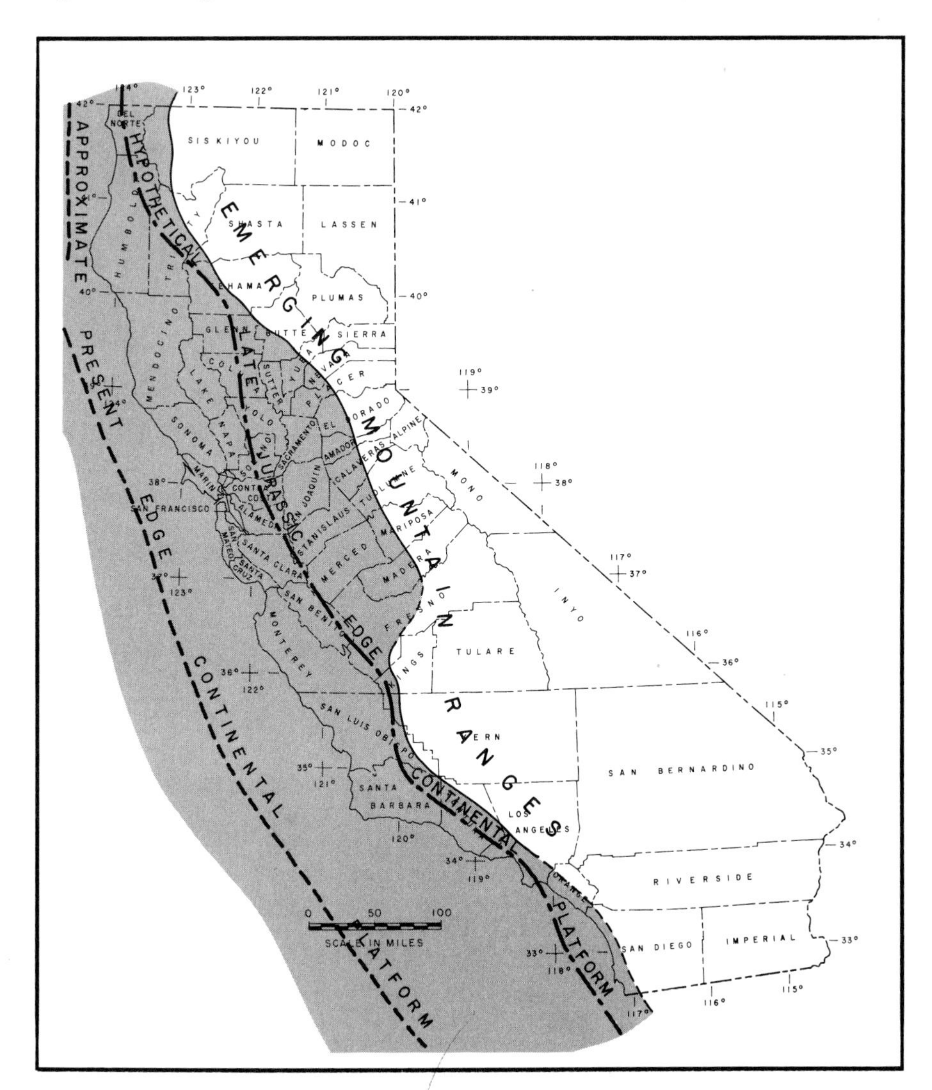


**FIGURE 8-1** *General areas of lands and seas in Late Jurassic time. The Nevadan orogeny was developing an extensive mountain system, including the Sierra Nevada, Klamath Mountains, and Peninsular Ranges. Beginning in the late Middle Jurassic, much of the Franciscan Formation was probably being deposited in troughs involved in a subduction zone—hence the attempt to show a "Late Jurassic edge of the continental platform."*

**FIGURE 8-2** *General areas of lands and seas in Late Cretaceous time. High and rising Sierra Nevada, Klamath Mountains, and various ranges in southern California were establishing the mountainous framework of the state. Map constructed from present-day outcrops; it does not include possible crustal-plate movements.*

section of 30,000 feet of metamorphosed pyroclastic (explosive volcanic) and sedimentary rocks of both marine and subaerial origin. Volcanic rocks in the lower part of this section have been radiometrically dated at 230 million to 265 million years old (Permian); but, 10,000 feet above the base of this section, early Jurassic marine fossils have been collected. Triassic rocks may be missing here, but in the southern Inyo Mountains not far away there are 1,800 feet of fossil-bearing marine shales and limestones of Triassic age.

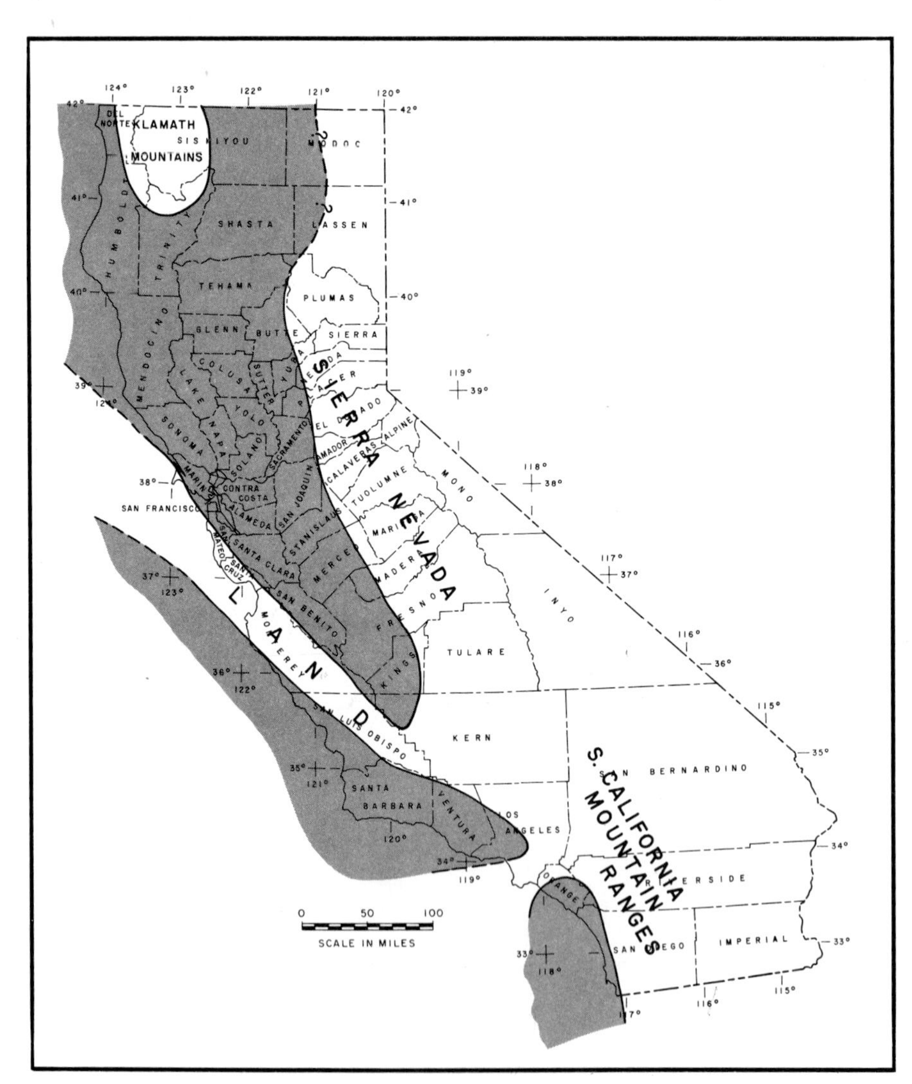

**8-3** *Looking east across Merced River up Yosemite Valley. El Capitan is on left, crest of Half Dome shows in lower center, Sentinel Rock is to Half Dome's right, and Bridalveil Fall is on near right. Flat floor of the valley has been built up by stream and lake sediments to a thickness of more than 1,000 feet since the Ice Age. Granitic rocks exposed in the walls of the valley are all Late Cretaceous in age—among the youngest in the Sierra Nevada batholith. (Mary Hill photo.)*

**8-4** *Serpentine outcrop, Tiburon Peninsula, Marin County, showing typical blocky jointing and rough weathered texture. (Salem Rich photo, California Division of Mines and Geology Bull. 183.)*

The most extensive and continuous outcrops of pre-Nevadan rocks are in the western foothill belt. The formations consist of metamorphosed clastic sedimentary and volcanic rocks including the Mariposa Slate, Logtown Ridge andesites, and sedimentary and volcanic rocks of the Cosumnes Formation. The few fossils show that both Triassic and Jurassic time are represented, but the rocks are mostly of early Late Jurassic age.

The Taylorsville region, at the northwestern end of the Sierra, includes about 800 feet of slate and 200 feet of the Hosselkus Limestone of Triassic age. The Mount Jura section in this region, which is the classic section of the marine Jurassic in California, consists of 13,000 feet of fossiliferous marine shale, mudstone, limestone, conglomerate, tuff, volcanic breccia and agglomerate, and sandstones full of volcanic particles. Shallow seas covered this area for the entire period.

In southern California, only widely scattered remnants of Jurassic rocks of pre-Nevadan age are found. For the most part, they consist of metamorphosed clastic sedimentary rocks—originally shale, sandstone, and some conglomerate—and abundant volcanic materials in great variety. Ages are often extremely uncertain because of scarcity of fossils, metamorphism, and lack of a clear stratigraphic succession. Examples are the Sidewinder volcanics in the Barstow area, the Santa Monica Slate in the Santa Monica Mountains near Los Angeles, and the Julian Schist in the Peninsular Ranges. However, the Bedford Canyon Formation in the Santa Ana Mountains, which consists of metamorphosed siltstone, graywacke, and marble, is well dated by marine fossils as early Late Jurassic in age; and the Santiago Peak volcanics of the Peninsular Ranges, near San Diego, are latest Jurassic.

In southern Death Valley, carbonate rocks of Permian age are overlain by 8,000 feet of metasedimentary and volcanic rocks containing early Triassic marine molluscs and corals. These materials were deposited in a northwesterly trending trough that may have been continuous with the Sierran geosyncline.

Pre-Nevadan formations are missing in the Coast Ranges (except in the far north adjacent to the Klamath Mountain province). There are several

**8-5** *Late Jurassic shale, Blue Canyon, Tehama County, typical of the Great Valley Series exposed on the eastern slopes of the Coast Ranges, western margin of the Sacramento Valley. (Olaf P. Jenkins photo.)*

**8-6** *"Tombstone rocks" of the pre-Nevadan middle Jurassic Logtown Ridge metamorphosed andesite formation in the Mother Lode belt, western foothills of the Sierra Nevada. (Sarah Ann Davis photo.)*

possible explanations: (1) they are present but have not been recognized because of lack of fossils and metamorphism; (2) the area was above sea level and was being eroded (not likely, because the seas of this time probably spread inland from the west); or, perhaps most likely, (3) the area of the Coast Ranges was deep sea floor, which we have not recognized in rock exposures.

Major elements in the structural framework of the state are certainly the Sierra Nevada and the Coast Ranges. We shall now look more closely at their rock formations and structure and draw up a reasonable sequence of events in the building of each which is consistent with what is known of geologic facts and processes.

## GRANITIC ROCKS AND THE BUILDING OF THE SIERRA NEVADA

A glance at the geologic map shows that the dominant feature of the Sierra Nevada is a great mass of granitic rock—the Sierra Nevada batholith. A batholith is a large mass of granitic rock—measurable in terms of hundreds of miles in length and up to a hundred or more miles in width—formed from the slow cooling of molten magma several miles below the surface. The southwestern portion of the Sierran batholith turns westward, around the southern end of the Great Valley, into the Coast Ranges. It continues northward into the Klamath Mountains and southward across the Transverse Ranges into the Peninsular Ranges.

How was the granite batholith formed and what are its relationships to the mountain ranges it occupies?

We shall base our discussion of these questions on a series of generalized west-east cross sections that illustrate reconstruction of several stages in Sierra Nevada history (Figure 8-8).

### present-day sierra nevada

Examine Figure 8-8d, which is a diagrammatic representation of the present-day Sierra Nevada. Measurements of gravity and studies of the passage of earthquake waves through the Sierra and underlying crust and mantle show that there is a Sierran root: peridotite (or serpentine) of the upper mantle and the overlying basaltic layer at the bottom of the earth's crust are bowed down to depths of about 30 miles (50 kilometers) compared with 10 to 15 miles under the Great Valley and slightly more in the Great Basin.

We have already learned that Paleozoic miogeosynclinal limestone-rich rocks lie unconformably on Precambrian formations east of the Sierra Nevada and that Paleozoic eugeosynclinal rocks (Calaveras Formation), rich in volcanics, are exposed in the western foothills. Mesozoic sedimentary and volcanic rocks lie on top of the Paleozoic.

The diagram shows that Paleozoic strata dip (tilt) generally from both east and west *toward the axis of the mountain range,* which is occupied by various types of granitic rocks (shown in check patterns). Thus, the essential structure of the Sierra is that of a huge complex syncline (synclinorium), more than 100 miles across, that has been intruded in its axial region by granite batholiths. Oldest rock in the *undisturbed* "sediments" which flank the western part of the range, and lie unconformably on the Paleozoic, earlier Mesozoic and plutonic rocks are Late Cretaceous marine sandstone, conglomerate, and shale.

## history of the sierra nevada in the light of plate-tectonics theory

Referring to Figures 8-8a and 6-5, the diagrams model a section across the coast of California as it may have been in early Jurassic time, about 150 million years ago. Here is oceanic crust being thrust under the American continent with new lava generated and volcanism along an island arc. Perhaps the Logtown Ridge Formation (p. 127, Figure 8-6) includes andesitic rocks formed in that island arc. Note that the ancient Paleozoic (Calaveras Formation, for example, p. 113) and Precambrian rocks have been deformed, intruded by old Triassic granitic rocks, and all overlain by young marine Mesozoic sedimentary strata up to 150 million years old.

By 140 million years ago (Figure 8-8b), pressures between the subducting oceanic plate and the overriding continental plate had become so intense that all older rock formations were thrust-faulted, folded, and elevated in the *Nevadan orogeny*. This orogeny was most intense in the present areas of the western Sierra Nevada

**8-7** *Northern face of Stanton Peak, High Sierra in Tuolumne County. Weathering and glacial erosion have emphasized the very prominent jointing, or sheeting, dipping to the right, in the Soldier Lake Granodiorite. Talus, accumulated since glacial times, lies at the base of the peak. (Charles W. Chesterman photo.)*

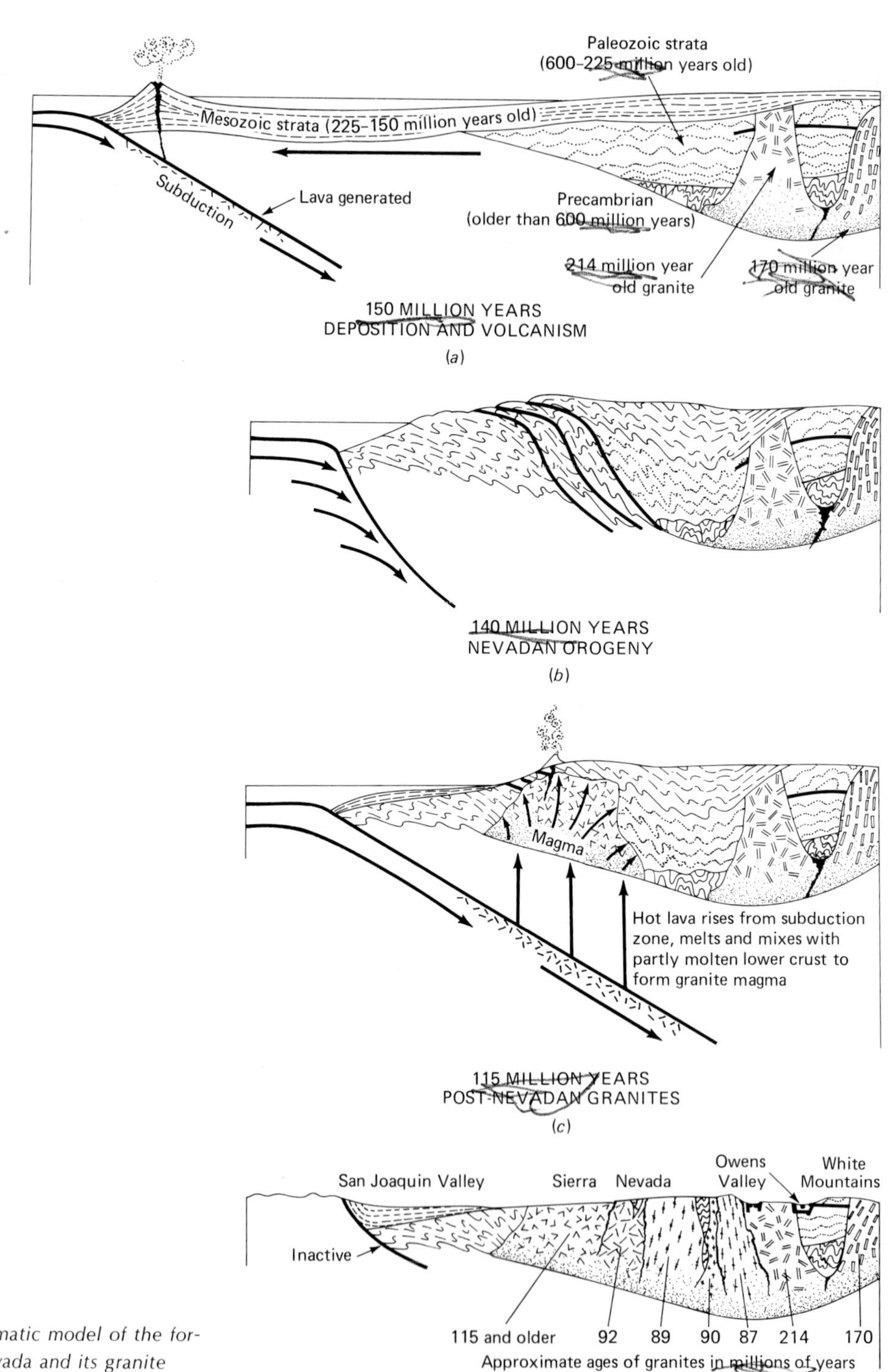


**FIGURE 8-8** *A diagrammatic model of the formation of the Sierra Nevada and its granite batholith consistent with plate tectonics theory. (After Bateman,* California Geology, *California Div. of Mines and Geology, January, 1974.)*

and Great Valley; it did extend into the Coast Range area, but folding was lower and more gentle. (Note the eastward-dipping, overlapping thrust slabs.) At this time, there was probably initial thrusting on the great Coast Range thrust, which dips eastward along the eastern flanks of the Coast Ranges.

In Figure 8-8c, we see that subduction has brought oceanic crust far beneath the American continental block generating magma that includes that formed by melting of continental granitic crust. In the western foothills of the Sierra we find granitic rocks of 115 to 140 million years old, with younger and younger granitic rocks eastward toward the heart of the high Sierra Nevada just west of Owens Valley.

The Sierra Nevada block of the present day (Figure 8-8d) is the result of 140 million years of uplift, westward tilting, and erosion. The old, inactive surface expression of the subduction zone is the Coast Range thrust seen on the west side of the Great Valley, where late Jurassic to late Cretaceous strata of the Great Valley Series are thrust over Franciscan rocks in the Coast Ranges.

## FRANCISCAN ROCKS AND THE BUILDING OF THE COAST RANGES AND CONTINENTAL MARGIN

The Franciscan Formation, a heterogeneous unit of eugeosynclinal sedimentary and volcanic rocks, consists predominantly of massively bedded graywacke with interbedded dark shale, minor chert and limestone, altered volcanic rock (greenstone), and various metamorphic rocks including the green chlorite-actinolite schists and the blue glaucophane schists. These rocks have all been intruded by, and are interbedded with, sheetlike masses of peridotite—mostly serpentinized and probably originating in the upper mantle—which are prevalent in many parts of the San Andreas and related fault zones. Rocks of the Franciscan Formation crop out very widely in all parts of the Coast Ranges, *except between the San Andreas and Nacimiento fault zones,* an area of continental granitic crust called Salinia.

The Franciscan Formation is perhaps as much as 50,000 feet thick, but no top or base has been observed; the sediments and volcanics were probably deposited directly on oceanic basalt, peridotite, and serpentine of the bottom of the crust and in the upper mantle. Based on considerations of the pressure-temperature conditions of formation of the blue schists and the minerals they contain—glaucophane, lawsonite, jadeite—geologists now postulate that these rocks indicate crystallization at low temperatures (less than 300° C), but high pressures (5,000 to 9,000 atmospheres). The rocks must therefore have reached a depth of 70,000 or more feet, through accumulation and downwarping so rapid that normal temperature conditions could not be established. Uplift must have been equally rapid. This could happen in a subduction zone.

We can visualize turbidity-current deposition and slumps of sediments down the continental slope, accompanied by submarine volcanism, into a deep eugeosynclinal trough at the base of the continental slope. Fossils are scarce in Franciscan rocks, but those found show Late Jurassic to early Late Cretaceous ages. The growing Sierra Nevada and Klamath Mountains were undoubtedly major sources of the sediments that make up the Franciscan Formation.

**8-9** *Colored Franciscan chert outcrop, Wolfback Ridge, Marin County. Average bedding layer in the photo is about 2 inches thick. This silica rock is extremely hard and brittle. The intense folding has obviously resulted in pervasive fracturing and minor faulting—particularly at the crests of the folds—but the material must have been in a semiplastic state under pressure when such tight folding took place (associated with pillow basalt and serpentine). (Mary Hill photo.)*

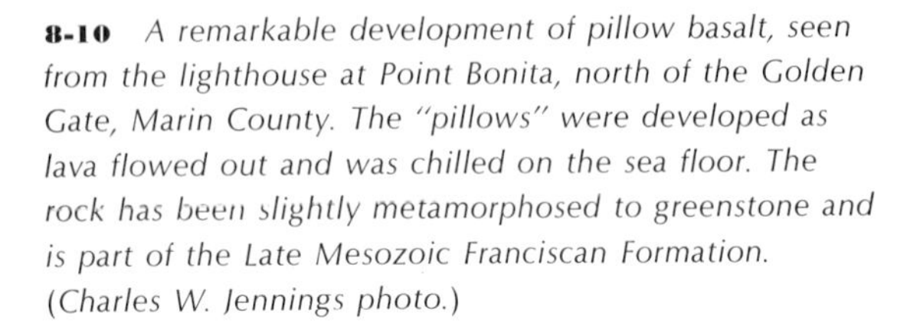

**8-10** *A remarkable development of pillow basalt, seen from the lighthouse at Point Bonita, north of the Golden Gate, Marin County. The "pillows" were developed as lava flowed out and was chilled on the sea floor. The rock has been slightly metamorphosed to greenstone and is part of the Late Mesozoic Franciscan Formation. (Charles W. Jennings photo.)*

Granitic rocks of the Coast Ranges seem to have been formed essentially contemporaneously with those of the Sierra Nevada and also with the entirely different Franciscan rocks. Radiometric ages of the quartz diorites and quartz monzonites of the Coast Ranges are mostly Late Cretaceous, but some are Late Jurassic. In the Coast Ranges, wherever the granitic rocks are found in contact with those of the Franciscan Formation that contact is a fault—the San Andreas fault, for example. Why isn't the granite ever found intruding Franciscan rocks? We have no good answer; this remains one of the problems connected with the Coast Ranges (but relate this to plate tectonic theory).

While Franciscan rocks were forming in great troughs at the bases of the continental slopes and granitic magmas were solidifying, a third major rock group was developing in relatively shallow seas on the shelves and slopes. About 30,000 feet of unmetamorphosed shelf-facies to deep-sea fan sandstone, shale, siltstone, and minor conglomerate and limestone of Late Jurassic to Late Cretaceous age crop out in the Coast Ranges. They are found, in some places, thrust faulted over the granitic rocks and the Franciscan Formation, but the thickest, most continuous section lies on the eastern flanks of the Diablo and Mendocino Ranges above the Coast Range thrust. The Late Jurassic part of this section is predominantly dark shale; the Lower and Upper Cretaceous rocks are mainly graywacke and arkose (granite sand) with minor shale and conglomerate.

## mountain building and westward extension of the continental platform

Profound faulting, roughly parallel to Coast Range structures, took place in the subduction zone as rocks of the Great Valley Series in the continental block were thrust many miles westward over oceanic-plate rocks of the Franciscan Formation (serpentine, basalt, chert, graywacke, and shale) in mid-Jurassic to early Cenozoic time. Subsequent folding, uplift, and erosion have left only remnants of the Coast Range thrust, except along the east side of the Coast Ranges, where it is well exposed. Over the

**8-11** *Typically weathered, eroded, highly sheared serpentine, west flank of San Benito Mountain, Diablo Range. (G. B. Oakeshott photo.)*

**8-12** *Tightly folded Franciscan blueschist on Santa Catalina Island. (Edgar H. Bailey photo, California Division of Mines and Geology Bull. 183.)*

**8-13** *Block of Franciscan gneiss, consisting of green pyroxene, red garnet, blue glaucophane, green epidote, and veinlets of lawsonite. Such isolated rock masses, common in the Franciscan Formation, may have been formed in the upper mantle, brought up by tectonic forces and left in their present position by erosion. (Edgar H. Bailey photo, California Division of Mines and Geology Bull. 183.)*

**8-14** *Vertically folded Upper Cretaceous strata at Point San Pedro, San Mateo County. Lighter-colored marine sandstone beds are alternating with dark shale. The folding took place during the Coast Range orogeny—late Miocene to Pleistocene time. (Mary Hill photo, California Division of Mines and Geology.)*

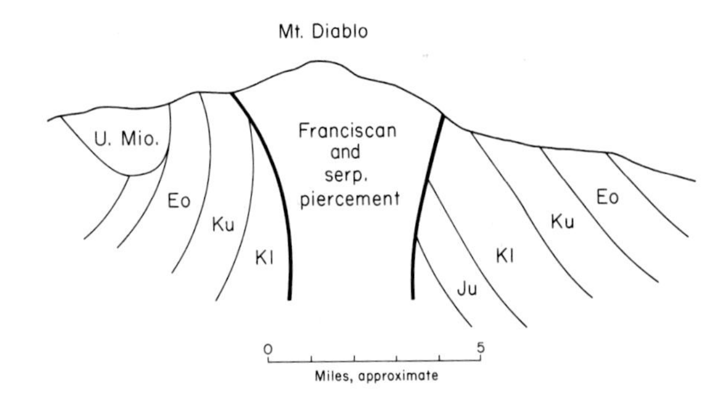


**FIGURE 8-15** *Mount Diablo, at the north end of the Diablo Range, consists of eroded Franciscan rocks and serpentine thrust upward as a piercement that has cut across younger rocks; or, are these steep faults an intensely folded segment of the Coast Range thrust?*

thrust the Great Valley Series is folded but relatively little disturbed; under the thrust the rocks of the Franciscan Formation are a tectonic jumble in many places, and some of its rocks are metamorphosed.

In early Cenozoic time—we think about mid-Eocene to Oligocene time—the regime was profoundly changed. Subduction and thrust faulting virtually ceased, and the great, right-lateral, strike-slip San Andreas fault system began to develop. Geologists differ in interpretation, but most believe cumulative right-lateral displacement totals 160 to 310 miles since middle Eocene time (see San Andreas fault, pp. 182–197). Along with the strike-slip faulting, folding and uplift have occurred, intermittently, in the Coast Ranges during Cenozoic time, culminating in major Coast Range orogeny in late Pliocene-Pleistocene time.

By the processes described here in the building of the Sierra Nevada and Coast Ranges, the American plate has been extended many miles westward and our present landscape of California has been built.

## CHARACTERISTIC MESOZOIC LIFE AND FOSSILS

The Mesozoic Era is best known for the dramatic and fantastic development of the reptiles. It has been deservedly called the Age of Reptiles. Great reptiles roamed the lands of the earth, swam in the seas, and flew in the air, in countless numbers during the Mesozoic Era, dominating life for more than 125 million years. Perhaps the most unusual of the animals that lived in the Age of Reptiles were the dinosaurs—literally "terrible lizards." Although they were the largest land animals that ever lived, the dinosaurs were exceeded in total bulk later by some of the modern whales. Roaming the low swampy lands and river floodplains over the site of the present Rocky Mountains were such beasts as *Tyrannosaurus rex,* the greatest land carnivore of all time; heavy armored forms like *Stegosaurus* and *Triceratops;* and bizarre forms like *Anatosaurus,* a duck-billed dinosaur. In the air were pterodactyls, with wingspreads up to 27 feet—more numerous than the birds of their time (Jurassic and Cretaceous Periods). A few of these creatures are shown in Figures 8-16 through 8-22.

### fossil reptiles from california

Few of the giant reptiles got as far west as California, for the Mesozoic environment was not favorable to their living. Land areas in California were very limited and such as did exist were largely mountainous—not the swampy river floodplains favored by the dinosaurs. In the seas, the eugeosynclinal troughs which were rapidly filling with turbidity sediments were unfavorable to preservation of bones of marine reptiles. Thus, it might be expected that Mesozoic fossil reptiles would be extremely rare in California.

The best chance of finding dinosaur bones might be in shallow marine shelf and brackish-water deposits just off a low, swampy coast. Just such conditions existed close to the end of Cretaceous time in the area of the present hills of the Diablo Range west of the San Joaquin Valley. There, in western Fresno County, about forty years ago a startling discovery was made—twenty-seven vertebrae of the duckbill dinosaur *Hadrosaurus*! The bones were excavated from shale of the very late Cretaceous Moreno Formation. The same formation, from Fresno, Stanislaus, and San Luis Obispo Counties, has yielded bones of mosasaurs—marine lizards several feet long—and plesiosaurs—strange, long-necked marine reptiles up to 40 feet in length, looking like the legendary "sea monsters." Farther north, in the same range in San Joaquin County, the snout of an ichthyosaur—a fishlike marine reptile 10 to 15 feet long—was found in chert of the Franciscan Formation, probably of Late Jurassic age.

The whole "race" of dinosaurs—on land, in the sea, and in the air—became extinct at the end of the Cretaceous Period all over the world, including California; none carried over into the Cenozoic Era. *Why,* is one of the geological problems for which we have no completely satisfactory answers.

**FIGURE 8-16** Brontosaurus—*the thunder lizard. He was the largest of the swamp-living, plant-eating dinosaurs; they grew to be from 50 to 85 feet long and weighed as much as 50 tons.* Brontosaurus *is best known in Cretaceous rocks in the Rocky Mountain area; not found in California. (Scratchboard drawing by Peter Oakeshott.)*

**FIGURE 8-17** Allosaurus. *One of the most vicious beasts of prey of all time, predecessor of the giant* Tyrannosaurus, Allosaurus *stood 20 feet high. His bones are found in the Late Jurassic rocks of the Rocky Mountain province, not in California. (Scratchboard drawing by Peter Oakeshott.)*

### other vertebrates

The first birds and the first true mammals developed in the Jurassic Period by evolution of branches of the reptiles, but both remained small and inconspicuous during the Mesozoic Era. Neither are known from Mesozoic rocks in California. Amphibians, which had developed rapidly and become prominent in the late Paleozoic Era, declined and became inconspicuous during the Mesozoic.

Fish were very numerous and highly developed in Mesozoic seas. They were distinctly more modern in Cretaceous time than they were in the Paleozoic. Fish with complete bony skeletons and well-formed fins and tails are abundant as fossils in some of the shale formations, for example, the late Cretaceous Moreno Formation of the Diablo Range. Some of the thinly laminated shale when split along its bedding planes exposes fish skeletons or impressions of the skeletons.

### invertebrates

Mesozoic rocks in California—particularly the miogeosynclinal types—contain abundant invertebrate fossil remains.

The most important shelled invertebrates were the molluscs: the coiled, gastropods, the bivalved pelecypods, and the coiled, many-chambered cephalopods. All are well represented in California. By Cretaceous time there were entire

**FIGURE 8-18** Stegosaurus—*the plate-bearing dinosaur of Late Jurassic age. He was a vegetarian and grew to be 30 feet long and 10 feet tall. His bones are found in the Rocky Mountains, not in California. (Scratchboard drawing by Peter Oakeshott.)*

reef structures composed of oysters (*Ostrea*) and similar shells. Large conical, twisted shells of the pelecypods known as rudistids have been found in Cretaceous strata along the coast of northern California. The pelecypods *Buchia piochii* (shown in the drawing at the beginning of Chapter 4) and *Buchia crassicollis* are, respectively, characteristic of late Jurassic and early Cretaceous beds. *Buchia concentrica* occurs in the late Jurassic Galice Formation in the northern Coast Ranges.

The ammonites—flat, coiled, external-shelled cephalopods which had begun to develop back in Devonian time—became exceedingly complex and abundant in the Mesozoic Era. In terms of invertebrate life, the Mesozoic Era might be called the Age of Ammonites. Their sutures (junctions of the shell partitions with the inner wall of the shell) were developed in great variety and complexity. The abundance, variety, complexity, and rapid changes of the ammonites make them the most useful index fossils of the era. Ammonite shells are known from less than 1 inch across to a diameter of 6 feet. They occur as fossils in most Mesozoic strata in California, but are most prolific in shelf-facies rocks of the Cretaceous Period. Extraordinarily, ammonites became extinct all over the world at the close of the Cretaceous Period.

Beginning with the Cretaceous Period, microscopic single-celled, chambered shells of the smaller foraminifera (much smaller than Paleozoic

**FIGURE 8-19** Anatosaurus—*the duck-billed dinosaur. Its bones have been found in the Late Cretaceous swamp sediments of the Diablo Range of California. The duckbill grew 15 to 20 feet tall and was highly specialized for scooping up large quantities of vegetable muck from swamps. (Scratchboard drawing by Peter Oakeshott.)*

**FIGURE 8-20** Triceratops—*the horned dinosaur. A vegetation eater, he was as much as 30 feet long and had a skull up to 8 feet long from the parrot beak to the base of the hood. Not found as a fossil in California. (Scratchboard drawing by Peter Oakeshott.)*

fusulines) become important for age determination and correlation. These calcium carbonate shells are found in great numbers in shelf-facies rock formations and in some Cretaceous marine shale beds in California; but they also are abundant in some of the limestone lenses in the Franciscan Formation.

### plants

California has no record of the forests of giant ferns, horsetails, and club mosses that lived in eastern North America in late Paleozoic time—particularly during the Pennsylvanian Period. Also, California lacks the record of cycads, conifers, and gingkos, which were so typical of Mesozoic environments elsewhere. The reason, of course, is the same as for this state's scarcity of dinosaur remains: *there just wasn't much land above sea level* and only very scant unfossiliferous land sediments remain.

**FIGURE 8-21** Plesiosaurus *and* Mosasaurus—*marine dinosaurs. They are known from Triassic to Cretaceous time.* Mosasaurus *(bottom) is known to a maximum length of 50 feet. These two lived in shallow late Cretaceous seas in the Diablo Range area, southern Coast Ranges of California. (Scratchboard drawing by Peter Oakeshott.)*

**FIGURE 8-22** Pteranodon—*the flying reptile. A batlike creature that reached a maximum wing spread of 27 feet. In the sea below is* Ichthyosaurus, *which grew to a maximum length of about 40 feet. Both* Pteranodon *and* Ichthyosaurus *lived from Mid-Triassic to Late Cretaceous time.* Ichthyosaurus *is known from Late Cretaceous rocks in the Diablo Range of California. (Scratchboard drawing by Peter Oakeshott.)*

**FIGURE 8-23** *Corals in Triassic Hosselkus limestone, Klamath Mountains. (Scratchboard drawing by Peter Oakeshott; slightly larger than natural size.)*

A very important plant preserved in fine-grained Cretaceous marine shale in the California Coast Ranges is the diatom. Diatoms are microscopic, single-celled plants that have lived in enormous numbers in both fresh and salt water from Cretaceous time on. They are so numerous and afford such an important source of food for marine animals that they have been called the grass of the sea.

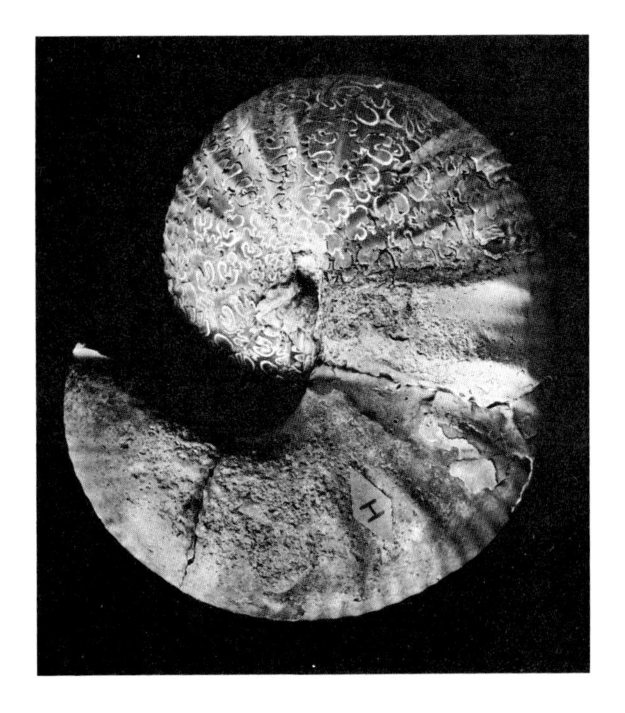


**8-24** *An ammonite—Scaphites—characteristic of Late Cretaceous time. A few inches across. (Smithsonian Institution photo.)*

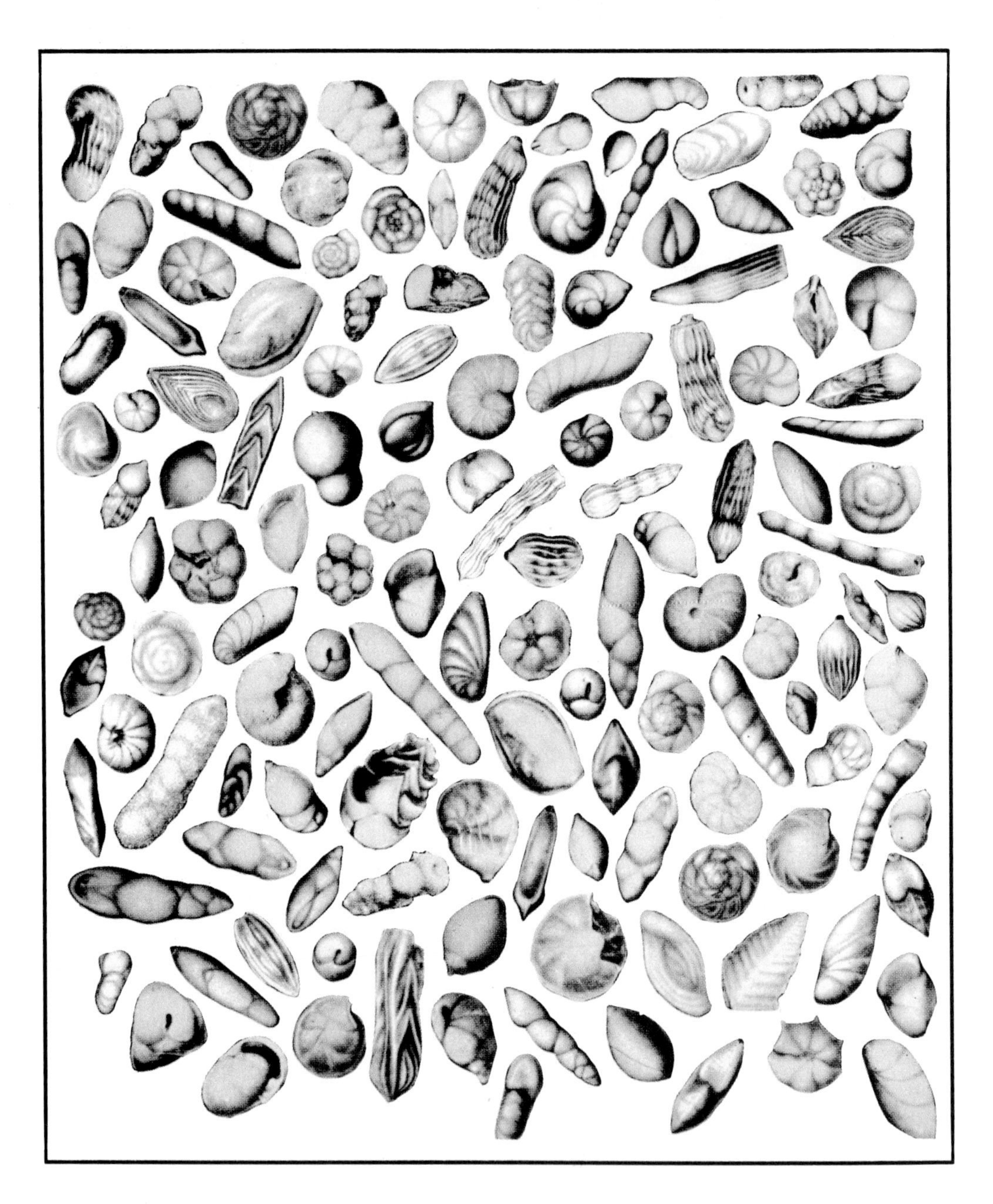

**FIGURE 8-25** *Foraminifera from Upper Cretaceous marine shale beds on the Stanford University campus. The largest of these is about one millimeter long. (From sketches by Perfecto M. Mary, California Division of Mines and Geology Special Report 66.)*

# 9 modern life and land emerge in the cenozoic era

One of the first signs of the emergence of modern life was in the vegetation. Cast of the delicate leaf of a broad-leaf maple preserved in Late Cenozoic sandstone.

■ The era of modern life, the Cenozoic (*ceno* means "recent," *zo* means "life"), dawned in the land that is California about 65 million years ago. For the first time in 1.8 billion years of "California" history we see a land that is recognizable, for the geologic framework of today had been built by the close of the Mesozoic Era.

An ancestral Sierra Nevada had emerged from the ancient geosyncline; never again did the seas cross its site. The Klamath Range was also "permanently" mountainous, as were most of the Peninsular Ranges. The Mojave Desert, Basin Ranges, and parts of the Transverse Ranges were probably also areas of uplift and high-standing lands. All of the positive lands were being eroded during Cretaceous and Cenozoic time; and as the lands rose higher and mountain building was continued and renewed from time to time, great masses of sediment were carried to the western seas or were deposited in inland basins. The action was in the Coast Ranges, the western Transverse Ranges, and the northwestern end of the Peninsular Ranges during Cenozoic times and we shall focus particularly on these areas as we consider that era. *The scene has shifted westward and southwestward during geologic time.*

## TYPES AND DISTRIBUTION OF CENOZOIC ROCKS

Being youngest, rocks of the Cenozoic Era generally overlie those of all older eras and are less likely to have been destroyed by erosion. Three major groups of Cenozoic rock formations are shown on the 1:2,500,000 colored geologic map of California (available from the U.S. Geological Survey or the California Division of Mines and Geology): (1) volcanic rocks, (2) marine sedimentary rocks, and (3) nonmarine sedimentary rocks and alluvial deposits.

Volcanic rocks are distributed literally all over the state. They range in composition from basalt to rhyolite and in age from Eocene to Recent (Holocene). The largest single area, by far, is the Modoc Plateau and Cascade Mountains in the northeastern corner of the state, but volcanic rocks are especially extensive throughout the Great Basin and Mojave Desert. There are also extensive fields of volcanic rocks in the Sonoma County–Clear Lake area, north of San Francisco Bay. The smallest amount of Cenozoic volcanic rock is in the Klamath Mountains and adjacent parts of the northern Coast Ranges.

Marine sedimentary rocks of Cenozoic age crop out most widely in the Coast Ranges, in the western Transverse Ranges, and along the coast of the Peninsular Ranges. Thousands of wells drilled for oil, gas, and water show that they also extensively underlie the alluvium of the Great

**FIGURE 9-1** *General area of lands and seas during Paleocene time. The lands were largely low hilly areas; the high mountain ranges resulting from Late Jurassic and Cretaceous orogenies had been greatly reduced by erosion.*

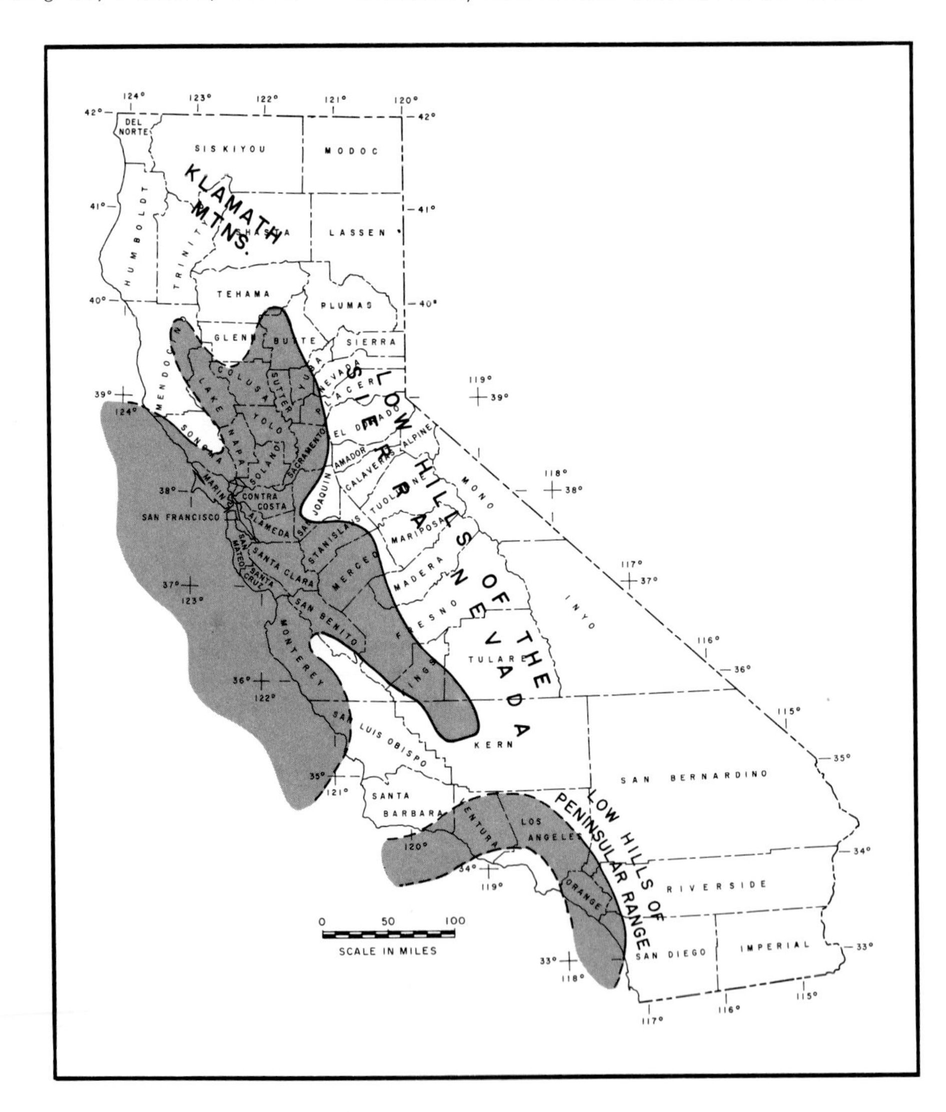

Valley. They have been found as much as 100 miles inland—at Sutter Buttes in the Sacramento Valley, east of Bakersfield in the foothills of the southern Sierra Nevada, and in the Orocopia Mountains northeast of the Salton Sea.

By mapping outcrops of the Cenozoic marine formations and inferring connections where rocks have been removed by erosion, geologists have outlined the shallow, broader seaways and complex, often narrow, arms of the seas that ebbed and flowed over the western margins of

**FIGURE 9-2** *General areas of lands and seas in late Eocene time. The lands were still low, but seas were shallow and low islands developed in the Coast and Transverse Range areas.*

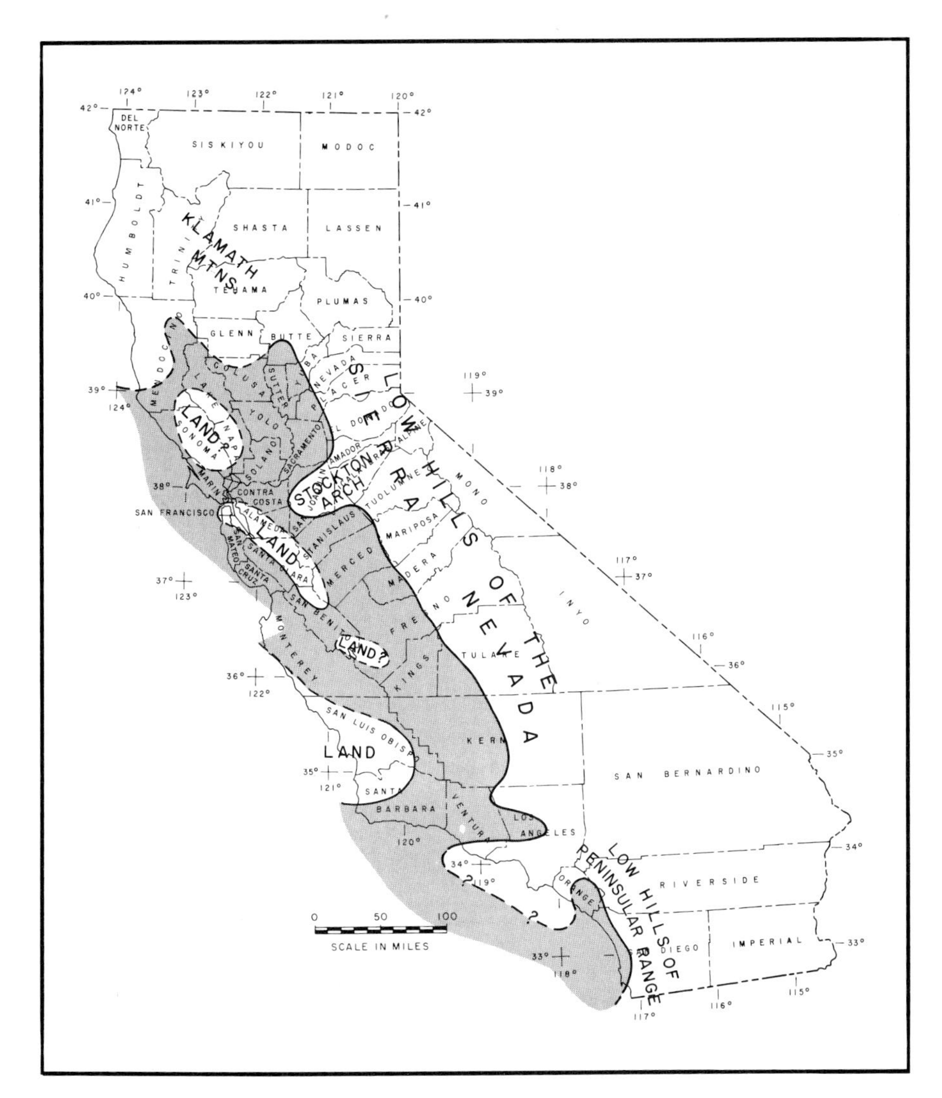

California during this latest era of geologic history (see Figures 9-1 through 9-5, which show the distribution of lands and seas). Although narrow, some of the marine troughs in the southern Coast Ranges and western Transverse Ranges subsided rapidly by faulting and folding and received many thousands of feet of marine and continental (land-laid) sediments. These maps (Figures 9-1 to 9-5) have not been adjusted for crustal-plate movements.

Alluvial deposits—mostly Quaternary in age—are found in basins and on plains throughout the

**FIGURE 9-3** *General areas of lands and seas in late Oligocene and early Miocene time. Narrow, shallow coastal seas existed across the Coast Ranges area and into the San Joaquin Valley. The northern Coast Ranges and Peninsular Ranges were emerging.*

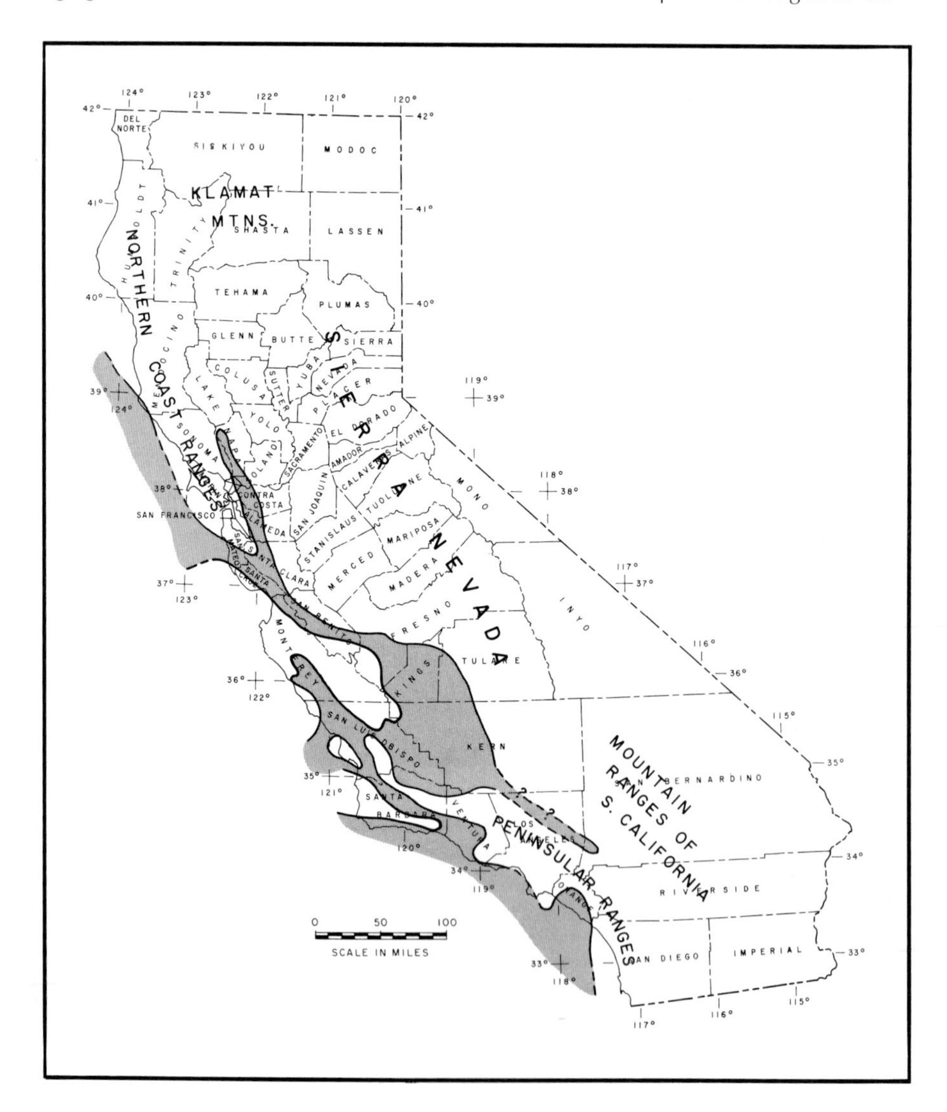

state, locally covering entire areas, as in the Great Valley. Wherever they occur, they tend to cover and obscure rocks of older age and thereby make interpretation of older geologic history more difficult. For the most part, Quaternary alluvium is little consolidated. Alluvium and nonmarine sedimentary rocks of Tertiary age are widespread in present-day inland basins, as in Death Valley and the Mono Basin. They are also found in presently mountainous areas where former basins existed, where they may be thoroughly consolidated, folded, and faulted.

**FIGURE 9-4** *General areas of lands and seas in late Miocene time. Broad seas extended across much of the Coast Ranges area into the San Joaquin Valley. Further uplift in Coast Ranges.*

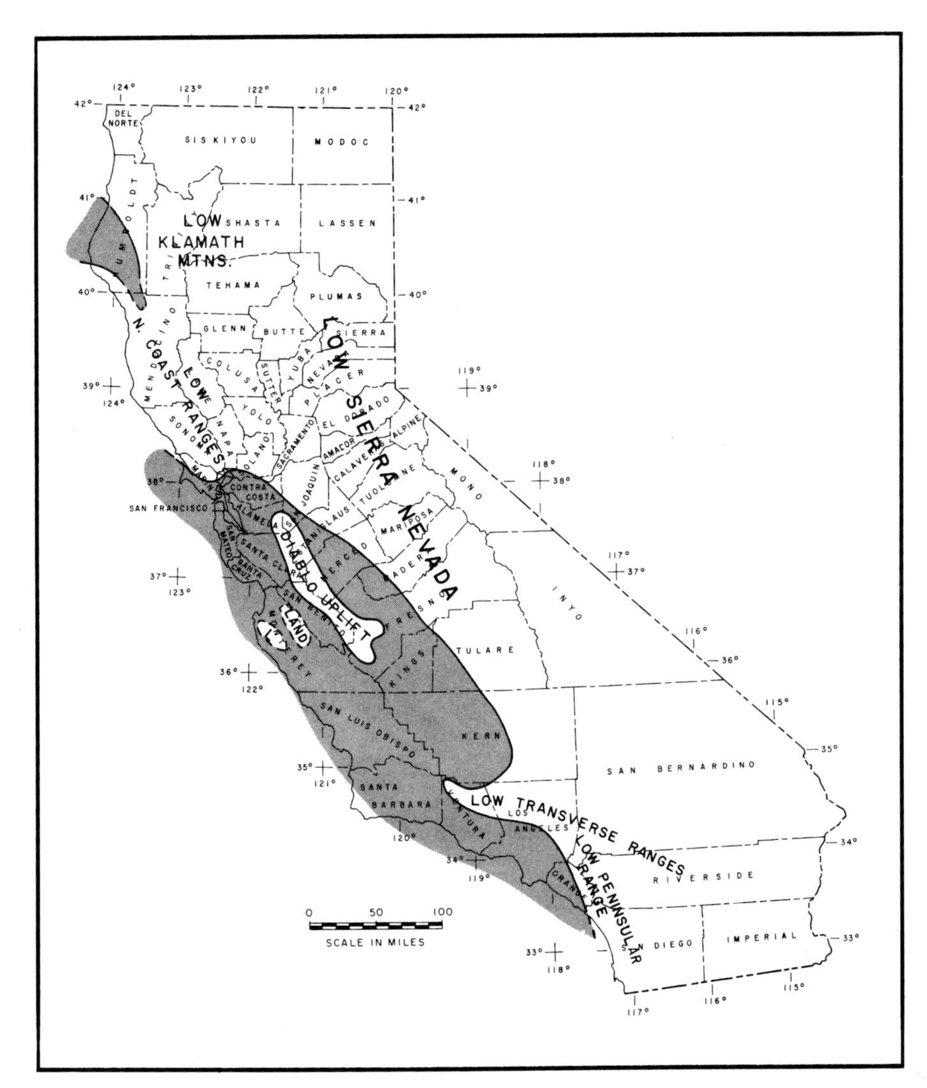

**FIGURE 9-5** *General areas of lands and seas in late Pliocene time. Seas were very restricted and close to the present-day coastal margins; except that a shallow sea still extended across the southern Coast Ranges area into the southwestern San Joaquin Valley. High mountains were similar to those of the present, except the Coast Ranges.*

Marine as well as nonmarine sedimentary rocks of the Cenozoic Era are largely *clastic*—conglomerate, sandstone, siltstone, and mudstone. Limestone occurs, but it makes up only a small fraction of the combined, total thickness of the Cenozoic rocks. Some sediments of largely organic and chemical origin—diatomaceous shale, silicious shale, and chert—are important types in the Coast Ranges.

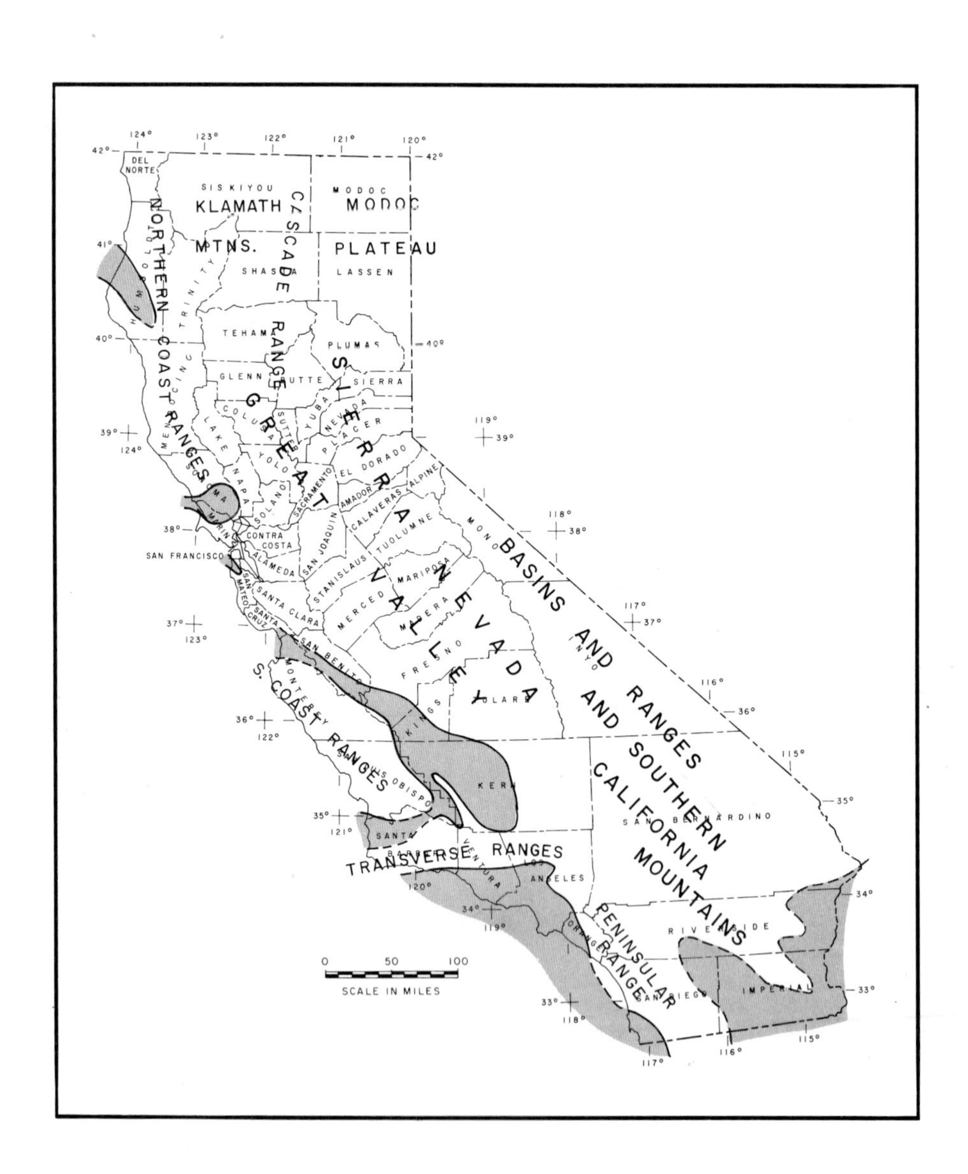

## EARLY TERTIARY ROCKS AND HISTORY

For convenience of discussion, "early Tertiary" is taken as the first 40 million years of the Tertiary Period, including the Paleocene, Eocene, and Oligocene Epochs. What was California like in the early Tertiary?

The Sierra Nevada had been reduced from high-standing Cretaceous mountains to a gently sloping upland by Paleocene time. There were broad, slowly moving streams carrying gravels (gold-bearing) to the Great Valley. Paleocene and early Eocene seas came from the west into the southern Sacramento Valley (Figures 9-1 and 9-2). Into these shallow seas poured clastic sediments—sand, gravel, silt, and clay—now consolidated as the Paleocene Martinez and Meganos Formations, and the early Eocene Capay Formation; all these are well known because they are important for gas production.

Paleocene marine sedimentary rocks over the state are quite similar to those of the Upper Cretaceous Period and are widely distributed through the Coast Ranges, San Joaquin Valley, and northern Peninsular Ranges (Santa Ana Mountains). However, Paleocene and Eocene seas were not as widespread as those of Late Cretaceous time.

Marine shale and sandstone were deposited in a basin occupying the San Joaquin Valley in Eocene time. In the low western foothills of the Sierra, the beautifully clean white sands and clays of the Ione Formation were deposited on floodplains and in shallow lakes and lagoons on the landward margins of mid-Eocene seas.

The Oligocene Epoch was a time of quite localized seas, which extended particularly over parts of the Santa Cruz Mountains, San Francisco Bay area, and the western portion of the San Joaquin Valley (Figure 9-3). Oligocene climates were seasonal and semiarid to arid. Extensive land-laid conglomerate and coarse sandstone, including red beds due to oxidized iron minerals, are widely distributed in the southern Coast Ranges and Transverse Ranges. Volcanic rocks thousands of feet thick are interbedded with coarse nonmarine sediments in the Transverse Ranges as the thick Vasquez Formation. The Oligocene Epoch is also represented by coarse land-laid sedimentary rocks in the Klamath Mountains, the Death Valley region, and Mojave Desert.

## LATE TERTIARY ROCKS AND HISTORY

Late Oligocene environments continued, for the most part, into early Miocene time, but with gradual broadening of the seas. Warm, shallow arms of the sea extended irregularly in numerous channels through coastal California from the northwestern end of the Santa Ana Mountains in southern California to Point Arena in northern California (Figure 9-3); a broad embayment covered the greater part of the southern San Joaquin Valley. This time and environment, characterized by a distinctive group of fossils, has been called the Vaqueros Stage.

During the late Tertiary, the principal seaways continued to be in the Coast Ranges. Middle Miocene seas were more widespread than those of the early Miocene. Large volumes of volcanic tuff, breccia, agglomerate, and andesite flows are found in middle Miocene rocks of the southern Coast and Transverse Ranges. Shallow seas spread and again reached a maximum in late Miocene time. The most widespread late Tertiary formation is the middle to late Miocene Monterey Formation, which is characterized by abundant siliceous shales and chert beds. Along with similar rocks, it extends from Point Arena to southern coastal California as far south as Oceanside on the western flank of the Peninsular Ranges. In southern California, rocks of the age of the Monterey Formation are more sandy than in the north—in the Santa Monica Mountains and the Puente Hills, for example.

About 20,000 feet of sediments ranging in age from Pliocene to Recent fill the Salton Trough. These are chiefly alluvial fan, floodplain, stream, and lake deposits, but the Pliocene Imperial Formation was deposited in a shallow sea that came in from the Gulf of California. In late Pliocene time shallow sea covered most of Imperial Valley and extended across eastern Imperial and Riverside Counties (Figure 9-5).

**9-6** *Table Mountain and San Joaquin River, Fresno and Madera Counties, western Sierra Nevada. Table Mountain is a bare, flat-topped ridge looping from the center toward lower left of the photo. It consists of the remnants of a Pliocene basalt flow which filled an ancient river channel. Quite resistant to erosion, it has been left as a curved table as much as 1,500 feet above the present level of the San Joaquin River (meandering across the lower part of the picture) and flowing toward the right (west). (Vertical aerial photo, U.S. Geological Survey, Prof. Paper 590, 1968; scale approximately 1 inch equals 4,085 feet; south is at top.)*

In the Sierra Nevada in late Oligocene to Miocene time, extensive masses of rhyolite tuff and gravels of the Valley Springs Formation blanketed the western slopes and filled stream valleys. This volcanic episode was followed, in the Pliocene, by eruptions of andesitic volcanic agglomerates, which formed mudflows and, with conglomerate and sandstone, buried the northern Sierra under 1,500 to 4,000 feet of debris of volcanic origin. This blanket, known as the Mehrten Formation, was thickest near its source at the crest of the Sierra Nevada and literally filled many deep valleys on the western slopes.

The western part of the Transverse Ranges, which received great thicknesses of mostly marine Cretaceous and lower Tertiary deposits, continued as a trough of deposition called the Ventura Basin during later Cenozoic time. More than 40,000 feet of Cenozoic sediments accumulated in the Ventura Basin.

Seaways were greatly restricted in California in Pliocene time; and by the end of the Pliocene Epoch, seas had left the San Joaquin Valley. As the Coast Ranges began to be elevated, Pliocene marine sands, muds, and some tuff were deposited in narrow, shallow embayments throughout the Coast Ranges as far north as the Eureka Basin. At least 15,000 feet of late Miocene and Pliocene marine muds and sands were deposited in the Los Angeles Basin; the kinds of bottom-dwelling foraminiferal fossils show that some sediments were deposited in water as deep as 5,000 to 6,000 feet.

Floods of gravel and coarse sand deposited in the channels, deltas, and floodplains of streams almost covered the site of the southern Coast Ranges, parts of the western Transverse Ranges, much of the Great Valley, and southern margins of the Klamath Mountains in late Pliocene and early Pleistocene time. The great coarsening and increased amount of continental sediments may be interpreted as a reflection of strong orogenic activity of that time.

Pleistocene seas encroached to only a minor extent across coastal California; they were broadest and the sediments were thickest in the near-coastal part of the Los Angeles Basin.

## BUILDING OF THE COAST RANGES

Shallow, early Paleocene seas advanced across the western margin of the state over the eroded surfaces of the older rocks, in most places. For instance, coarse sandstone and conglomerate of the Paleocene Carmelo Formation at famous Point Lobos State Reserve south of Monterey lie on the eroded surface of Late Cretaceous granitic rock. At Point San Pedro, about 12 miles south of San Francisco, similar Paleocene strata lie with a slight angular unconformity on Upper Cretaceous strata. Such occurrences are evidence of uplift and some folding at the close of the Cretaceous Period. On the eastern flanks of the Diablo Range, however, marine sedimentation seems to have been locally continuous from Cretaceous into Paleocene time.

The Paleocene and Eocene epochs covered 25 million years of crustal quiet. Then, gradually, intermittent crustal unrest began. From late Eocene to early middle Miocene time, seaways became more restricted and parts of the Coast Ranges were elevated above sea level. For example, in the Coast Ranges, Oligocene formations crop out only in the San Francisco Bay area and southern ranges; probably the northern Coast Ranges were above sea level during Oligocene time. In the Santa Cruz, Santa Lucia, and Diablo Ranges, shallow-water marine sandstone, shale, conglomerate, and local tuff beds represent deposits in rather restricted bays and channels. Southward from the San Francisco Bay area, the Oligocene strata become more continental and in the southern Coast Ranges are mostly land-laid.

In the later part of the early Tertiary, crustal movements began in the Coast Ranges. In different portions quite different events took place. Details of geologic history during the Cenozoic Era were strikingly different in different ranges within the Coast Range system. Marine embayments, high-standing lands, folded and faulted ranges of hills, active volcanos, erosion, and deposition all went on locally at the same moment in history. Numerous unconformities between upper Tertiary rock formations are one type of evidence of crustal activity.

The enormous volumes of volcanic materials—tuff, breccia, agglomerate, rhyolitic to andesitic and basaltic flows, and plugs—that were extruded during middle Miocene time preceded and accompanied local crustal movements. After advances of the seas in the late Miocene, that epoch closed with important uplift and folding in many of the Coast Ranges. Pliocene beds in many places, but not everywhere, are unconformable on Miocene formations.

Culmination of the Coast Range orogeny came during late Pliocene to mid-Pleistocene time. Abundant evidence for this is found in the essentially flat-lying attitude of widespread late Pleistocene and Holocene sediments that lie with angular unconformities on all older formations. These relationships not only prevail throughout the Coast Range province but also extend into the Transverse Ranges, the Los Angeles Basin and northern part of the Peninsular Ranges. Relationships between the land-laid sedimentary and volcanic rock formations of the Mojave Desert also show this orogeny, although it is not as well dated there as in the Coast Ranges.

This major mountain-building epoch did not affect all regions at precisely the same time but progressed in waves through the Coast Ranges. For example, in the Coast Ranges just north of the San Francisco Bay area, the late Pliocene marine Merced Formation and Sonoma volcanics lie over wide areas with only gentle dips on older formations, although locally steeply folded strata are found. Perhaps here the most intense orogenic episode was in the middle or earlier Pliocene time. On the other hand, in the Ventura Basin part of the Transverse Ranges, lower Pleistocene strata that have been folded and faulted into vertical

**9-7** *Tilted conglomerate of the Paleocene Carmelo Formation lying on the eroded surface of granitic rocks (in background) at Point Lobos State Park, Monterey County. (Oliver E. Bowen photo.)*

positions are unconformably overlain by flat-lying upper Pleistocene beds; one of the most violent mountain-building pulses was clearly in the middle Pleistocene.

Folding, faulting, and uplift have gone on contemporaneously in the Coast Ranges; sometimes one process or another has seemed to predominate in building a given range. Folding and faulting have been so complex and discontinuous that the axes of individual folds can seldom be traced more than a few miles. Trends of the folds are generally parallel to the major faults and thus tend to strike a little more westerly than the trend of the Coast Ranges as a whole.

As shown on the large geologic map, three major north-northwest-trending fault zones dominate the structural pattern of the Coast Ranges as a whole. The most westerly of these is the Nacimiento fault zone which separates the western coastal block of Franciscan basement rocks from the granitic block east of that fault zone.

The great San Andreas fault strikes about 35 degrees west of north obliquely across the Coast Ranges and Coast Range structures through Shelter Cove in coastal Humboldt County, 650 miles south to the Gulf of California. It extends far beyond the Coast Range province to cross the Transverse Ranges, forms the southwestern boundary of the Mojave block, and extends into the Salton Trough. North of Point Arena the fault extends northward, mostly offshore, to probable junction with the west-trending Pacific Ocean Mendocino fault zone. Like the Nacimiento fault, the San Andreas fault separates Franciscan rocks from granitic basement. Movements on the San Andreas fault during the Quaternary Period have been largely horizontal right-lateral (east block moving relatively south); older displacements, probably first accompanying uplift and initial mountain building at the close of Jurassic time, perhaps were predominantly vertical. Geologists are strongly divided on this point, however; some maintain that there has been cumulative right-lateral displacement of several hundred miles since Late Mesozoic time.

Prominent branches of the San Andreas fault system in the San Francisco Bay area are the Pilarcitos, Hayward, and Calaveras faults. South of the Tehachapi Mountains, the San Gabriel, San Jacinto, Elsinore, Banning, and Imperial faults are all part of the San Andreas system. All these faults presently are active except the Pilarcitos and San Gabriel.

A third important fault zone, probably related to the Nacimiento and San Andreas zones, is the South Fork Mountain fault, which separates Franciscan rocks of the northern Coast Ranges on the west from Mesozoic and older crystalline rocks of the Klamath Mountains on the east. This fault very probably extends southward under young rock formations of the western Great Valley to form a similar boundary between the Coast Range Franciscan basement and the Sierra Nevada granitic rocks. The South Fork Mountain–West Valley fault is an old inactive one, probably dating back to the Nevadan orogeny of Late Jurassic time. The fault boundary between the Franciscan and Sierran granitic areas may well be related to thrusting of the oceanic crust beneath the continental margin in Late Mesozoic time.

Apparently of major importance in the western mountain structures are elongate masses of peridotite and serpentine. They are essentially sill-like bodies that intruded and are interbedded with older rocks, contemporaneously with the Late Jurassic and Mid-to-Late Cretaceous orogenic episodes.

Peridotite, serpentine, basalt, and their alteration products are part of the "ophiolite" group of rocks derived from the upper mantle and oceanic crust. Ophiolites make up a large part of oceanic subducted plates. Such rocks are most clearly exposed underlying the Great Valley Late Jurassic-Cretaceous sequence on the eastern flanks of the Coast Ranges.

**9-8** *Aerial-oblique view of western Marin County looking northwest. The San Andreas fault zone extends across the width of the trough lying between Bolinas Bay in the foreground and Tomales Bay in the upper left. Granite lies west of the fault and rocks of the Franciscan Formation are on the east. (Robert E. Wallace photo, U.S. Geological Survey.)*

## OROGENY IN OTHER AREAS

From latest Cretaceous, during the Cenozoic Era, and into modern times, the Sierra Nevada, Klamath Mountains, and southern Peninsular Ranges behaved as resistant blocks; they were not strongly folded internally. However, they were uplifted and tilted from time to time with some internal faulting. Both the Sierra Nevada and Peninsular Ranges were tilted westward and profoundly faulted along their eastern borders. The Sierra Nevada, for example, was a broad, gentle, westward-sloping upland. In the northern Sierra, volcanism took place from Eocene to Holocene time. Uplift at irregular intervals during the Cenozoic Era rejuvenated the Sierra and caused deep and extensive erosion. Thousands of feet of vertical fault movements took place on the eastern margin of the Sierra Nevada to accomodate the westward tilting of the block. The most recent uplift and westward tilting occurred after volcanic rocks, dated radiometrically at 3 million years, were erupted. Uplift and faulting continue to the present day.

## CALIFORNIA'S ICE AGE

The most striking event in the Pleistocene history of North America is certainly the development of thick continental glaciers, which covered much of the northern part of the continent. Although much less important and widespread, during the great Pleistocene glaciation California had its own Ice Age. No continental glaciers developed, but alpine, or mountain-valley glaciers, left unmistakable evidence of their former presence in the U-shaped valleys, horn-shaped peaks, serrated (saw-toothed) ridges, and glacially eroded basins. Most of the dramatically sharp and beautiful scenery of the High Sierra is the result of glacial erosion. Rock surfaces polished and grooved, or striated, by the great eroding power of moving ice give clear evidence of the passage of ancient glaciers. Lower down, on the eastern slopes particularly, are the striking effects of deposition by glaciers: long ridges of glacial till (fragmented rock deposited by ice) left along the sides of valleys as lateral moraines and remnants of terminal moraines across the valleys, marking the ends of former glaciers. The shaping of landforms by glaciers was discussed briefly in Chapter 3.

Such unique and apparent signs have made it rather easy for geologists and naturalists to mark the areas of former glaciation. But as always in efforts to reconstruct geologic history—even that as recent as the Ice Age—there are complications. Four great continental ice sheets advanced and retreated over northern North America during the Pleistocene Epoch, each one leaving new records and partly erasing the older. There were probably four major advances of alpine glaciers at about the same time in the high mountainous areas of California; but, if these did occur, they are hard to correlate with each other and with glacial epochs of the continental interior. Latest studies in the high southern Sierra Nevada show a succession of eleven recognizable glacial tills. Each new epoch of glaciation partly obscured the older by erosion and by deposition of later till over older.

In attempting to understand glacial succession in California, geologists have studied weathered zones and soils between successive till layers, the extent and depth of weathering of rock fragments in the tills (granite boulders of the oldest tills are deeply disintegrated by weathering), and the state of preservation of glacial striations and polish on the rock surfaces. But the most useful recent tool has been radiometric dating. Some of the glacial deposits are overlain or underlain by volcanic deposits that have been dated.

North from inland southern California, U.S. Highway 395 runs across the Mojave Desert, east of the Sierra Nevada, through Reno, Nevada, back across the Modoc Plateau in California, and into Oregon. For more than 200 miles, from Owens Lake to Reno, this highway offers an incomparable view of glacial features of the Sierra Nevada, from the high serrated crests and cirques (valley-head glacial basins) to moraines and glacial deposits on the lower slopes. North of Bishop, highway cuts expose detailed features of the glacial moraines crossed. Fine paved highways extend a few miles

**9-9** *View of the High Sierra Nevada southerly, across Kaweah Basin, Sequoia National Park. Olancha Peak on the horizon, upper left. The big cirque, near the center of the photo, is occupied by a living rock glacier. Outside and around the rock glacier are lateral and terminal moraines of the last ice glacier to occupy the cirque. The trimline, or sharp break, in the lower slope probably represents the outer limit of the Tioga glacial stage—the last major glacial epoch in the Sierra. Note the broad unglaciated surface in the background in contrast to the sharp glaciated ridges and basins in the middle ground and foreground. (John L. Burnett photo.)*

west of Highway 395 at Independence, Big Pine, Bishop, Rock Creek, and many other points into the best examples of glacial features in California. The road west of Big Pine, in Owens Valley, is the access to Palisade Glacier, California's largest living glacier.

About 2,000 feet above the floor of McGee Creek (crossed by Highway 395 west of Bishop) is the McGee Till, about 2.5 million years old. In the same general area is the younger Sherwin Till, older than the 700,000-year-old overlying Bishop Tuff. *The McGee and Sherwin are the oldest radiometrically dated Quaternary glacial deposits in the world.* Several thousand feet of uplift of the Sierra Nevada appear to have occurred between McGee and Sherwin time.

**9-10** *McGee Creek moraines at Crowley Lake, eastern Sierra Nevada. Part of Crowley Lake appears in lower left corner of photo. Highway 395 crosses from center left to lower right. McGee Creek turns in a broad curve in the lower right quadrant and has built a small delta in the lake. Hilton Creek runs through the center of the photo into Crowley Lake. McGee Creek runs between two sharply defined curving, lateral moraines. Careful examination reveals two small faults that parallel Highway 395 in the lower right quadrant of the photo; one is within 1,000 feet of the highway, while the other is about 2 miles to the upper right. (Vertical aerial photo, U.S. Geological Survey, Prof. Paper 590, 1968; scale approximately 1 inch equals 5,652 feet;* south is at top.)

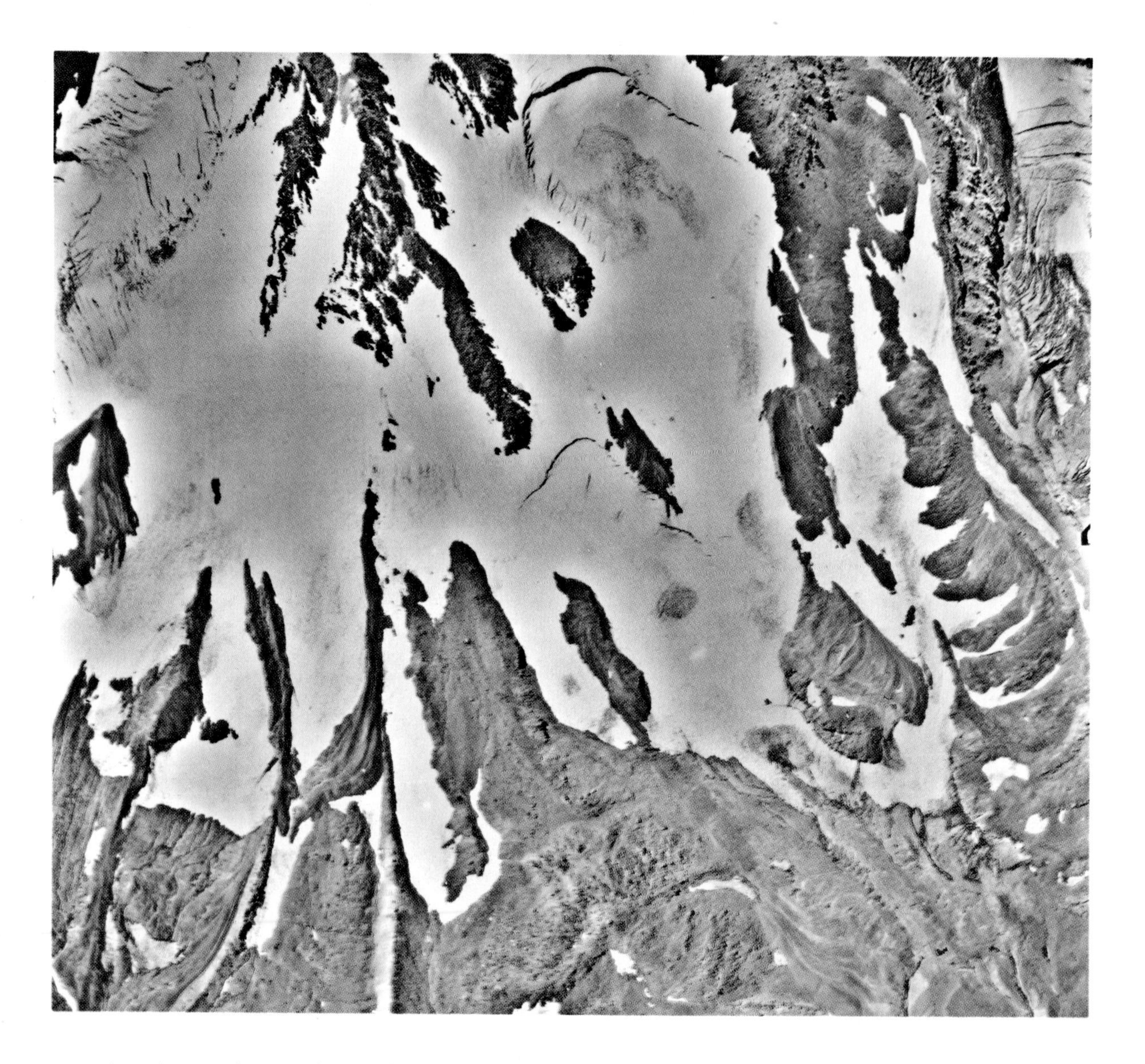

**9-11** *Summit cone of Mount Shasta, Siskiyou County; top near center. The summit (elevation 14,162 feet) is marked by a central platform with rocky crags of volcanic rock. Slopes, dropping off at an angle of about 35 degrees, are covered with rubble. Glaciers are best developed on the north and east. Note the numerous longitudinal and transverse crevasses in the ice and snow, and the arc-shaped crevasses at the heads of glaciers. Volcanic rubble has been rearranged by the ice to develop small lateral moraines. (Vertical aerial photo, U.S. Geological Survey, Prof. Paper 590, 1968; scale approximately 1 inch equals 2,361 feet;* south is at top.)

In the Truckee–Lake Tahoe area, tills and other features of glaciation are also well exposed. Here is also an old till—younger than Sherwin but older than a volcanic tuff dated at 370,000 years.

Above, and younger than all these, is a succession of tills—given names like Tahoe, Tenaya, and Tioga, after the geographic localities Lake Tahoe, Lake Tenaya and Tioga Pass—which are younger than volcanic rocks that have been dated by potassium-argon means at 60,000 to 90,000 years old. These are believed to correlate with the well-known Wisconsin Stage of glaciation of the continental interior—the last major advance of the ice over North America.

At the maximum extent of glaciation, the Sierra Nevada was largely covered by ice (except for the steepest slopes, which could not hold ice) along 270 miles of its crest, with a width of about 30 miles. Ice extended down west-slope canyons to elevations as low as 3,000 feet and depths as great as 6,000 feet in valleys like Yosemite. Because of lower precipitation on the eastern side of the range, ice did not extend to nearly as low elevations. There was also ice in the White Mountains, east of Owens Valley. Elsewhere in the state, Mount Shasta, the Trinity Alps, and Lassen Peak had alpine glaciers. The farthest-south glaciers in California were above elevation 8,700 feet on the slopes of San Gorgonio Mountain in the San Bernardino Mountains.

Existing glaciers in California are much fewer and cover much smaller areas than in the Pleistocene. John Muir reported that there were sixty-five glaciers in the Sierra in the 1870s; today glacial geologists count about sixty. Palisade Glacier, one of the largest, is only a mile long. Mount Shasta has several small glaciers. All California's living glaciers are now above 10,000 feet elevation and most show signs of gradual decrease in size as we move into a warmer and drier climatic cycle.

Glaciers are fed by snow and are maintained where the delicate seasonal balance of snowfall and temperature are such that the annual snowfall does not melt but accumulates over the years.

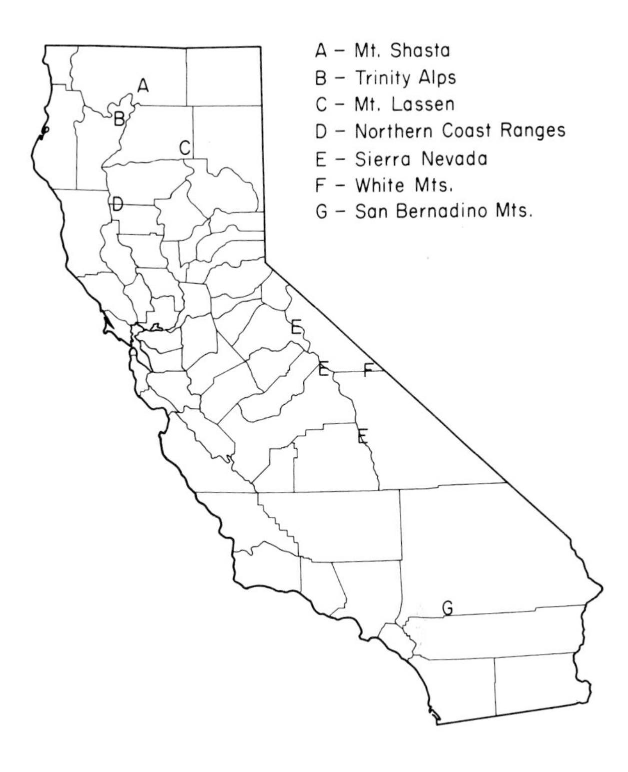


**FIGURE 9-12** *Areas glaciated during the Ice Age. (California Division of Mines and Geology,* Mineral Information *Service, March, 1964.)*

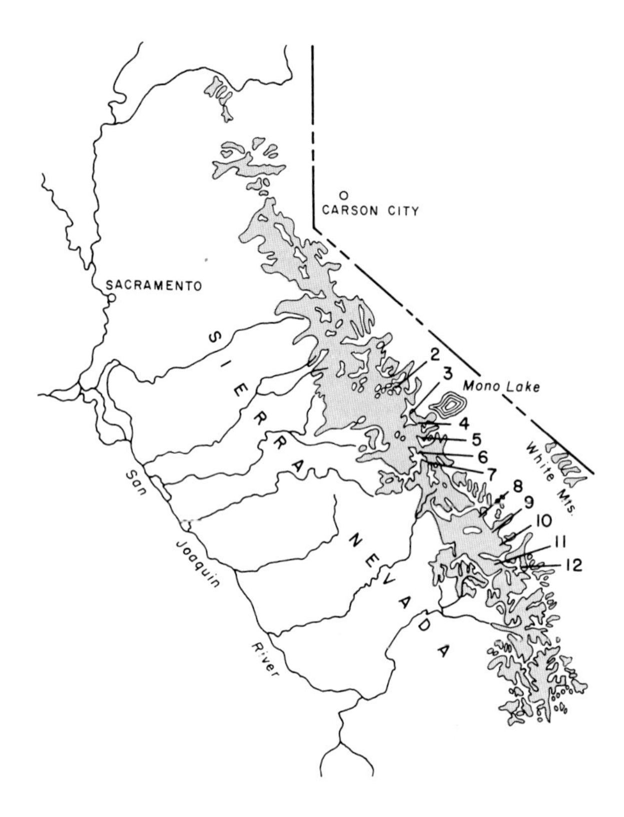


**FIGURE 9-13** *Existing glaciers in the Sierra Nevada. (2) Sawtooth Ridge—many small glaciers; (3) Mount Conness—Conness glacier and others to the southwest; (4) Mount Dana—Dana glacier; (5) Kuna Peak, Koip Peak—several glaciers; (6) Mount Lyell—Lyell glacier and other small glaciers to the northwest and south; (7) Ritter Range—many small glaciers; (8) Mount Abbot—several glaciers to the northwest and southwest; (9) Mount Humphreys—several glaciers to the northwest; (10) Glacier Divide, Mount Darwin, Mount Haeckel, Mount Powell—Goethe glacier, Darwin glacier, and many others to the northwest and southeast; some are large in size; (11) Mount Goddard, Black Divide—many glaciers; (12) The Palisades, Mount Bolton Brown, Split Mountain—Palisade Glacier, Middle Palisade Glacier, and many other glaciers.* (California Division of Mines and Geology, Mineral Information Service, *March, 1964.*)

Today, California is emerging from its last Ice Age, but we have no clues as to whether this is to be a short interglacial epoch or is to be a much more permanent change.

## LIFE FORMS AND THEIR FOSSIL REMAINS

The varying locations of lands and seas, thick marine and nonmarine sedimentary rock formations, and variable climates of Tertiary and Pleistocene times account for the rich fauna (animals) and flora (plants) preserved as fossils in the Cenozoic rocks of California.

### cenozoic land animals

A spectacular record of the animal life that lived in California during the Cenozoic Era—the last 65 million years of geologic time—has been preserved in rock strata of this Era in various parts of the state. Mammals completely took the center of the stage and shortly dominated life of the earth.

In 1952, a paleontologist, working in the El Paso Mountains just north of the Garlock fault on the northwestern edge of the Mojave Desert, found a few scattered crocodile bones and teeth in land-laid clay, sand, and gravel beds of the Paleocene Goler Formation. A couple of years later, working in the same locality, he found a fragment of a jaw and teeth of a small condylarth (a primitive, generalized mammal about the size of a small dog). This is the oldest known mammal in California and the oldest west of central Utah.

Late Eocene beds of the Rose Canyon Shale, just north of San Diego, have yielded fossil remains of lemur-monkeys, insectivores, primitive rodents, and a rhinoceros. In Ventura County, beds of the Sespe Formation contain assemblages of mammals ranging in age from late Eocene to early Miocene. The earliest animals include lemur-monkeys, doglike carnivores, opossums, and rodents; the later beds contain tree squirrels, a small rhinoceros, small deer, and an early bear-dog. Lemurs and monkeys—especially rare as fossils—belong to the same order of mammals as man and were certainly the most intelligent creatures of their time.

One of the famous early Tertiary mammal localities is in the early Oligocene Titus Canyon Formation of northern Death Valley. Here the small browsing horse *Mesohippus,* tapirs, dogs, rodents, rhinos, and tiny deer roamed a wet, lush, wooded countryside startlingly different from that of today.

Several wonderful vertebrate faunas of early Miocene to early Pliocene age have been found only a few miles from Los Angeles in the Mint Canyon area at the eastern end of the Ventura Basin on the northern slopes of the western San Gabriel Mountains. Bones in the early Miocene Tick Canyon Formation include a hawk (the oldest well-identified bird in California), pocket mice and rabbits, *Parahippus* (a medium-sized browsing horse), camels, and an oreodont (an extinct mammal looking something like a small sheep, but with some of the characteristics of pig, deer, and camel). Overlying beds of the Mint Canyon Formation (lower Pliocene on the scale of the vertebrate paleontologists but upper Miocene in relation to marine invertebrates) contain bones of three grazing horses, mastodon elephants, rhinoceros, dogs, peccaries, camels, and pronghorn antelope. The climate was mild, subtropical, and semiarid. A rich late Miocene fauna, including early saber-toothed cats, as well as horses, camels, rodents, dogs, deer, and birds, has been found in the Barstow Formation near Barstow in the Mojave Desert.

Oldest fossils of land animals in the north have been found east of the Berkeley Hills in the Mount Diablo area. Here are fossils, dated on the border of Miocene-Pliocene time, that include horse teeth, a mastodon elephant, camels, and an oreodont. Most famous locality is Black Hawk Ranch on the southern slopes of Mount Diablo where, in addition to the above, are remains of squirrels, a gray fox, a ring-tailed cat, lizards, and beavers. Several stages of later Pliocene faunas are also known.

The best early Pleistocene fauna in western North America has been collected in recent years from the Irvington gravels northeast of the Hayward fault in Fremont. In these beds is the first appearance of the Old World mammoth elephant. Other forms are deer, camels, the true horse *Equus,* many rodents, saber-toothed cats, a large wolf, coyote, and a flightless goose.

In the heart of the city of Los Angeles in the expensive Wilshire Boulevard district is 23-acre Hancock Park. This is one of the most remarkable parks in the world, for here is preserved, labeled, and explained one of the most famous concentrations of fossil mammals and birds ever found. This is Rancho La Brea (*brea* is Spanish for "tar"), a region of natural tar seeps that bring heavy oil to the surface from petroleum-bearing

**9-14** *The 14-million-year-old* Paleoparadoxia*—a one-ton, 9-foot-long mammal similar to a sea lion that frequented upper Miocene seas—is shown in a sitting position on land. Discovered in 1964 in an excavation at Stanford University on the San Francisco Peninsula, this rare specimen is the first of its kind found in North America. Its skeleton was found in excavating for the linear accelerator. (U.S. Geological Survey photo of a sketch by Charles A. Repenning.)*

**9-15** *The Charles R. Knight mural of the Rancho La Brea scene, on view in Hancock Hall at Los Angeles County Museum of Natural History. Giant ground sloths, wolves, horses, camels, and a saber-tooth tiger in a Pleistocene scene in the city of Los Angeles. The snow-capped San Gabriel Mountains are in the background, with tar pits in foreground. (Los Angeles County Museum of Natural History photo.)*

**9-16** *From the Charles R. Knight mural of the late Pleistocene Rancho La Brea scene in western Los Angeles. Giant wolves, saber-tooth tigers, and huge ground sloth at the tar pits. Giant vultures sit overhead; Imperial elephants and bison may be seen in the background. (Los Angeles County Museum of Natural History photo.)*

beds below. In addition to 50 species of mammals and 110 bird species, snakes, turtles, toads, molluscs, "bugs and beetles," and plants make up the assemblage of life trapped in tar. The seeps are still active and are capable of attracting today's life; consequently here is preserved a sequence of animals transitional from late Pleistocene to Holocene age.

Some of the more striking extinct forms are bison, wolf, bears, the saber-toothed tiger *Smilodon* (official state fossil), mammoths, mastodons, *Equus,* a western camel, and the giant ground sloth *Megalonyx.* Presence of bison particularly distinguishes this late Pleistocene Rancholabrean fauna from the early Pleistocene Irvingtonian fauna.

In the widespread lands of prehistoric Cenozoic California there were numerous less conspicuous land animals—vertebrates and invertebrates—which we lack space to discuss here. However, what is known of early man, the highest form of all mammals?

How old is man in California? Was he a contemporary of the mammoth elephant, ground sloth, camel, and saber-tooth cat, whose remains are so abundantly represented at Rancho La Brea?

We do not know, positively, the answer to either of these questions. With his higher intelligence, man was not so easily trapped in tar pits or so casually buried in sediments. At most, prehistoric man was never very numerous in California. His bones are among the rarest of all fossils, and the artifacts of most ancient man—stone tools, obsidian arrowheads and spearheads, bone tools, grinding stones, and the like—are difficult to date; the most primitive artifacts are even difficult to distinguish from the same materials "accidentally" fashioned by weathering and erosion!

Some of the oldest authentic dates for prehistoric man in the state have come from southern California. Extending from Santa Barbara to San Diego, there are a number of radiocarbon-dated sites from 7,000 to 4,000 years ago. Shell from kitchen middens (refuse heaps from ancient

**9-17** *Howard Ball life-size restoration of Pleistocene female mammoth elephant caught in asphalt lake at Rancho La Brea, Hancock Park, in Los Angeles. (Photo by Armando Solis, Los Angeles County Museum of Natural History.)*

habitations) on Santa Rosa Island have given a date of 7,350 years for what archaeologists have called the dune dweller culture. Near La Jolla, on the San Diego coast, a site was occupied during the span of time from 5,460 to 7,380 years ago. The culture represented has yielded grinding stones, scraper planes, flake scrapers, choppers, pebble hammerstones, and bone awls and punches. Inland, old lake shore lines and caves have most frequently yielded artifacts of ancient man. In nearby Nevada and Arizona, such sites have proven the existence of man as long ago as 11,000 years, and evidence is good for 12,000 years ago.

There has been some discussion about the "Laguna lady," whose skull, dug up at Laguna Beach in the early 1930s, has been dated by radiocarbon as 17,000 years old. A shadow of doubt on this determination is cast by the fact that the history of what has happened to the skull (and even whether it is truly the one originally found) is somewhat obscure.

Returning to our initial questions on earliest man in California: Yes, it seems pretty well established that man did live in southern California as a contemporary of the last of the great mammals. Concerning evidence for the oldest man in the state, geologists must defer to the opinions of leading anthropologists. Robert F. Heizer, professor of anthropology at the University of California, has judged that the evidence is good for man's presence about the 10,000 B.C. time level and that evidence from 9,000 B.C. is secure.

**9-18** *Life-size restorations of Rancho La Brea Pleistocene mammals on view at Hancock Park in Los Angeles. Created by sculptor Howard Ball. Two saber-tooth cats on left, baby mammoth and adult male on bank, female mammoth in the asphalt lake. (Photo by Armando Solis, Los Angeles County Museum of Natural History.)*

## characteristic marine fossils of california

Marine fossils of the Cenozoic Era from the Paleocene Epoch through the Pleistocene Epoch have been extensively used by the petroleum industry for determining ages and for correlating formations. Those life forms that were very abundant and geographically widespread and that varied markedly and rapidly with time and environment (temperature, depth, and clarity of water) make the most useful guide, marker, or index fossils.

The tiny single-celled, calcium carbonate shells of the foraminifera best fulfill all of these criteria. These fossil shells occur in great numbers and variety in almost all fine-grained marine sediments. The whole Tertiary Period has been broken down into thin time zones based on evolution of the foraminifera.

The *diatoms* (microscopic, single-celled, silica-secreting plants) are not so useful for stratigraphic correlation, but they existed in enormous numbers in both marine and lake waters. Beds of diatomaceous shale up to several thousand feet thick are known in the upper Miocene–lower Pliocene of the Coast Ranges.

Other invertebrate fossils of greatest use in paleontology, more or less in order of their usefulness, are pelecypods, gastropods, corals, bryozoans, crustaceans, and echinoids. Some of the best-known and most striking marker fossils in California rocks are large specimens of the pelecypod *Venericardia* in the Eocene; giant thick-shelled *Pectens* and *Dosinia* from Miocene to Holocene age; the spirally coiled, tower-shaped gastropod *Turritella,* characteristic of the early and late Tertiary; the huge oyster (up to 18 inches long) *Ostrea titan* in upper Miocene sediments; and the echinoid *Dendraster,* the genus to which all living sand dollars belong, known only from Pliocene to Holocene time.

Holocene species of these invertebrates are unknown in the Eocene of California, but still-existing species of the Holocene begin to appear in rocks of the Oligocene Epoch. In lower to

**9-19** *The Pleistocene condorlike vulture* Teratornis *from Rancho La Brea. (Los Angeles County Museum of Natural History photo.)*

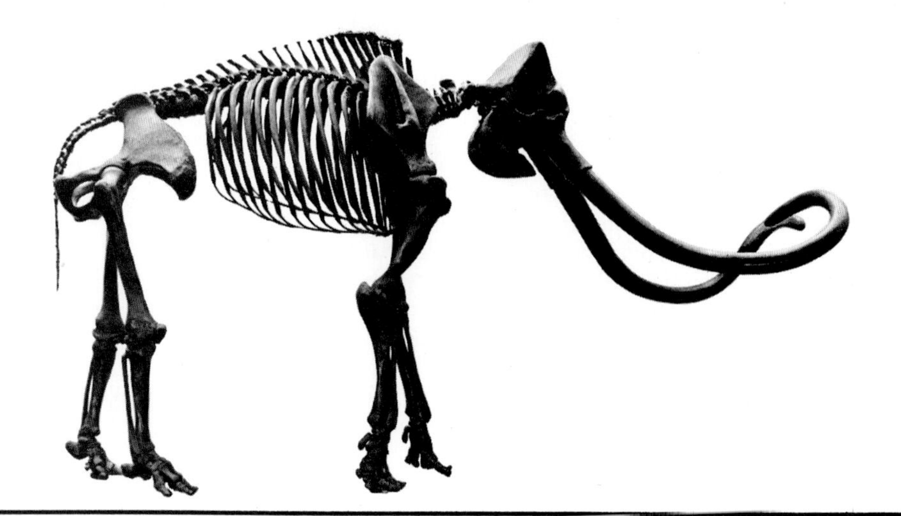

**9-20** *Imperial mammoth skeleton from Rancho La Brea. (Los Angeles County Museum of Natural History photo.)*

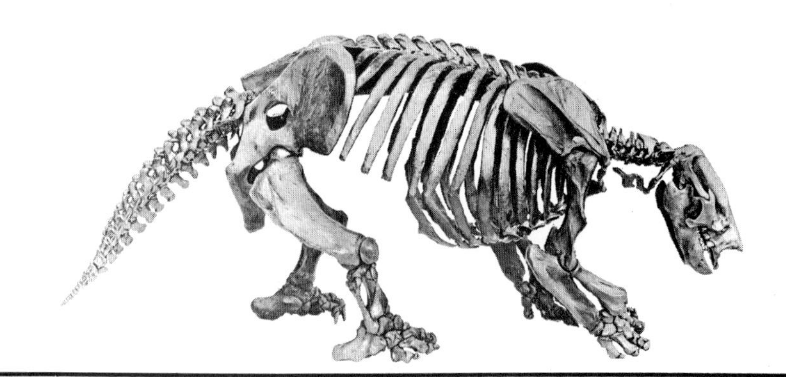

**9-21** *The massive skeleton of the giant ground sloth* Paramylodon *from Rancho La Brea. (Los Angeles County Museum of Natural History photo.)*

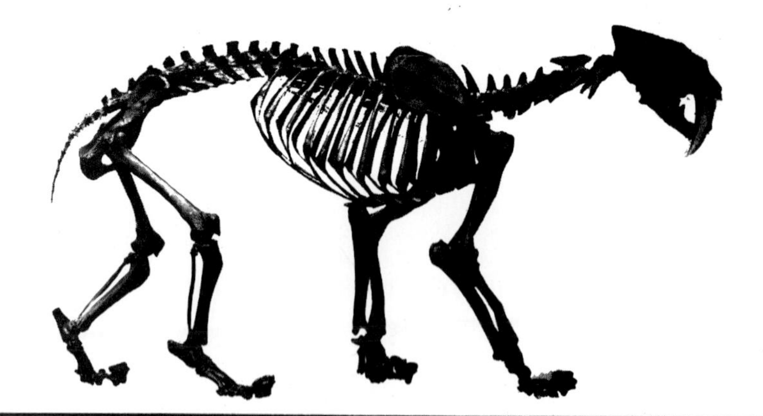

**9-22** *Skeleton of the sabertooth tiger* Smilodon *from Rancho La Brea. (Los Angeles County Museum of Natural History photo.)*

middle Miocene time there appeared more than twenty species that are still living; of the Pliocene species at least a hundred are still living; and in the late Pleistocene there existed at least 95 percent of the forms that are found today in waters of nearby regions.

Many of the forms were adapted to rather narrow temperature ranges and are therefore good indicators of past climates. The marine faunas of California indicate a gradual cooling from early to late Tertiary time. Cooling becomes most pronounced in the late Pliocene and early Pleistocene Epochs. Warming followed in late Pleistocene time. For example, in the San Pedro beds in south coastal California, the cool-water form *Pecten caurinus* occurs in early Pleistocene sands; however, warming of the climate again in the late Pleistocene is shown by the presence of the warm-water *Pecten vogdesi*.

We cannot leave this brief sketch of the marine fossils of California without mentioning the unusual occurrence of great numbers of fossil fish skeletons and impressions found in diatomaceous shale of late Miocene to early Pliocene age. Particularly in the diatomaceous shale quarries worked at Lompoc on the Santa Barbara coast, marvelously perfect specimens have been found between the thin laminations of diatomite. The forms are quite similar to living fishes, although of different species.

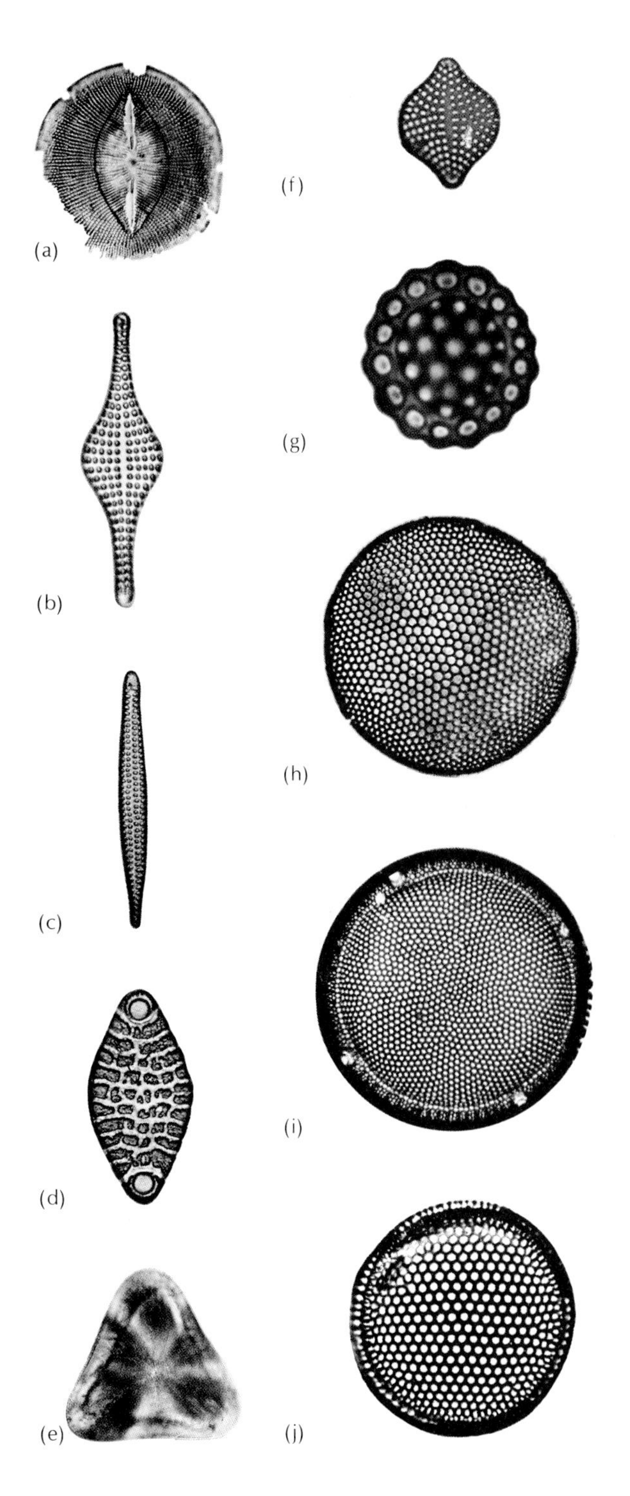


**9-23** *Middle Miocene diatoms from Sharktooth Hill, Kern County, except figure* (d), *which is from the middle Miocene, Santa Cruz Island. (Photomicrographs by G Dallas Hanna, California Academy of Sciences.)* (a) Raphidodiscus marylandicus *Christian, length 0.0555 millimeter.* (b) Rhaphoneis elegans *Pantacseh and Grunow, length 0.0668 millimeter.* (c) Sceptroneis caduceus *Ehrenberg, length 0.1466 millimeter.* (d) Xystothaia hustedti *Hanna, length 0.0714 millimeter.* (e) Cymatogonia amblyceros (*Ehrenberg*), *length of one side 0.090 millimeter.* (f) Rhaphoneis obesula *Hanna, length 0.020 millimeter.* (g) Macrora stella (*Azpéitia*), *diameter 0.0152 millimeter.* (h) Rattrayella inconspicuua *Rattray, diameter 0.0966 millimeter.* (i) Eupodiscus antiquus *Cox, diameter 0.1426 millimeter.* (j) Cascinodiscus lineatus *Ehrenberg, diameter 0.0412 millimeter.*

## plants

Nothing tells us more specifically about the natural environment and local conditions of a past age than the fossil flora. The oldest land plants in California are known from leaf impressions of tree ferns, palms, and large-leafed evergreens found near Elsinore, Riverside County, in beds of the Paleocene Martinez Stage. These plants do not grow wild in the state today; they show clearly that at that time southern California had a tropical savanna climate marked by heavy summer rains. In the San Francisco Bay area, leaves of the fan palm, magnolia, and avocado have been found in coal beds of the middle Eocene Tesla Formation. These, also, show a tropical climate.

Far north in Trinity County, silt layers interbedded with gold-bearing gravels of the Weaverville Formation have yielded fossil remains of Oligocene swamp cypress, walnut, laurel, and the oldest California remains of the redwood *Metasequoia* (Photograph 9-27), native only in China today. The Weaverville flora shows a warm, humid climate such as in the Gulf states today.

Interestingly, Miocene floras show the beginning of cooler temperate to semiarid climates. A Miocene flora from northeastern California includes alder, chestnut, maple, hickory, pine, and poplar. The living representatives of these trees are in forests of cool, humid, temperate regions, where most of the rainfall comes in the summer—quite different from the winter-rainfall maximum of California today. In Kern County, fossil oak leaves show a more arid climate.

A late Miocene-early Pliocene flora of the Berkeley Hills in the San Francisco Bay area includes cherry, alder, sycamore, poplar, and willow leaves, with swamp cypress near the coasts.

**9-24** (Above) *The Tertiary pelecypod* Trigonia. *About 2 inches across. (Smithsonian Institution photo.)*

**9-25** *Sand dollar* Scutella, *an echinoid from Tertiary time. About 2 inches across. (Smithsonian Institution photo.)*

**9-26** *Fish skeletons*—Xyne grex—*in upper Miocene diatomite, Lompoc, Santa Barbara County. (California Division of Mines and Geology,* Mineral Information Service, *February, 1964.)*

**9-27** Metasequoia, *the "Dawn redwood," found in the Miocene of northern California and Oregon. (Smithsonian Institution photo.)*

These begin to somewhat resemble present-day forms. During the Pliocene, climates cooled notably and became less humid; the forests became almost wholly modern in their appearance. At the Petrified Forest on the highway from Santa Rosa to Calistoga, north of San Francisco Bay, are beautiful petrified trunks of the huge redwoods (*Sequoia*) buried in sediments of the Sonoma volcanics. Here, also, are seeds and leaves of fir, spruce, Douglas fir, and hemlock. The tuff containing these fossil plants has been radiometrically dated as 3.5 million years—near the Pliocene-Pleistocene boundary.

Near the end of the Pliocene Epoch, considerable cooling took place; and during parts of Pleistocene time, with glaciation in the Sierra Nevada and Trinity Alps, climates were cooler and more humid than today. A rare occurrence of the coast redwood (*Sequoia sempervirens*) has been

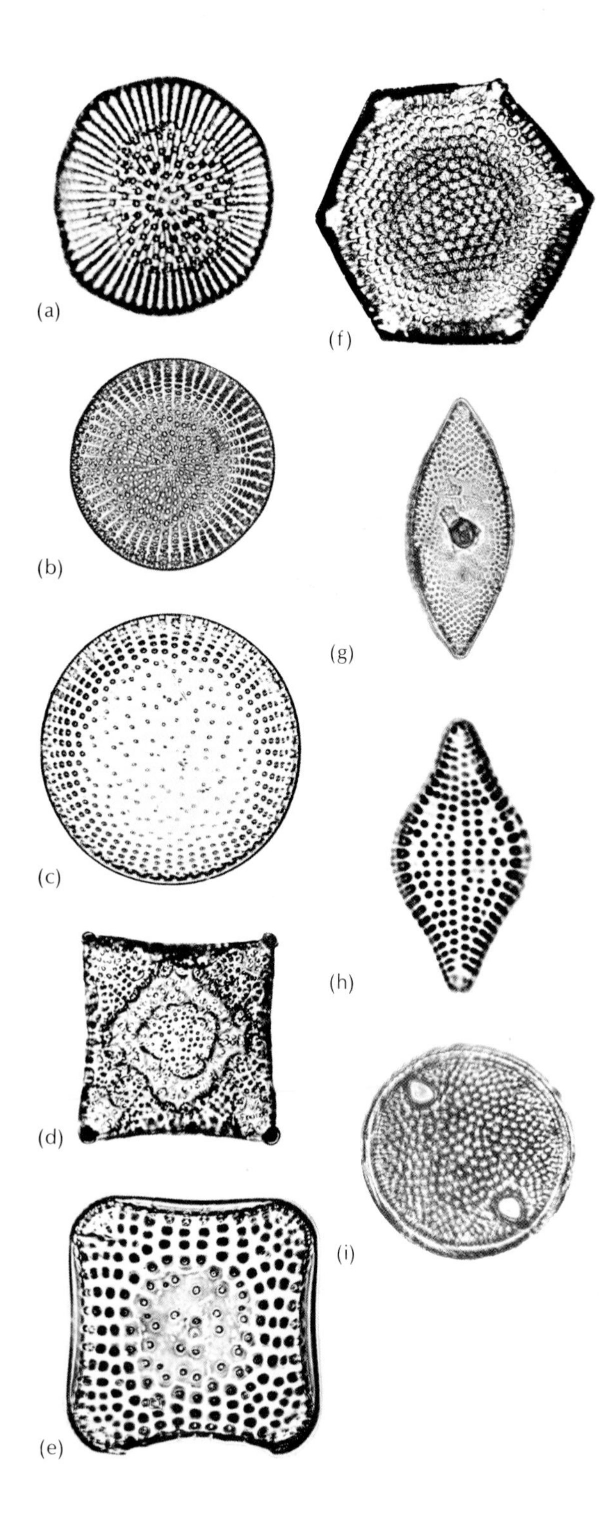


**9-28** *Diatoms from middle Miocene of Santa Cruz Island. (Photomicrographs by G Dallas Hanna, California Academy of Sciences.)* (a) Stictodiscus *sp, diameter 0.0727 millimeter.* (b) Stictodiscus hardmanianus *Greville, diameter 0.0818 millimeter.* (c) Stictodiscus *sp, diameter 0.1273 millimeter.* (d) Triceratium, *length of one side 0.0636 millimeter.* (e) Triceratium, *length of one side 0.0545 millimeter.* (f) Triceratium, *diameter 0.210 millimeter.* (g) Rutilaria, *length 0.100 millimeter.* (h) Rhaphoneis amphiceros *Ehrenberg, length 0.0364 millimeter.* (i) Pseudauliscus, *diameter 0.050 millimeter.*

**9-29** *Photomicrograph of* Triceratium *sp, a middle Miocene diatom from Santa Cruz Island. Length of one side is 0.0636 millimeter. (G D. Hanna photo.)*

found in Pleistocene deposits near Carpenteria on the Santa Barbara coast. The farthest-south living redwoods are in the Big Sur country, about 20 miles south of Carmel.

Flowering plants and grasses had appeared in the Cretaceous Period. They developed rapidly during the early Tertiary and were no doubt a major factor in the rise of mammals to dominance in the Tertiary period.

## GEOTHERMAL ENERGY

A unique development, for the United States, is found at The Geysers in Sonoma County about 90 miles north of San Francisco. In the past few years, power companies have begun to use natural steam from wells drilled into this hot-spring area to produce electrical power. By 1968, this district was producing 82,000 kilowatts of electricity; enough to supply the power needs of a city of 90,000 people. By 1976, production was over 500,000 kilowatts.

At The Geysers, wells up to a few thousand feet deep are producing steam from intensely faulted and sheared rocks of the Late Mesozoic Franciscan Formation. Although the steam comes directly from fractured Franciscan rocks, the water appears to be normal groundwater; and the source of heat is probably still-hot Late Cenozoic volcanic rocks at depth. Such rocks crop out extensively at the surface only a few miles away.

California's extensive areas of Late Quaternary volcanic rocks and numerous hot springs suggest a considerable potential for geothermal energy.

The great advantage of geothermal steam as a source of energy is its cheapness as compared with that of electrical energy from falling water and from steam plants where a source of heat must be used such as burning coal, oil, gas, or the decay of radioactive minerals (nuclear energy). It takes a million barrels of oil to generate 100,000 kilowatts of electricity.

**9-30** *Kettleman Hills Middle Dome, July 1953. Gently dipping marine sandstone and mudstone of the early Pliocene Etchegoin Formation exposed in the eroded core of the anticline. Southwestern part of the San Joaquin Valley appears on the horizon in upper left. (Charles W. Jennings photo, California Division of Mines and Geology.)*

**9-31** *Casa Diablo Hot Springs, Highway 395, Mono County. This is an area being investigated for possible development of electrical power from natural-steam wells. (Mary Hill photo.)*

*The building of mountains by volcanic action continues in dynamic California. The sharp, snow-capped 10,453-foot Lassen Peak looms on the horizon across Fall River Valley and the lesser volcanic peaks and domes of Lassen Volcanic National Park.*

# 10 dynamic california today— an uneasy crust and changing landforms

■ Geologists have traced several epochs of mountain building in California, with scant records of some even as far back as the Late Precambrian Era. Evidence of orogeny and granitic intrusion at or near the close of the Paleozoic Era has been found in widely different California localities—in the Klamath Mountains and in the Sierra Nevada, Peninsular, and Transverse Ranges. The great Nevadan orogeny (Late Jurassic) formed a mountain system extending north from California into Oregon and northeast into Idaho, including the Sierra Nevada and Klamath Mountains, and through the Peninsular Ranges into Baja California. Important mountain building also took place in the Coast Ranges at that time. Again, in the early Late Cretaceous Period, orogeny was renewed in the same general area.

Finally, late Miocene to late Pleistocene time brought the Coast Range orogeny, with emphasis on the building of the Coast and Transverse Ranges. Uplift and mountain building also spread far into the Mojave Desert and Great Basin. Renewed broad uplift and westward tilting of the Sierra Nevada and Peninsular Ranges reflected Plio-Pleistocene orogeny in those areas.

Has mountain building ended in California? Have uplift and depression, folding, volcanic activity, and earth movement ceased?

Quite the contrary. Geologists have evidence for believing that all these mountain-building forces are operating today; perhaps even at rates comparable to those during major orogenies of past ages. It is not likely that the mountain ranges of California were ever higher and more continuous or that the relief between mountain crests and valley basins was ever greater than it is today. We shall now examine some features and evidences of geology in action in California as seen today.

## VERTICAL UPLIFT AND SUBSIDENCE

Quaternary history of the Coast and Transverse Ranges has been dominated by warping and faulting. Evidence of these activities is particularly clear along the coast, where sea level serves as a reference plane for the latest movements. Marine-cut terraces at many different levels, up to about 1,300 feet above sea level, occur discontinuously along the coast; offshore topographic maps show that similar terraces

**10-1** *View north across Helen Lake at 10,453-foot Lassen Peak, Lassen Volcanic National Park, Shasta County. This composite volcano must be considered active, as it last had a series of eruptions in the period 1914 to 1917. (Mary Hill photo.)*

extend out onto the continental shelf to several hundred feet below sea level. Each terrace was formed by wave and current erosion very close to a former sea level. Often the terrace deposits contain sea shells, and often later sediments from the higher lands have sloughed down over their surfaces.

The higher (older) terraces are deformed more than the low terraces, but even the oldest of these have been dated radiometrically as only late Pleistocene. Progressively greater deformation of the older terraces suggests that deformation has been more or less continuous throughout latest geologic time.

**10-2** *Looking north across terraces of the Palos Verdes Hills and the Los Angeles Basin. This remarkable succession of thirteen recognized marine terraces is probably due to a combination of uplift of the land and fluctuations of sea level during advances and retreats of the ice during Quaternary time. The youngest folded rock formation underlying the flat terraces is early Pleistocene in age. (John S. Shelton photo.)*

Although marine terraces extend all along California's coast, none are more impressive than the giant's stairway of thirteen wave-cut terraces (the highest at 1,300 feet elevation) along the coast of the Palos Verdes Hills. Early-day geologists pointed out that fossils in even the oldest of these terraces are all similar to living forms and thus the succession of uplifts has been extremely recent. The older of the Palos Verdes terraces have been unmistakably warped.

Inland, a great many stream-cut terraces have been left at different levels along the sides of downcutting streams. However, terraces may be formed whenever the downcutting power of a stream is increased, from whatever cause; not all terraces result directly from uplift. Lack of fossils makes these stream terraces difficult to date. Indeed, one of the most difficult problems geologists have is how to correlate terraces of all types, marine and nonmarine.

Beside uplift (or subsidence) and warping, we must not neglect the fact that the level of the sea itself fluctuated with the advance and retreat of

the great continental glaciers during the Pleistocene. Sea level dropped as much as 350 feet during the last advance of glaciers, when large amounts of water were trapped in continental glaciers, and rose again when the ice melted.

Thus, geologists are faced with two great problems: (1) relating the times of terrace cutting with advances and retreats of the continental glaciers, and (2) determining to what extent the terraces are the result of crustal movements as opposed to change in sea level.

The United States Coast and Geodetic Survey is responsible for networks of topographic leveling all over the country. In California, it has run about 50,000 miles of precise levels, including releveling to determine vertical changes in elevation. Both uplift and subsidence have been found at various times since 1906.

Some of the most detailed releveling has been done in cooperation with the state of California because of ground subsidence in the San Joaquin Valley along the route of the great aqueduct that takes water from northern to southern California. Maximum subsidence was found to be 22.9 feet, from 1943 to 1964, at a point 10 miles southwest of the little town of Mendota on the

**10-3** *The eucalyptus tree growing on a terrace at Santa Barbara Point, Santa Barbara County, shows by its exposed root system that the sea cliff has been rapidly receding landward due to wave erosion at its base. (Robert M. Norris photo, California Division of Mines and Geology,* Mineral Information Service, *June, 1968.)*

**10-4** *Marine terraces, San Clemente Island, Los Angeles County. Gently dipping Miocene volcanic rocks form the island, which rises nearly 2,000 feet above sea level. The southwestern slope, which is shown by the photo, is a spectacular flight of Pleistocene marine terraces. A rocky cliff forms the coastline. The terraces are discontinuous. The risers of the steps are steep slopes or cliffs 20 to 100 feet high; the treads are gentle slopes from a few feet to 1,000 feet wide. Streams have deeply cut the slope since Pleistocene time. Note the curving patterns of waves in the upper right.* (*Vertical aerial photo, U.S. Geological Survey, Prof. Paper 590, 1968; scale approximately 1 inch equals 2,330 feet;* south is at top.)

western side of the San Joaquin Valley. In the center of the valley, near Delano, about 30 miles north of Bakersfield, subsidence was 11.4 feet during the period from 1930 to 1964. At the southern end of San Francisco Bay the land surface sank 11.2 feet from 1912 to 1963.

Extensive research on causes of these examples of subsidence has shown that it is due to removal of groundwater by pumping for irrigation and other water uses and is not related to mountain building. Groundwater fills the pore spaces in soil and rock formations below the water table; when this support is removed by pumping, compaction and subsidence take place. The removal of oil and gas may have the same effect. Subsidence, due to removal of oil and gas in the Wilmington area of the Los Angeles Harbor, literally formed a sinkhole in the land surface over an elliptical area of about 10 square miles and with a maximum sinking of 28 feet. What this did to harbor installations and the problems involved in the "rise" of sea level can readily be imagined! A few years ago, the state of California developed a regular program of replacement of the gas and oil removed; the program involves injection, under pressure, of tremendous amounts of seawater (about a million

**FIGURE 10-5** *Land subsidence from 1934 to 1967 in the Santa Clara Valley due primarily to pumping of fresh water for irrigation. (U.S. Geological Survey diagram.)*

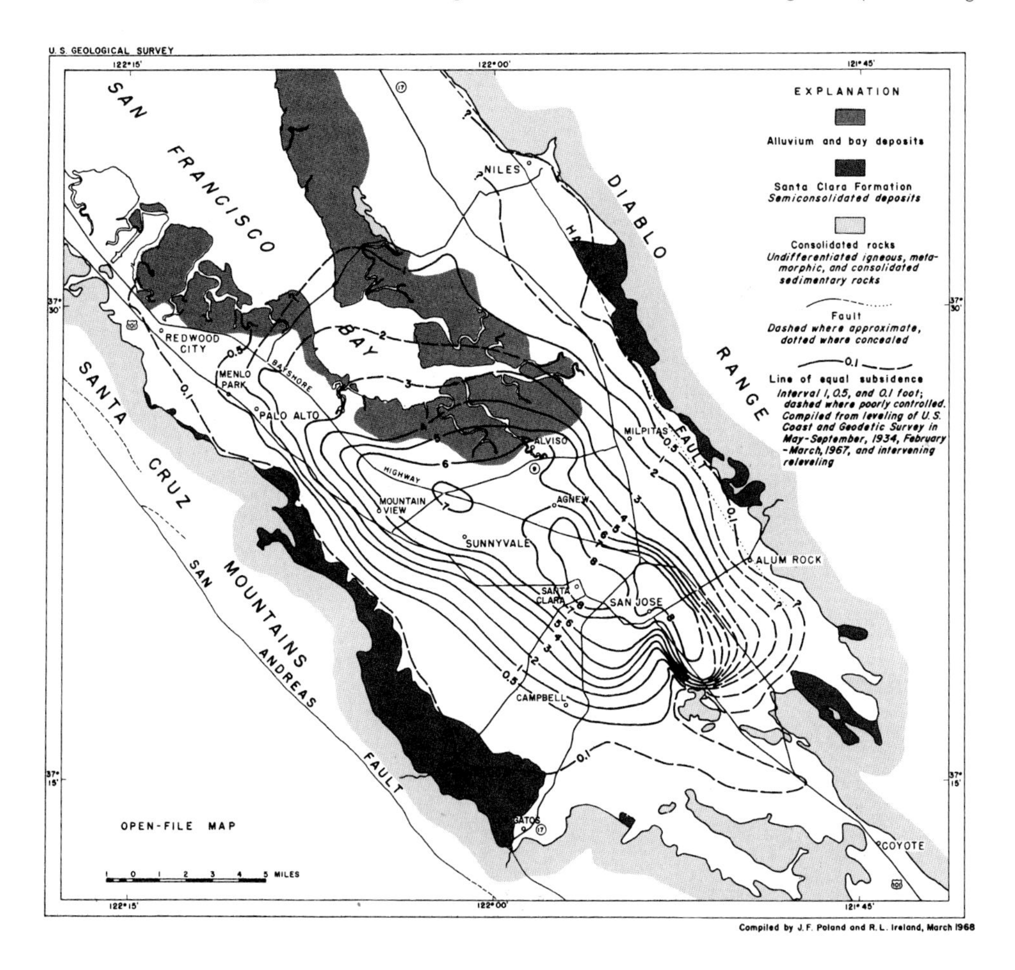

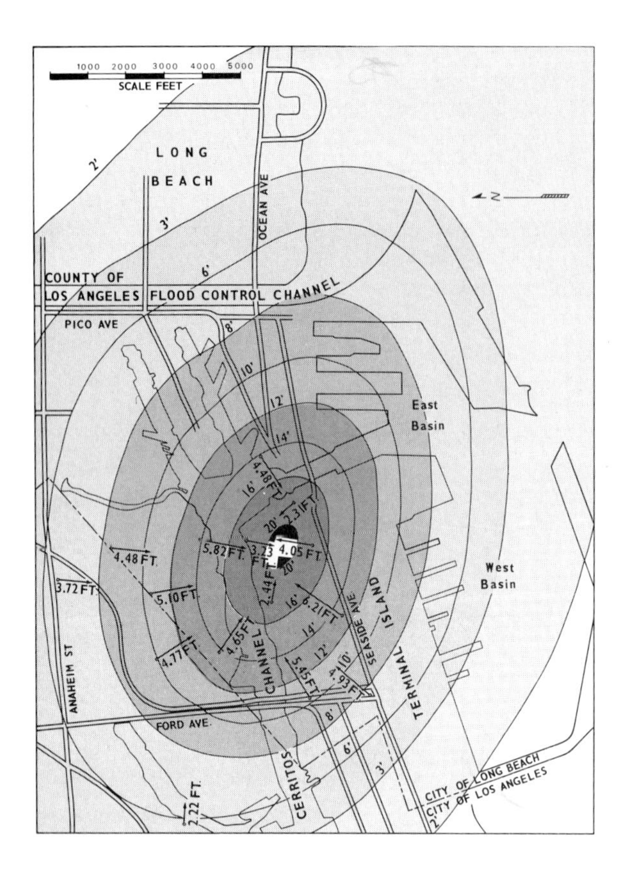


**FIGURE 10-6** *Land subsidence in the Wilmington area, Los Angeles County, due to pumping of oil and gas, finally reached about 28 feet. In the past few years, this has been stopped by pumping 1 million barrels per day of seawater back into the formation to replace the oil and gas withdrawn. (Pacific Fire Rating Bureau.)*

barrels a day in 1968) into the reservoir rocks of the Wilmington field. Under this subsidence-abatement program, regulated by the State Division of Oil and Gas, subsidence has been virtually stopped, although removal of oil and gas is going on at an accelerated rate.

Subsidence by removal of fluids from underground is just one instance—but a spectacular one—of geologic effects being brought about by man's own interference with natural processes.

Subsidence is not all man-induced by any means; releveling has shown that vertical movements—both subsidence and uplift—are also tectonic, that is, are due to natural deformational processes. For example, a rise of 0.21 feet (relative to a reference network) took place between 1948 and 1963 in the Alum Rock Park area east of San Jose. In the vicinity of Lebec (on State Highway 99 in the Tehachapi Mountains), releveling in 1965 showed an upheaval of 0.82 foot in seventeen years, or about 5 feet per century. This is about fifty times as fast as the 1,000 feet per million years that has been established for parts of the Sierra Nevada! Of course, there is no reason to expect that any such rate of local uplift will continue. Uplifts of 2 to 26 millimeters in the year 1964 to 1965 were measured at fifteen locations in southern California near the San Andreas and other major faults.

The "Palmdale bulge," along the San Andreas fault in southern California, has amounted to an uplift of about 1 foot in 15 years. It continued in 1976, but a year later it seemed to be subsiding! Is this an earthquake precursor? We do not know.

## UPLIFT AND TILTING OF THE SIERRA NEVADA

Uplift and gentle tilting of the Sierra Nevada took place intermittently, in stages, in late Cenozoic time. How fast did such uplift take place? With the aid of numbers of radiometric dates on volcanic rocks formed in the Sierra Nevada throughout Cenozoic time, geologists are able to say that rates of uplift were very different at different times and at different places. One such figure shows an average of 1,000 feet of uplift per

**10-7** *Sierra Nevada and Owens Valley near Lone Pine, Inyo County. The extremely steep eastern slope of the Sierra Nevada here consists of granitic rocks deeply incised by small streams, which run eastward (left) to the Owens River in Owens Valley. The smooth-appearing left two-thirds of the photo consists of coalesced alluvial fans made up of coarse fragments of granitic rock. A special feature of the photo is the geologically recent Independence fault which sharply separates the alluvial fan from the granitic base of the mountains. This scarp shows that the mountain block has been uplifted, but note also that the course of Shephard Creek (largest stream running right to left across the center of the photo) shows a strong suggestion of right-lateral offset of about 1,300 feet. (Vertical aerial photo, U.S. Geological Survey, Prof. Paper 590, 1968; scale approximately 1 inch equals 5,342 feet;* south *is at top.)*

million years for the past 10 million years. First-order releveling across the Sierra Nevada via Interstate Highway 80 has shown that the High Sierra has been uplifted by about 4 inches (10 centimeters) during the last 30 years (California Division of Mines and Geology, *California Geology,* March 1977). Uplift gave added power of erosion to the westward-flowing streams, so that each episode of uplift and tilting was recorded by the cutting of deeply entrenched canyons and by the formation of erosion surfaces—gently sloping summits and terraces. Remnants of several of these canyons and erosion surfaces remain. Almost continuous volcanism in parts of the Sierra has furnished a radioactive time clock. From time to time in the late Cenozoic Era, lava, volcanic mudflows, and ash filled the valleys and changed drainage courses. Erosion followed, and later volcanics covered the eroded surfaces. Dozens of radiometric dates of these volcanic rocks have enabled geologists to work out the closely related sequences of uplift and tilting, erosion, volcanic activity, and renewed faulting. For example, by this means we know that the intensive faulting of the eastern side of the Sierra (accompanied by uplift and westward tilting) began about 2.5 million years ago.

## THE SAN ANDREAS FAULT AND EARTHQUAKE HISTORY

The San Andreas fault is California's most spectacular and most widely known structural feature. Few specific geologic features on earth have received more public attention. Sound reasons for this are found in the series of historic earthquakes that have originated in movements in the San Andreas fault zone and in continuing surface displacements, both accompanied and unaccompanied by earthquakes. This active fault is of tremendous engineering significance, for no engineering structure can cross it without jeopardy and all major structures within its potential area of seismicity must incorporate earthquake-resistant design features. In 1963, a proposal for a large nuclear power plant installation on Bodega Head, north of San Francisco, was abandoned because of public controversy over the dangers of renewed movements and earthquakes on this nearby fault. Expensive design features have been incorporated into the state's plan to transport some of northern California's excess of water to water-deficient southern California, in order to ensure uninterrupted service across the fault in the event of fault movements and earthquakes in the Tehachapi Mountains area.

Geologists and seismologists the world over have directed their attention to the San Andreas fault because of (1) the great San Francisco earthquake of 1906 and many lesser shocks that have originated in the fault zone; (2) development of the "elastic rebound" theory of earthquakes by H. F. Reid; (3) striking geologic effects of former movements and continuing surface movements in the fault zone; and (4) postulated horizontal displacements of hundreds of miles—east block moving south. Observing the 1906 fault displacements in the ground surface, Reid recognized that stresses built up in the elastic rocks of the earth's crust could cause strain (distortion and bending) to the point of rupture. It was this sudden rupture that caused the earthquake!

### location and extent of the fault

The San Andreas fault strikes (bears) northwesterly in a nearly straight line in the northern Coast Ranges province and extends southeastward for a total length of about 650 miles from Shelter Cove across geologic structures and lineation of the Coast Ranges at a low angle, then south across the Transverse Ranges and into the Salton Trough. Latest movement in the fault zone has thus been clearly later than major structural features of those provinces. This recent movement may, however, be an expression of renewed activity along an older fault zone that existed before the geologic provinces now in existence were fully outlined.

The long northwesterly trend of the fault zone is interrupted in three places (see the 1:2,500,000-scale colored geologic map of California available from the U.S. Geological Survey or the California Division of Mines and Geology): (1) At Cape Mendocino, where it turns abruptly westward to enter the Mendocino fracture zone; (2) at the southern end of the Coast Ranges where they merge into the Transverse Ranges and where the fault turns to strike east into the complex knot of major faults in the Tehachapi Mountains; and (3) in the San Bernardino Mountains where the main San Andreas (Mission Creek) fault appears to turn southward into the Salton Trough.

## earthquake history

The earliest California earthquake noted in written records was felt by explorer Gaspar de Portolá and his party in 1769 while they were camped on the Santa Ana River about 30 miles southeast of Los Angeles. The earliest seismographs in use in California—also the earliest in the United States—were installed by the University of California at Lick Observatory on Mount Hamilton and at the University at Berkeley in 1887. The earliest seisomograms of a major California earthquake are those of the San Francisco earthquake of 1906, which was recorded at seven California stations as well as elsewhere throughout the world.

What is meant by a large earthquake or a small earthquake? How can the "strength" of earthquakes be compared in different places over the earth?

In 1935, Dr. Charles F. Richter, of the California Institute of Technology, devised a means of comparing the total energy of earthquakes expressed in terms of a figure now called the Richter magnitude, or simply magnitude (M). The maximum trace amplitude of an earthquake shock wave is measured in thousandths of a millimeter, from seismograms of a certain standard seismograph at a standard distance from the epicenter. The logarithm of the amplitude is then determined. Constants have been worked out to make the figures comparable for other seismometers at other distances. On this Richter scale, magnitude $M = 2$ is the smallest earthquake felt. Earthquakes of magnitude 4.5 to 5 cause small local damage, and those of magnitude 5.5 to 6 may cause considerable damage. Magnitudes of 7 or more are associated with "major" earthquakes, and those of 7.75 and over, with "great" earthquakes. The earthquake at Long Beach in 1933, with $M = 6.3$, was a "moderate" earthquake (but a very damaging one); the one at Arvin-Tehachapi in 1952, at $M = 7.7$, was a major earthquake; and the San Francisco earthquake of 1906, at $M = 8.25$, was a great earthquake. Local size or strength has long been measured by an intensity scale based on how the earthquake is felt and its apparent damage. The commonest intensity scale in use is the Modified Mercalli.

One of the Bay area's largest earthquakes centered on the Hayward fault (within the San Andreas fault zone) in the East Bay on June 10, 1836. Surface faulting (ground rupture) took place at the base of the Berkeley Hills from Mission San Jose to San Pablo. On October 21, 1868, another large earthquake was centered on the Hayward fault, with surface faulting for about 20 miles—from Warm Springs to San Leandro at the present southeastern boundary of Oakland. Maximum right-lateral (east block moving south) offset was about 3 feet.

In June of 1838 a strong earthquake originating on the San Andreas fault was accompanied by surface rupturing from Santa Clara, near San Jose, almost to San Francisco. This damaged the Presidio at San Francisco and the missions at San Jose, Santa Clara, and San Francisco. Another strong earthquake centered on the San Andreas fault in the Santa Cruz Mountains on October 8, 1865. This was accompanied by ground cracks, landslides, and dust clouds; buildings were damaged in San Francisco and at the New Almaden mercury mine, which is only a few miles east of the active part of the fault. On April 24, 1890, a strong earthquake damaged Watsonville, Hollister, and Gilroy, south of San Jose, and closed an important highway and railroad pass by landslides.

The famous San Francisco earthquake, 5:12 A.M. local time April 18, 1906, was among California's two or three strongest. Visible surface faulting occurred from San Juan Bautista to Point Arena, about 200 miles north, where the San Andreas fault enters the ocean. At the same time, surface faulting also occurred 75 miles north of Point Arena at Shelter Cove in Humboldt County,

(a)

**10-8** (a) *San Andreas fault zone west of southern San Joaquin Valley. (Aerial oblique courtesy of Malcolm Clark, U.S. Geological Survey.)* (b) *Index to aerial oblique photo of San Andreas fault zone. (Malcolm Clark, U.S. Geological Survey.)*

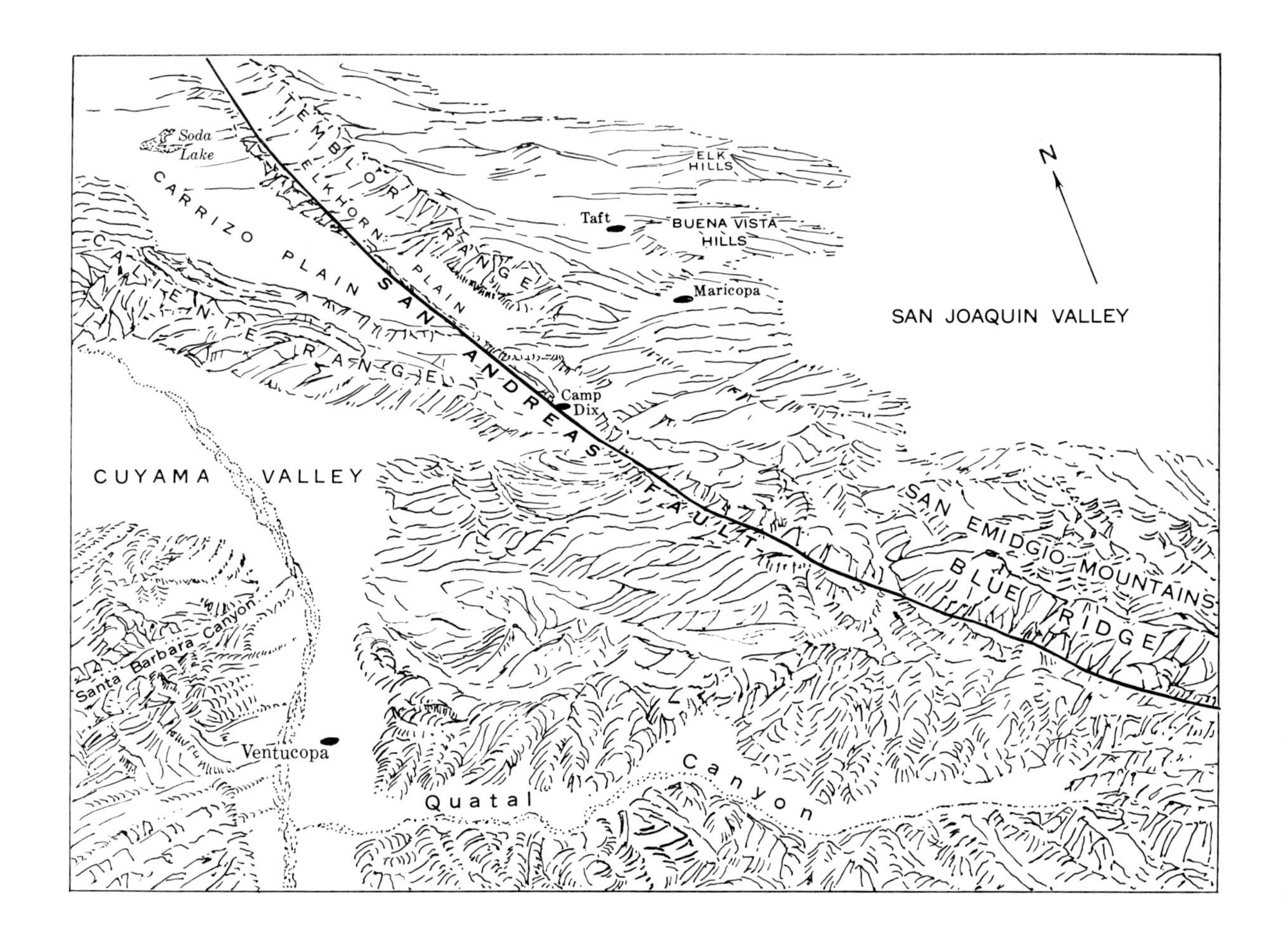
Soda Lake
TEMBLOR RANGE
ELK HILLS
N
ELKHORN PLAIN
CARRIZO PLAIN
Taft
BUENA VISTA HILLS
CALIENTE RANGE
SAN ANDREAS FAULT
Maricopa
SAN JOAQUIN VALLEY
Camp Dix
CUYAMA VALLEY
SAN EMIDGIO MOUNTAINS
BLUE RIDGE
Santa Barbara Canyon
Ventucopa
Quatal Canyon

(b)

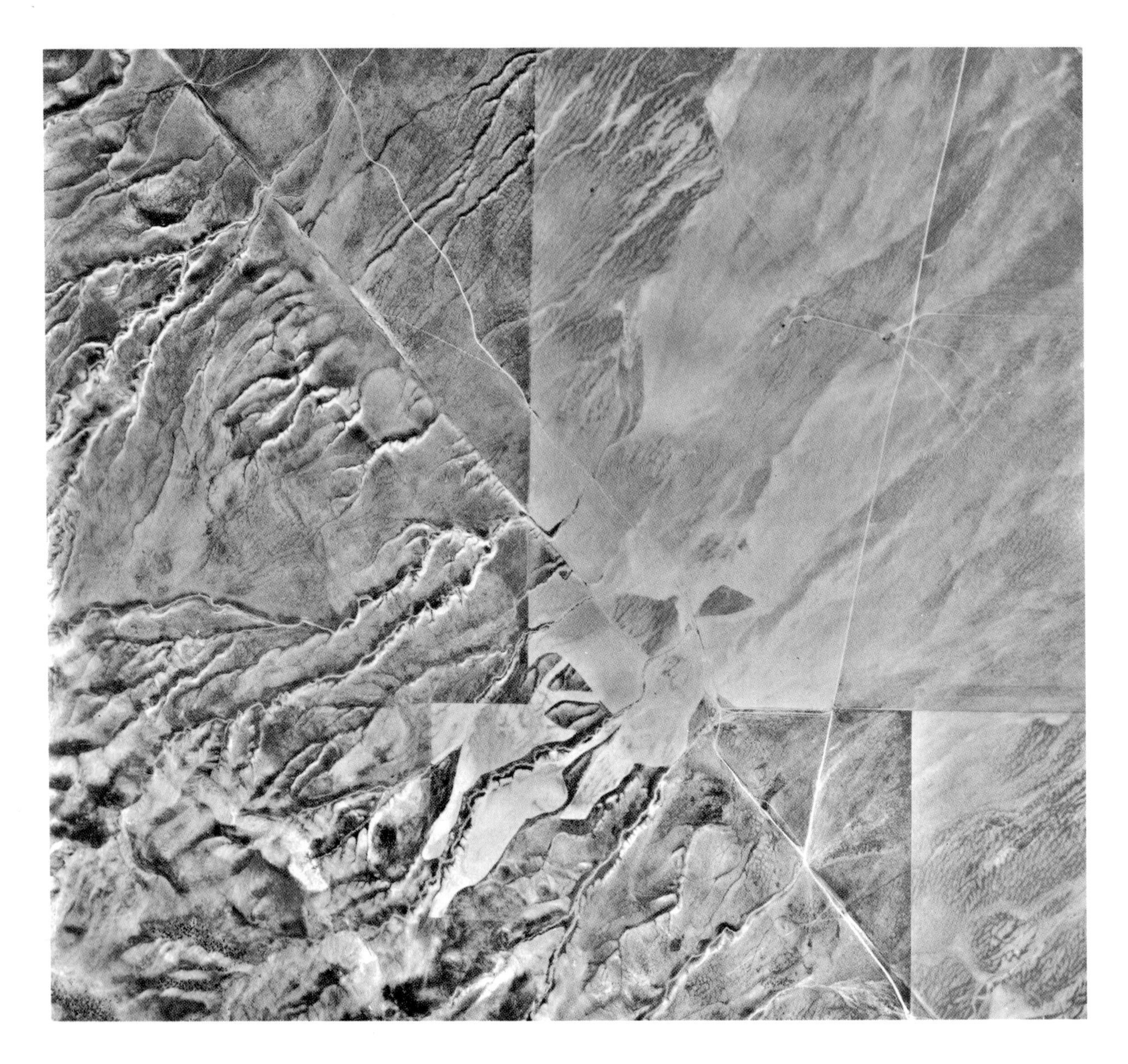

**10-9** *San Andreas fault, Carrizo Plain, San Luis Obispo County. The fault appears as a line running diagonally across the photo from upper left to lower right. Trenches, scarps, and offset drainage define the trace of the fault. Note that there is an older fault trace parallel to the latest. Streams have cut deeply into the low hills on the northeast (left); they run across the fault and deposit sediment on the plains to the right. Fence lines and roads should not be confused with the fault trace. An offset stream at the center of the photo shows a right-lateral horizontal component of displacement amounting to about 500 feet. Other streams—probably younger—show less apparent offset. (Vertical aerial photo. U.S. Geological Survey, Prof. Paper 590, 1968; scale approximately 1 inch equals 2,300 feet; south is at top.)*

probably along an extension of the San Andreas fault. The 1906 scarp viewed at Shelter Cove in 1963 clearly shows upthrow of 6 to 8 feet on the east side; there was no evidence of a horizontal component of displacement. However, offset of a line of old trees and an old fence viewed east of Point Arena in 1963 gave clear evidence of right-lateral displacement on the order of about 14 feet. The epicenter of the earthquake was just west of the Golden Gate, off San Francisco. At the south end of Tomales Bay a road was offset a reported 20 feet in a right-lateral sense. Magnitude of the earthquake is generally computed from the 1906 seismograms at about 8.25 on the Richter scale. Damage has been estimated at from $350 million to $1 billion in 1906 dollars, and about 700 people were killed. A large part of the loss resulted from tremendous fires in San Francisco, where gas from broken mains fed the flames and ruptures in water lines prevented effective fire fighting. Most extensive ground breaking in the city was near the waterfront in areas of natural bay mud and artificial fill.

Another of California's great earthquakes, comparable in magnitude to the San Francisco 1906 earthquake, was caused by displacement on a segment of the San Andreas fault extending through the southern part of the Coast Ranges province and southward across the Transverse Ranges. This Fort Tejon earthquake of January 9, 1857, probably was centered in the region between Fort Tejon in the Tehachapi Mountains and the Carrizo Plain in the southern Coast Ranges. Surface faulting extended for 200 to 275 miles from Cholame Valley, along the northeastern side of the Carrizo Plain, through Tejon Pass, Elizabeth Lake, Cajon Pass, and along the southern side of the San Bernardino Mountains. Accounts of this earthquake are unsatisfactory and inconclusive, but horizontal surface displacement may have amounted to as much as 30 feet in a right-lateral sense (east block moving south).

Among California's greatest earthquakes since the explorations of the Portolá party was that in Owens Valley on March 26, 1872. At Lone Pine, 23 of the 250 inhabitants were killed and 52 of their 59 adobe houses were destroyed. The shock was felt from Mount Shasta to San Diego. Surface faulting at the eastern foot of the Sierra Nevada produced scarps with a maximum net vertical displacement of about 13 feet and horizontal right-lateral offset of about 16 feet. Surface faulting extended for perhaps 100 miles. This fault, of course, has no direct relationship to the San Andreas fault, but is probably related to a zone of active faults in west-central Nevada and adjacent eastern California. Pleistocene and recent uplifts of the Sierra Nevada have occurred in part along this fault zone.

There have been, in historic times, two truly great earthquakes (Fort Tejon and San Francisco) originating on the San Andreas fault, each accompanied by more than 200 miles of surface ruptures: one in the south and one in the north. The southern segment of the San Andreas fault zone—south of the Tehachapi Mountains—has been quiet since 1857 on the San Andreas fault proper, but recently very active on the closely related San Jacinto, Elsinore, Inglewood, and Imperial faults. In the northern segment, that marked by surface rupture in 1906, many earthquakes have originated on the San Andreas fault and its auxiliary faults in the East Bay—the Hayward and Calaveras faults. However, since 1906 no earthquakes have originated on the most northerly segment from Marin County to

(a)

**10-10** (a) *High aerial oblique view of intersection of San Andreas, Garlock, and Big Pine faults, Tehachapi Mountain area.* (*Photo courtesy of Malcolm Clark, U.S. Geological Survey.*) (b) *Index to aerial oblique photo of intersection of San Andreas, Garlock, and Big Pine faults.* (*Malcolm Clark, U.S. Geological Survey.*)

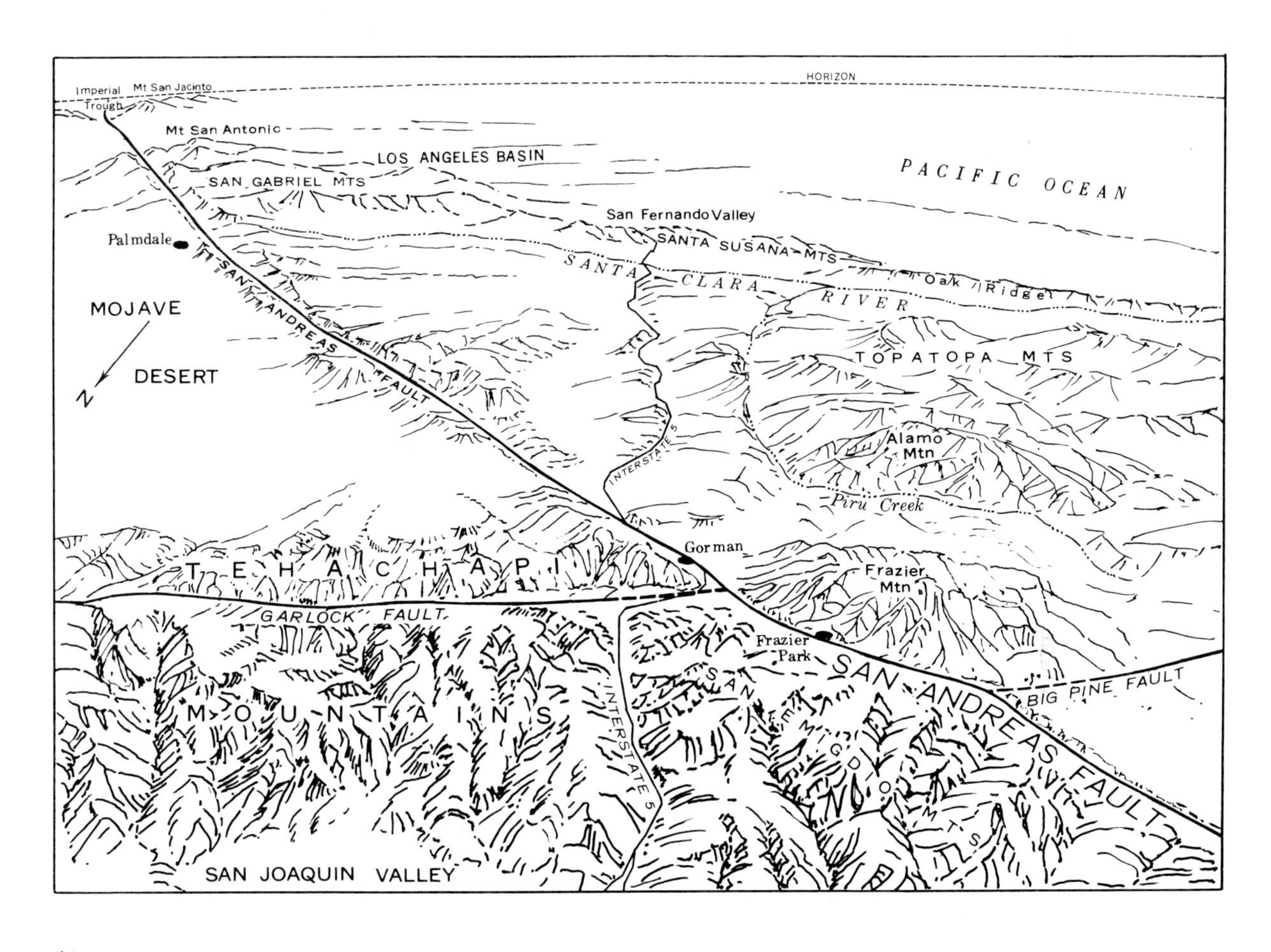
HORIZON
Imperial Trough
Mt San Jacinto
Mt San Antonio
LOS ANGELES BASIN
PACIFIC OCEAN
SAN GABRIEL MTS
San Fernando Valley
SANTA SUSANA MTS
Palmdale
SANTA CLARA RIVER
Oak Ridge
MOJAVE
DESERT
N
SAN ANDREAS FAULT
TOPATOPA MTS
INTERSTATE 5
Alamo Mtn
Piru Creek
Gorman
TEHACHAPI
Frazier Mtn
GARLOCK FAULT
Frazier Park
SAN ANDREAS FAULT
BIG PINE FAULT
MOUNTAINS
SAN EMIGDIO MTS
INTERSTATE 5
SAN JOAQUIN VALLEY

(b)

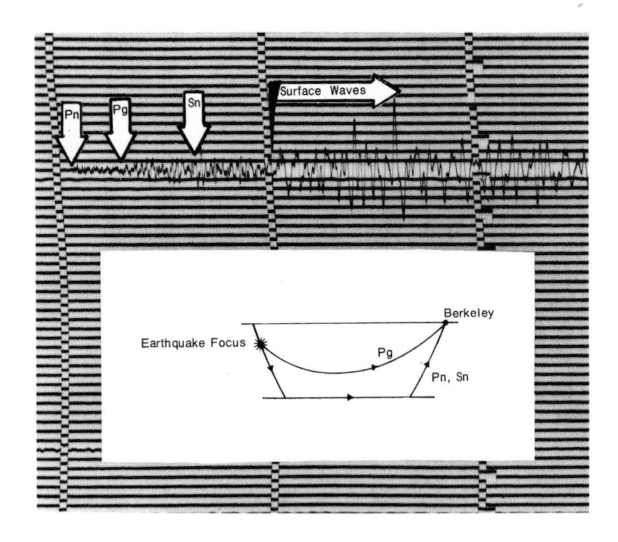


**FIGURE 10-11** *The record (seismogram) of a typical California earthquake. At 01:42:19.5 GCT (1 hour, 42 minutes, and 19.5 seconds after midnight Greenwich Civil Time) an earthquake took place off the northern California coast at latitude 40.1° N and longitude 124.4° W, close to the northern end of the San Andreas fault zone. Depth of focus was about 10 kilometers (6 miles) and Richter magnitude computed from the seismogram at the Berkeley Seismograph Station was between 5.8 and 6.0 (BRK 5.8–6.0). The record was made by a Wood-Anderson seismograph which responded to the east-west component of the ground motion. Distance between vertical broken lines on the seismogram represents one minute of time. For example, it was about one minute after the earthquake occurred that the surface waves began to appear on the seismogram. The seismologist has labeled the first appearance of four kinds of waves:* Pn *and* Pg*—fast-moving push-pull waves;* Sn*—slower waves of distortion, and surface waves. The small inset shows diagrammatically the paths of the P and S waves from the earthquake focus through the crust and mantle to Berkeley. The earthquake was felt as far away as Klamath Falls, Oregon. Chimneys were cracked and people were frightened in the nearby small towns of Petrolia and Honeydew. Several aftershocks. occurred. (Seismogram courtesy of Director Bruce A. Bolt, University of California Seismograph Stations, Berkeley.)*

Humboldt County. The strongest earthquake in the Bay area since 1906 was the San Francisco earthquake of March 22, 1957, of Richter magnitude 5.3. It originated at shallow depth near Mussel Rock, off the coast a few miles south of San Francisco; there was no surface faulting. No lives were lost, but minor damage to many homes in the Westlake-Daly City district totaled about $1 million. It may be that the San Andreas fault is "locked" temporarily in these two inactive segments and that stresses which are building up will eventually be relieved by sudden displacements and resultant earthquakes.

## landforms in the fault zone

Extensive activity along the San Andreas fault zone in Quaternary time has developed a linear depression, marked by all the features of a classic rift valley, extending the entire length of the fault and encompassing a width from a few hundred feet to more than $1\frac{1}{2}$ miles. Rift valley features are particularly well expressed in the San Francisco Bay area, in the arid Carrizo Plain, and along the northern side of the San Gabriel Mountains in southern California. Within the rift zone, fault gouge and breccia always occur as well as a disorganized jumble of fault-brecciated rocks from both sides of the fault, the result of hundreds of repeated ruptures on different fault planes during late Pleistocene and Holocene time. Features of the rift valleys have resulted from (1) repeated, discontinuous fault ruptures on the surface, often with the development of minor graben (downdropped fault blocks), horsts (uplifted fault blocks), and pressure ridges; (2) landsliding, triggered by earthquake waves and surface faulting; and (3) erosion of brecciated, readily weathered rock. Within the rift-valley troughs, it is common to find Pliocene to Holocene sediments.

**10-12** *Offset dam road, San Andreas fault, Crystal Springs Lakes on San Francisco Peninsula, 1906. Movement was largely horizontal, in a right-lateral sense (fault block opposite the observer moved to the right). (Photo from Marshall Moxam, in California Division of Mines and Geology collection.)*

**10-13** *Typical street scene in San Francisco the morning after the earthquake of April 18, 1906. Because of broken gas mains and water lines fires raged unchecked through the city. (Photo by W. E. Worden, in California Division of Mines and Geology collection.)*

North of San Francisco, across Marin County, the fault follows a remarkably straight course approximately 35 degrees west of north. The most prominent features are Bolinas Bay and the long, linear Tomales Bay, which lie in portions of the rift valley drowned by rising sea waters following the Pleistocene glacial epoch. Between these bays, the rift zone is a steep-sided trough, in places as deep as 1,500 feet, with its lower levels characterized by a remarkable succession of minor alternating ridges and gullies parallel to the general trend of the fault zone. Surfaces of the ridges and gullies are spotted by irregular hummocks and hollows; many of the hollows are undrained and have developed sag ponds, which are common along the San Andreas rift. Geologically recent adjustment of the drainage in the rift zone leaves little positive evidence of the amount and direction of recent displacement, except for that which took place in 1906. Offset lines of trees in this area still show the 13- to 14-foot horizontal right slip of 1906; just south of Point Arena trees also show a similar amount of offset. In the long stretch north of Fort Ross to a

**10-14** *Wagon road to Alpine, 5 miles west of Stanford University. This is not the San Andreas fault proper but is ground rupturing within the broad fault zone due to lurching and irregular soil compaction. San Francisco earthquake, 1906. (J. C. Branner photo collection.)*

**10-15** *Lone Pine, Inyo County, and Sierra Nevada fault scarps formed in the Owens Valley earthquake in 1872. The low Alabama Hills are on the right (west); Highway 395 passes through the center of town; and the Owens River runs southeasterly across the photo into dry, saline Owens Lake (upper left). The great 1872 fault scarp is marked strongly along the base of the Alabama Hills; Diaz Lake is in a fault graben. Note two types of rock in the Alabama Hills: minutely jointed granite and the smoother, eroded surfaces of Jurassic-Triassic metamorphic rocks. Beach ridges of Owens Lake show in the upper left corner. (Vertical aerial photo, U.S. Geological Survey, Prof. Paper 590, 1968; scale approximately 1 inch equals 5,340 feet; south is at top.)*

point a few miles south of Point Arena, the broad expression of the rift zone is clear, but minor features within the zone have been obscured by erosion by the Gualala and Garcia Rivers and by the dense forest cover of the area.

**10-16** *Looking northeast at Bear Mountain. Mole-track fault scarp extending toward the northeast is the trace of the White Wolf fault, formed at the time of the major July 21, 1952, Arvin-Tehachapi earthquake. The White Wolf fault parallels the Garlock fault at the extreme southern end of the San Joaquin Valley. (Lauren A. Wright photo, California Division of Mines and Geology Bull. 171.)*

South of San Francisco, across San Mateo County, the San Andreas fault zone follows the same trend as to the north, but is less straight and is complicated by several subparallel faults. Near Mussel Rock, south of San Francisco, where the fault goes out to sea, are great landslides which obscure the fault trace. For a few miles to the southeast is a succession of sag ponds, notched ridges, and rift-valley lakes within a deeply trenched valley. The long, narrow San Andreas and Crystal Springs Lakes are natural lakes, which were enlarged many years ago by the artificial dams built to impound San Francisco's water supply. Similar rift-valley features mark the fault southward to the Tehachapi Mountains; because of the local aridity, they are particularly clear and striking in the Temblor Range area, in the Cholame Valley, and in the Carrizo Plain.

South of the Tehachapi Mountains, the striking rift-valley landforms continue, expressed as sag ponds (Elizabeth Lake and Palmdale Reservoir) along the northern margin of the San Gabriel Mountains, through the Cajon Pass, and along the southern side of the San Bernardino Mountains. None of the offsets and other rift-valley features in this southern segment of the great fault appears to be younger than 1857.

### displacement on the fault

In 1953, two well-known geologists—M. L. Hill and T. W. Dibblee, Jr.—advanced the possibility of cumulative horizontal right-lateral displacement of possibly 350 miles on the San Andreas fault since Jurassic time. This hypothesis has received wide acceptance among earth scientists, has intrigued geologists, and has been an important factor in stimulating work on the fault. The geologists compared rock types, fossils, and gradational changes in rock characteristics in attempting to match units across the fault. At the opposite extreme, other geologists have felt less confident about this "matching" of rock units and have stated that horizontal movement on the northern

**10-17** *Trace of White Wolf fault crossing fence, looking southwest toward Arvin. Slack wires show compression across the fault; the uphill side moved up and toward the right in a direction opposite to movements on the San Andreas fault. (Lauren A. Wright photo, California Division of Mines and Geology Bull. 171.)*

**10-18** *Cracking in water-saturated alluvium, White Wolf fault zone near Comanche Creek. (G. B. Oakeshott photo, California Division of Mines and Geology Bull. 171.)*

segment of the San Andreas fault has been less than 1 mile!

Some of the latest work by geologists, seismologists, and engineers on the San Andreas fault in the San Francisco Bay area and on the closely related Hayward fault in the East Bay shows that these faults are still active. In several places surface "creep," or slippage, is taking place. At the Almaden Winery a few miles south of Hollister, for example, creep occurs in spasms of movement of small fractions of an inch, separated by intervals of weeks or months. Average displacement, east block moving relatively south, is $\frac{1}{2}$ inch per year, wrenching apart a winery building by the movement. Several cases of well-substantiated right-lateral creep, on the order of $\frac{1}{8}$ to $\frac{1}{4}$ inch per year, have also been recognized along the 1868 trace of the Hayward fault from Irvington (Fremont) to the University of California stadium in Berkeley. Other active faults in the state are also showing creep movements. Frequent earthquakes, with epicenters on these faults, show that present-day movements are taking place.

## unsolved problems

South of the Tehachapi Mountains, right lateral horizontal displacement on the San Andreas fault system seems to have been a maximum of 160 miles. All formations from Precambrian to Oligocene appear to have been displaced by the same amount. This suggests that the San Andreas fault proper originated about Oligocene time.

Earlier—from late Jurassic to Eocene time—California coastal regions were involved in the complexities of a subduction zone, with the oceanic crust being subducted under the western edge of North America. By latest Cretaceous and early Tertiary times, the subduction zone and the East Pacific rise (zone of sea floor spreading) were being buried under the margin of the American plate. Plate tectonists view the San Andreas fault as a "transform" fault which separates the northerly drifting Pacific plate from the westerly drifting American plate, and which connects two segments of the East Pacific rise—the Salton Trough in the south and the Gorda Basin north of Cape Mendocino.

At the present day, we find great blocks of upper-mantle and oceanic-crust rocks (the Franciscan assemblage and its ophiolites) in western California which have been added to the North American continent. The present San Andreas fault lies within the oceanic province near its north end and far inland within the granitic continental province south of the Tehachapi Mountains. At its extreme south end, the fault appears to enter the Gulf of California and offset the East Pacific rise. Perhaps the San Andreas fault turns westward—at its north end—and merges into the Mendocino fracture zone. Across the southern and central Coast Ranges, the San Andreas fault does appear to mark the boundary between old oceanic crust (Franciscan and serpentine) and continental crust (granitic rocks). But here the granitic rocks lie west of the fault, perhaps because of large lateral displacement since late Eocene time.

Thus, great difficulties attend the concept of the San Andreas fault as a simple transform-fault boundary between the moving American and Pacific plates. Answers await the intensive work of geologists, seismologists, and other scientists of many disciplines.

## LANDSLIDES AND EARTHFLOWS

Literally thousands of landslides throughout the state, in all geologic provinces and in all its varied climates, are still another facet of California's dynamic landscape. In Chapter 3, landsliding was discussed as a part of mass wasting—a major process in changing the face of the land. Here, we shall take another look at landslides for their typical occurrence in areas of active crustal movements.

Landslides are, of course, closely associated with active faulting and earthquakes. More landslides per square mile appear along California's active fault zones than in any other areas of similar sizes. Faulting plays two major roles in landsliding: fault zones are the sites of crushed and fractured rock—rock of low strength which may slide readily; and faults cause earthquakes which, in turn, trigger slides and rock falls.

The 1906 San Francisco earthquake triggered many landslides over an area of approximately 13,000 square miles. One of the large ones was a slump at Cape Fortunas (near the north end of the San Andreas fault), which was about 1 mile long and ½ mile wide.

The Kern County earthquakes of 1952 triggered hundreds of slides, particularly along the White Wolf fault that caused the earthquake. However, as in any major earthquake, many slides and rock falls occurred more than 60 miles from the fault. Rock falls blocked the Angeles Crest Highway in the San Gabriel Mountains between Pasadena on the south side and Vincent on the north side of the range. For months after the main earthquake, slides were triggered on steep slopes by aftershocks.

Even in the small (Richter magnitude 5.3) San Francisco earthquake of 1957, significant damage was done by landslides that occurred in steep sea cliffs along the coast south of the city and buried State Highway 1. A segment of this highway was finally abandoned because of excessive cost of repair and maintenance.

### causes of landslides

Obviously, slides move down slope under the force of gravity. In addition, there must be loose or weathered rock, which has low strength.

The inherent, natural properties of rock are important in the formation and behavior of landslides. Obviously, clayey, schistose, or thoroughly fractured or jointed rock, particularly where natural planes of weakness have developed, would be likely to slide. Certain rock formations that are distributed widely in the state are notorious for inherent weakness and susceptibility to sliding.

The steeper the slope, with a rock of given type and structure, the more likely that sliding will occur. However, there are many examples of vertical or even overhanging cliffs, like some of the granite walls of Yosemite, that remain relatively stable. Mudflows, on the other hand, can occur in areas with as little as 1 degree of slope.

A major factor in landsliding is water, which adds weight, breaks down internal coherence, and

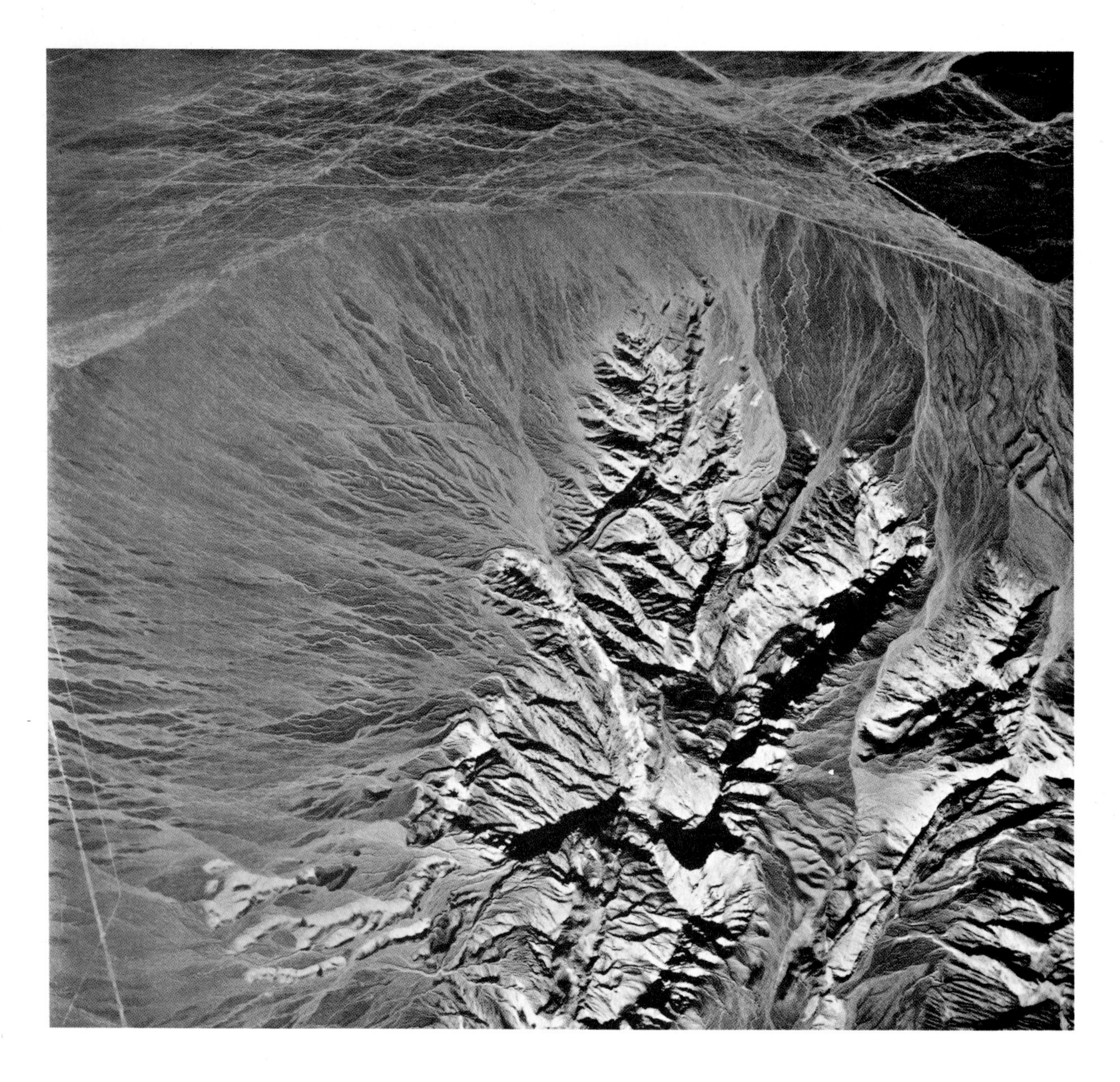

**10-19** *Landslides, east of Funeral Mountains, near Death Valley Junction, Inyo County. Tilted, stratified Paleozoic and Tertiary sedimentary formations form the mountainous area, which is surrounded by coalescing alluvial fans. The undisturbed desert pavement—dark because of desert varnish—is cut by numerous meandering and braided washes. Several landslides form light-colored, finger-shaped ridges that extend out over the desert pavement; good examples are in the lower left corner.* (*Vertical aerial photo. U.S. Geological Survey, Prof. Paper 590, 1968; scale approximately 1 inch equals 5,400 feet;* south is at top.)

lubricates movement. In California's seasonal climates, many more landslides occur during and after the wet winter season than at any other time. The amount of water present in landslide material varies between the widest extremes—from a rock fall with almost no water to a mudflow that may carry more water than solids!

### size and velocity of movement

How big may landslides get? And how fast can they move? All sizes and forms of slides occur. The prehistoric Blackhawk landslide (Figure 10-21) extends 5 miles onto the desert floor from the northern side of the San Bernardino Mountains in the Transverse Ranges. It consists of 370 million cubic yards of limestone fragments. Research strongly suggests that this slide moved at velocities as high as 200 to 300 miles per hour on a cushion of air! Another prehistoric slide at Tin Mountain, in the Death Valley country, has a calculated volume of 2,350 million cubic yards.

Judging from the effects produced and from some eyewitness accounts, velocities may get as high as 300 miles per hour. At the other extreme, many mass movements of rock and soil take place extremely slowly. Soil creep may take place down a slope at a fraction of a foot per year.

## LAKES AND RUNNING WATER

The principal elements in the state's drainage pattern were reviewed in Chapter 1, and landforms shaped by running water were discussed in Chapter 3. We shall now look briefly at California's lakes and streams in relation to crustal unrest and changing climates of the Quaternary Period.

Individual lakes are geologically short-lived, particularly so in a land of crustal unrest and rapidly changing landforms. Every lake exists only because of a depression of some sort and maintains itself only if the inflow of water is equal to, or greater than, the loss of water by evaporation and sinking. Once a lake is formed, its own destruction is started by filling of its basin by sediment and downcutting by erosion at its outlet. Its life may be prolonged if tectonic downwarping of its bed occurs.

Basins may be formed in a great variety of ways. The High Sierra is dotted with small lakes of glacial origin—basins scooped out of rock by glacial erosion or formed behind moraines that have dammed stream valleys. Donner Lake and Norden Lake near the crest of the Sierra on Interstate Highway 80 are of glacial origin. Nearby Lake Tahoe, the largest and deepest (1,645 feet) freshwater lake that touches California, is partly glacial, but its basin is probably mainly due to faulting, with an assist from Quaternary volcanic rocks that formed a dam across the Truckee River. The floor of Yosemite Valley consists largely of lake sediments formed in a lake behind an old dam made by glacial terminal moraines. The small gem of a lake called Mirror Lake, in Yosemite Valley, was formed by a rock fall across Tenaya Creek.

All of the state's large lakes lie in basins that are partly, or wholly, of structural origin. The largest lake of all—the saline Salton Sea—lies in a downwarped and faulted basin within the southern extension of the great San Andreas fault system. The lake was formed in 1905 when the lower Colorado River overflowed its banks and entered the Salton Sink. The "sea," whose shore line is now about 230 feet below sea level, is all that is left of Pleistocene Lake Coahuilla, a large freshwater lake that was formed by deltaic sediments of the Colorado River which dammed the northern end of the depression of the Gulf of California. The largest freshwater lake—Clear Lake, north of San Francisco Bay—lies in a basin produced by Quaternary crustal warping plus geologically late damming by volcanism.

Surprise Valley Lakes, Goose Lake, and Honey Lake in northeastern California, all lie in tilted fault troughs, as do many other lesser lakes. Eagle Lake is in a basin formed by faulting and Quaternary volcanic action. The tiny pond called Badwater lies at the bottom of Death Valley—a great downdropped fault block—which was filled by freshwater Lake Manly several hundred feet deep during the wettest, coolest stages of the Pleistocene Epoch.

Saline Mono Lake lies in a tremendous, nearly

**10-20** *Tunnel Road landslide of December 9, 1950, east side of Berkeley Hills, Contra Costa County. Slide occurred in clay and gravel of the Pliocene Orinda Formation after a heavy rainfall season. Slide mass was eventually stabilized by removing the water through horizontal drain holes.* (*Bill Young photo,* San Francisco Chronicle, *California Division of Mines and Geology,* Mineral Information Service, *October, 1967.*)

circular, downfaulted depression that was probably formed by crustal warping in combination with faulting. California's volcanic crater lakes are few and small; one example is a perfect little heart-shaped lake, about $\frac{1}{4}$ mile across, that lies in a volcanic crater on Paoha Island in the middle of Mono Lake.

Contemporary lakes of the Mojave Desert and Great Basin are all saline, or intermittently dry playas—such as Searles Lake, Owens Lake, Danby, Cadiz, and Bristol Lakes—because of aridity and very high rates of evaporation. During cool humid climates of the Pleistocene Epoch, freshwater lakes dotted the Mojave Desert and Great Basin. For example, glacial meltwater fed an Owens Lake, which was then fresh and overflowed into Searles Lake, thence into a large lake in Panamint Valley, and finally into Lake Manly in Death Valley. In the late Pleistocene (600,000 years ago), Lake Corcoran—a shallow freshwater lake—covered much of the central San Joaquin Valley. Today, Tulare and Buena Vista Lakes in the southern San Joaquin Valley exist as broad, shallow lakes only during wet years—such as the winter of 1968-1969!

Man has been making profound changes in California's landscape, greatly altering natural geologic processes. None has been more far-reaching than the building of dams, with consequent formation of artificial lakes. Among the largest are Shasta Lake on the Sacramento River, which contains 4.5 million acre-feet of

**10-21** *Blackhawk landslide, dated as 17,000 years old by the radiocarbon method, extends 5 miles out from the front of the San Bernardino Mountains in San Bernardino County. It is composed of fragments of Mississippian Furnace Limestone which came from high up in the range. (John J. Shelton photo, California Division of Mines and Geology,* Mineral Information Service, *October, 1967.)*

water; Oroville Reservoir on the Feather River, with 3.5 million acre-feet; Clair Engle Lake on the Trinity River; and San Luis Reservoir, formed by a dam across San Luis Creek on the northwestern side of the San Joaquin Valley. (An acre-foot is the amount of water required to cover an acre to a depth of 1 foot or 325,850 gallons of water). In California there are more than fifty such lakes that exceed 50,000 acre-feet of water storage and hundreds of lesser lakes.

It is a very obvious fact that streams and drainage patterns follow the slopes of the land. As far back as the latest Cretaceous Period, a pattern of river drainage was established westward down the slopes of the Sierra Nevada, with many smaller streams draining into the trunk streams—like the modern Sacramento and San Joaquin Rivers—which emptied into the ocean. This pattern has been persistent, lasting—with modifications—through all the vicissitudes of complex geologic changes and great crustal movements that were sometimes directly athwart the stream channels. The power of erosion by running water is indeed a great and wonderful thing!

## WATER—THE MOST IMPORTANT NATURAL RESOURCE OF THE LAND

No mineral resource is used in such large quantities, or is so essential to man, as water. Far more water is used in California (1977 population over 21 million) than in any other state.

An average of about 190 million acre-feet of water falls as precipitation on the state annually. About 71 million acre-feet of this runoff is in the major streams; the balance enters the ground and adds to groundwater in storage or is lost by evaporation and use by plants. The Great Valley has the largest developed groundwater reservoirs of any area of comparable size in the United States; these reservoir sands can store 600 million acre-feet. In the Sacramento Valley, there is a good balance between withdrawal of water and underground recharge; in the San Joaquin Valley, withdrawals are faster and greater than recharge, so ancient groundwater is literally being "mined." South of the Tehachapi Mountains, water must generally be mined or imported.

California has tremendous water problems. These arise because of (1) rapidly growing use of water, (2) the fact that 70 percent of the precipitation occurs in the northern third of the state while 77 percent of the water needs are in the southern two-thirds, and (3) the fact that nearly all the precipitation occurs from November to April. The state now uses about 25 million acre-feet of water annually, 90 percent of which is for irrigated agriculture. Because of its topography and drainage pattern, California gets little natural drainage from out of state. The Klamath River brings in some water from Oregon, and southern California annually uses about 4 million acre-feet from the Colorado River.

The people of California have been water conscious and have been engaged in water projects ever since the Spanish mission at San Diego first set up an irrigation system in 1770. The little pueblo of Los Angeles built a domestic water-supply system on the Los Angeles River in 1781. Extensive water systems were initiated by the gold miners, who, beginning about 1850, diverted streams in order to get at gold in stream channels and for use in washing gravel for gold recovery. Man has been building dams, diverting streams, and guiding them into new channels ever since. Few major streams in the state remain unaltered by man's interference with the normal course of geologic processes.

In late years, tremendous progress has been made in the efficient utilization of both surface water and groundwater in the state, and in equalizing the seasonal and regional imbalance of supply. So vast is this problem of redistribution that only the federal government and the state in cooperation can find the solutions and do the construction necessary.

Since 1963, the state has had under construction one of the world's greatest water redistribution projects. Known as the California State Water Project and estimated to cost about $2.8 billion, it is designed to take excess water from northern California for use in dry southern California. The project reaches from the upper Feather River in the northern Sierra Nevada to the Oroville Dam in Butte County and from the Sacramento–San Joaquin delta to the Perris Reservoir near Lake Mathews in Riverside County. The facilities include dams, reservoirs, canals, tunnels, and pumping plants; many are unprecedented in size. The major transmission structure is the California Aqueduct—444 miles long from the delta of the Sacramento-San Joaquin Rivers to Perris Reservoir in southern California. Great geological problems mark the aqueduct route; particularly, ground subsidence along the western side of the San Joaquin Valley and the active San Andreas fault, which the aqueduct must cross. In northern California, the Oroville Dam is the highest earthfill dam (747 feet) in the world and the highest dam of any kind in the United States. It is more than 1 mile long at the crest. Eventually, water for the project will come from as far north as Clair Engle Lake on the Trinity River and probably also from the Klamath River.

Finally, one of the great uses for water in California has been for electrical power. By 1969, development of hydroelectric power from falling water had just about reached the economic maximum in the state. The other major source of power has been steam from the burning of petroleum and natural gas. At The Geysers, electrical energy is being developed directly from natural steam, the only such source in the United States. The potential of geothermal energy in California is as yet unknown.

In the past few years, power from nuclear reactors has been rapidly coming to the fore and will be a major source of electrical energy before the end of the century. Nuclear energy is also offering the interesting possibility of furnishing the large amounts of power necessary to desalt seawater to furnish a freshwater supply to coastal communities in arid climates, such as those along the coastline from Santa Barbara to San Diego.

**10-22** *Tufa cones along the southern shore of Mono Lake. These consist principally of calcium carbonate deposited from warm spring waters. The water of Mono Lake is so highly alkaline that only the tiny brine shrimp live in it. (Charles W. Chesterman photo.)*

□ Where to go and what to see—geologically—is the theme of Part Three. California's very extensive major highways and secondary roads offer convenient access to most of the rock formations and geologic features of the state.

The national parks and monuments located in California are oriented around striking rock formations and geologic features which ideally reveal and depict geology typical of the provinces in which they are located. These are Lava Beds National Monument in the Modoc Plateau; Lassen Volcanic National Park in the southern Cascade Range; Yosemite National Park, Devils Postpile National Monument, Kings Canyon National Park and Sequoia National Park in the Sierra Nevada; Point Reyes National Seashore and Muir Woods National Monument in the Coast Ranges; the huge Death Valley National Monument in the Great Basin Province; Joshua Tree National Monument in the Mojave Desert; Channel Islands National Monuments on Anacapa Island (a westward extension of the Transverse Ranges onto the continental borderland) and Santa Barbara Island (a part of the Peninsular Ranges on the continental borderland); and Cabrillo National Monument at San Diego in the Peninsular Ranges. Many of the state and local parks—too numerous to mention individually—preserve interesting rock formations and structures for public view.

Reading of each chapter of Part Three should be preceded, first, by review of the appropriate section in Chapter 1 ("California's Mountains and Valleys"), for in that chapter is an introduction and orientation to each of the natural provinces; and second, by rereading the pertinent chapters and sections on geologic history in Part Two. The serious student will want to study the 1:2,500,000-scale colored geologic map of California (available from the U.S. Geological Survey or the California Division of Mines and Geology) and, when he goes into the field, will guide himself by the latest highway map. Details of the regional geology can best be followed along the highways by using the map sheets of the *Geologic Map of California* (scale: 1 inch equals 4 miles) published by the California Division of Mines and Geology.

California law does not permit parking or walking along the freeways; you must get off onto the frontage roads and local highways to photograph, examine, or sample the rock formations!

# III

# GEOLOGIC VIEWS AND JOURNEYS IN THE NATURAL PROVINCES

■ California's Coast Ranges extend for 600 miles in a northwesterly direction from the Transverse Ranges in the south to a point beyond the California-Oregon border. Their width varies from about 20 miles in the north to about 80 miles west of the northern Sacramento Valley. Their geology is extremely complex—in variety of rock formations as well as in structure. Interestingly, however, the oldest rocks are probably no older than late Paleozoic!

Broad features of Coast Range topography, history, and structure have been outlined in Chapters 1, 8, and 9; but the detailed geologic history of some of the separate ranges will be reviewed in this chapter. Geology of the Coast Ranges is best followed in detail on the *Geologic Map of California* (an index to the map sheets covering the Coast Ranges is given in Figure 11-1). To gain a still better view and understanding of Coast Range structural relationships, study also the two geologic structure sections across the southern Coast Ranges in Figure 11-2.

It was noted in Chapter 1 that the 30- to 40-degree west-of-north trend of Coast Range geologic structures is cut obliquely across by the irregular but more northerly trending coastline. Thus, by following highways near the coast, the traveler does get something of a geologic cross section and views a variety of rock formations as he cuts across individual ranges.

## HIGHWAY GEOLOGY

Where to go and what routes to take to best see the geology of the Coast Ranges? Literally dozens of paved highways running the length and breadth of the Coast Ranges expose scenic landscapes, fascinating rock sections, and striking geologic features. The Coast Highway—U.S. Highway 101—runs the entire length of the Coast Ranges from the Santa Ynez River at the northern margin of the Santa Ynez Mountains in the Transverse Ranges to the Oregon border. Only locally does it really follow the coast; generally it travels near-coastal valleys, like the Salinas Valley, which it follows for about 100 miles. U.S. 101 is fast; the 400 miles between San Francisco and Los Angeles is easily negotiable in eight hours; the nearly 400

*Cove at Elk on the Mendocino County coast. Geologically recent uplift has left the wave-cut terrace high above sea level. Vertical sea cliffs are rapidly retreating landward under the attack of ocean waves and currents; islets and sea stacks are left behind.*

# 11 the coast ranges and continental borderland

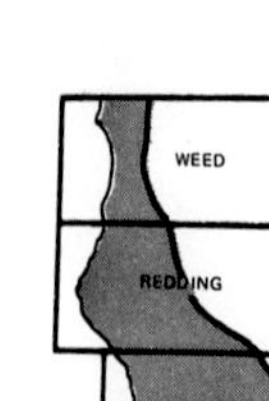

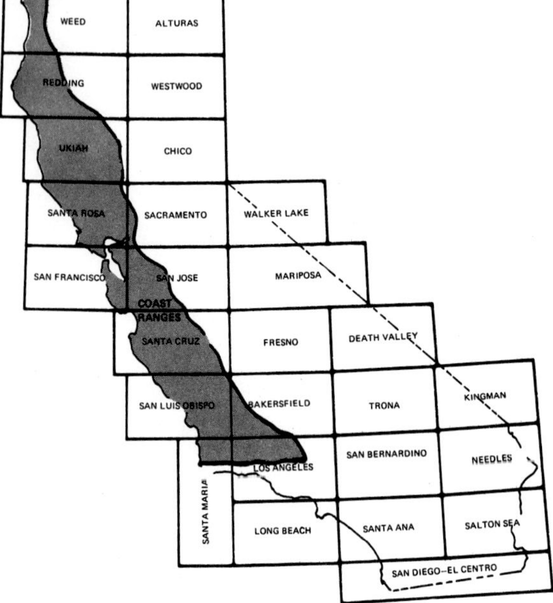


**FIGURE 11-1** *Index to map sheets* (Geologic Map of California) *covering the Coast Ranges province.* (*California Division of Mines and Geology.*)

miles from San Francisco to the Oregon border takes about nine hours. If one wants to enjoy the natural features and check the geology with the map sheets of the *Geologic Map* these trips take longer!

State Highway 1 is truly the coast route, clinging to the rugged cliffs and following along the coastal terraces (shown in Photograph 11-5) for almost the entire length of the Coast Ranges. With geologic structural features of the Coast Ranges trending slightly more westerly than the coast itself, a good sampling of the geology is seen. The enforced slower pace of this winding scenic highway also contributes to geologic observations!

Innumerable lesser highways and secondary roads cross the ranges from west to east, except for parts of the rugged Santa Lucia Range south of Monterey Bay, and the high Mendocino Range west of the northern Sacramento Valley, in both of which access is more difficult.

Since the southern Coast Ranges offer the greatest variety of rock formations and structures we shall follow the Coast Highway south.

## A GEOLOGIC TRIP ALONG U.S. 101 IN THE SOUTHERN COAST RANGES—GILROY TO BUELLTON

Gilroy lies just north of the northern boundary of the Santa Cruz map sheet in the southern part of the structural depression of San Francisco Bay and Santa Clara Valley between the foothills of the southeastern end of the Santa Cruz Mountains on the west and the rounded hills of the Diablo Range on the east.

### crossing the santa cruz mountains

About 6 miles south of Gilroy, U.S. Highway 101 begins to cross the low Santa Cruz Mountains at Sargent. High on the left (east of Sargent) is the Sargent landslide which took place in 1940.

Strongly tilted siltstone and sandstone beds in the roadcut are those of the marine lower Pliocene Purisima Formation; fossil shells can be found here. Four miles south of Sargent, the great San Andreas rift zone is crossed. Here it is marked by a rift valley and the fault can be easily seen on the highway because it marks the contact between Purisima sandstone and a dark granitic rock. For an interesting side trip, exit from freeway at the Watsonville off-ramp and follow a small secondary road for 4 miles northwest directly along the rift zone, by Anzar Lake, to Chittenden Pass. On the Chittenden Pass road is a striking exposure of the San Andreas fault contact between the Oligocene San Lorenzo shale on the northeast and granitic rock on the southwest. At this point, on the southeast side of the pass, is the huge quarry of the Granite Rock Company in which the granitic rock has been broken by movements in the fault zone.

Continuing from the fault zone, Highway 101 turns west for about 4 miles across bold outcrops of the tilted coarse sandstone beds of the marine Oligocene Pinecate Member of the San Lorenzo Formation. These are best seen from the southern segment of the divided highway, where a picnic spot has been set up in the rocks. The road very shortly passes out of the Oligocene outcrops and into the blanket of loosely cemented dark-reddish dune and floodplain sands of the Pleistocene Aromas Formation. A mile or so beyond the Aromas contact is an old quarry to the northwest of the highway in which the Aromas sand may be

**FIGURE 11-2** *Two geologic sections along northeast lines across the southern Coast Ranges: Top (Santa Cruz section)—from the coast at Gamboa Point (about latitude 36°) to the San Joaquin Valley near Dos Palos, Santa Cruz map sheet of* Geologic Map of California. *Bottom (San Luis Obispo section)—from the coast at Point Estero (about latitude 35°25′) to the San Joaquin Valley near Avenal, San Luis Obispo map sheet of* Geologic Map. *Explanation of symbols, from youngest to oldest:*

Qal—*modern alluvium*
Qf—*alluvial fan deposits*
Qt—*older Quaternary alluvium and terrace deposits*
Qc—*Pleistocene nonmarine sediments*
QP—*Plio-Pleistocene nonmarine sediments*
Pc—*Pliocene nonmarine sediments*
Pu—*Upper Pliocene marine sediments, western side San Joaquin Valley.*
Pml—*Middle and lower Pliocene marine sediments*
Muc—*Upper Miocene nonmarine sediments*
Mu—*Upper Miocene marine sedimentary rocks*
Mm—*Middle Miocene marine sedimentary rocks*
Ml—*Lower Miocene marine sedimentary rocks*
Mv—*Miocene volcanic rocks*
E—*Eocene marine sedimentary rocks*
Ep—*Paleocene marine sedimentary rocks*
Ku—*Upper Cretaceous marine sedimentary rocks*
Kl—*Lower Cretaceous marine sedimentary rocks*
Gr—*Granitic rocks, principally Late Cretaceous*
ub—*Ultrabasic rocks, principally serpentine*
KJf—*Upper Jurassic to Upper Cretaceous Franciscan Formation*
m—*Metamorphic rocks of the Sur Series*

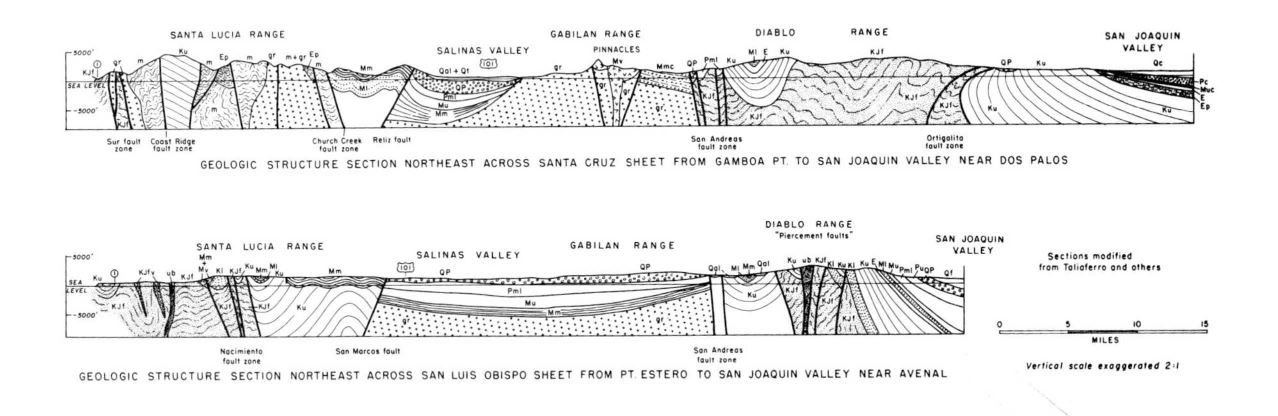

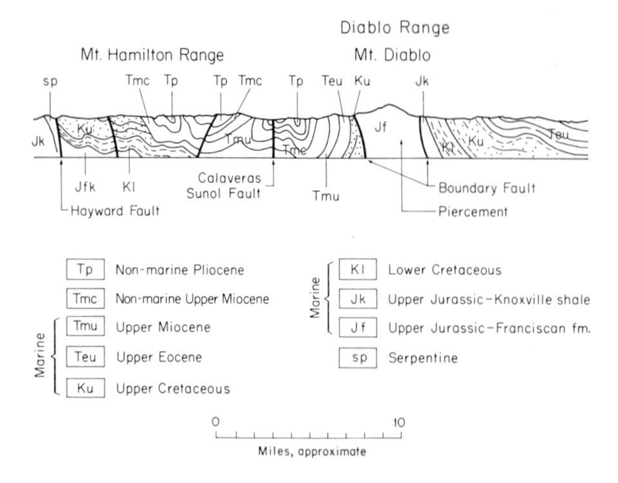


**FIGURE 11-3** *Generalized structure section across the Diablo Range from its western spur (the "Mount Hamilton Range") to the east side, showing Franciscan and associated rocks (Jf) squeezed upward along faults as a piercement at Mount Diablo.*

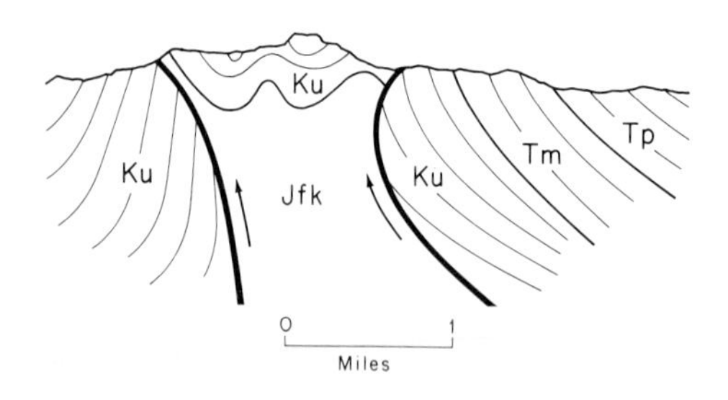


**FIGURE 11-4** *A small piercement similar to that in figure 11-3 in the Tent Hills in the southern Diablo Range. This one consists of serpentine only.*

seen lying directly on granitic rock. The highway then continues due south to Salinas over low hills made up of the eroded Aromas sand and still younger terrace deposits.

For several miles along this segment of the highway, the prominent white scar of the Natividad dolomite quarry of Kaiser Aluminum and Chemical Company can be seen a few miles to the east. East of the quarries, the steep, high peak at the north end of the Gabilan Range is Fremont Peak. Fremont Peak State Park is easily reached by a side trip through San Juan Bautista. The peak is about 11 miles from the mission at San Juan Bautista and is one of the best places to see exposures of the crystalline limestone, dolomite, quartzite, and schist of the Sur Series, which here strikes due west across the Gabilan Range. Barite was mined at one time on Fremont Peak. Special Report 56 of the California Division of Mines and Geology describes the geology of this area. Low on the northern slope of the peak, near San Juan Bautista, are the quarries and plant of the Ideal Cement Company, which utilizes several high-calcium limestone deposits in the Sur Series and uses Monterey Shale from Chittenden Pass for the necessary clayey fraction.

## salinas valley

For approximately the next 100 miles, Highway 101 follows southeast along the structural depression of the Salinas Valley between the steep, rugged Santa Lucia Range on the west and the low, subdued topography of the Gabilan Range on the east. In the lower part of the Salinas Valley, bedrock is downwarped several hundred feet below sea level. The ancient Pleistocene Salinas River may well have cut its canyon far to the west, heading into the submarine Monterey Canyon, so clearly shown by blue contours on the Santa Cruz map sheet of the *Geologic Map*. For many miles along the western side of Salinas Valley, at the base of the Santa Lucia Range, great elevated alluvial fans are a prominent feature.

## GEOLOGY OF THE SAN FRANCISCO URBAN AREA

About 40 air-line miles east of the city of San Francisco and the Golden Gate, the San Joaquin River and Sacramento River join, then run across a large delta and westward into San Francisco Bay. By late Pliocene time, the combined river had established approximately its present course to the Pacific Ocean along the route of Suisun Bay, Carquinez Strait, San Pablo Bay, and the Golden Gate.

The bay, with an area of about 400 square miles, is a highly irregular group of connected water bodies. Since man's occupation, area of the bay

**11-5** *Looking south along State Highway 1, south of Carmel, Monterey County. This is part of a wave-cut terrace, with the old shore line about 190 feet above sea level, which has been prominently developed along this part of the coast. Its history has probably involved complex late Quaternary uplift and fluctuations of sea level during and following the Ice Age. Very rapid present-day cutting by wave action and recession of the sea cliffs are obvious. At one point along the coast of southern California, not far away, the sea cliff is receding inland at a rate of more than 50 feet per century. (John S. Shelton photo.)*

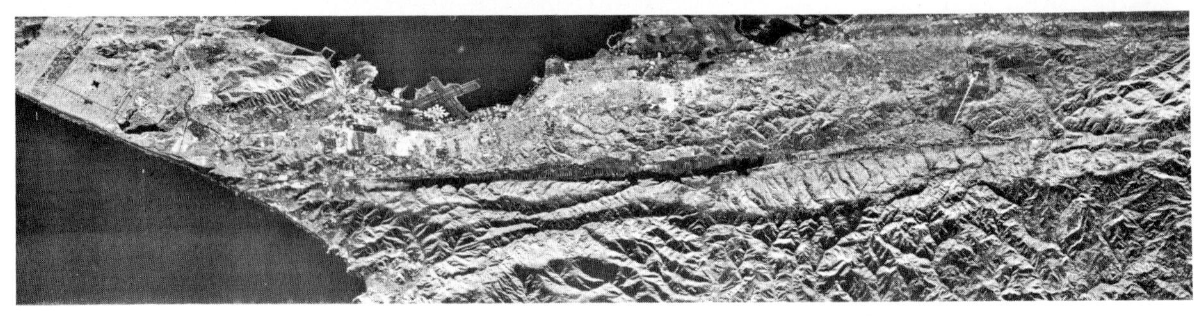

(a)

GOLDEN GATE BRIDGE
OAKLAND
OAKLAND AIRPORT
SAN FRANCISCO
PACIFIC OCEAN
SAN FRANCISCO BAY

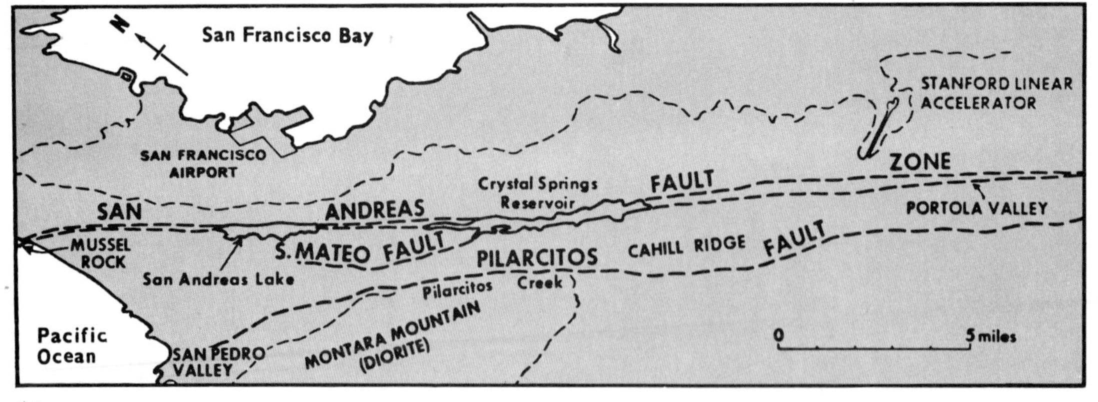
N
San Francisco Bay
STANFORD LINEAR ACCELERATOR
SAN FRANCISCO AIRPORT
ZONE
FAULT
Crystal Springs Reservoir
SAN
ANDREAS
PORTOLA VALLEY
MUSSEL ROCK
S. MATEO FAULT
PILARCITOS
CAHILL RIDGE
FAULT
San Andreas Lake
Pilarcitos
Creek
MONTARA MOUNTAIN (DIORITE)
Pacific Ocean
SAN PEDRO VALLEY
0
5 miles

(b)

has shrunk, as a result of filling and building along its margins, from its "original" area of some 700 square miles. Most of San Francisco Bay is less than 10 feet deep, but natural channels have been deepened and are kept open by dredging to depths of 30 feet or more so that ocean-going ships can go up the Sacramento River to Sacramento and up the San Joaquin River to Stockton. Greatest depth of the bay is at the mile-wide Golden Gate where bedrock is kept scoured by strong tidal-current action to a depth of 341 feet.

San Francisco Bay is a late Pliocene structural depression, which was drowned by rising seas in the interglacial stages of the Pleistocene Epoch; it marks a natural topographic separation between the northern and southern Coast Ranges (Illustration 11-6). This structural depression extends continuously southward into the Santa Clara and San Benito Valleys and fingers out to the north across San Pablo Bay into the Petaluma, Sonoma, and Napa Valleys. To the east of the bay are the Berkeley Hills; west of the bay are the low hills of the San Francisco and Marin Peninsulas. At the northeastern end of the bay, Carquinez Strait and Suisun Bay extend eastward into the Great Valley. San Francisco Bay opens to the Pacific Ocean through the Golden Gate, a channel eroded by the ancient Sacramento River during glacial epochs of the Pleistocene.

Complex geology of the bay area can be generalized by recognizing three geologically distinct blocks, separated by the San Andreas and Hayward faults: The Point Reyes–Montara, or western, block; the San Francisco–Marin, or bay block; and the Berkeley Hills, or eastern, block. The striking San Andreas rift zone generally bisects Marin and San Mateo Counties but passes west of the city of San Francisco, between it and the bare, granitic Farallon Islands. North of the Golden Gate, the fault separates the Marin Peninsula on the east from Point Reyes Peninsula (30 miles northwest of San Francisco) on the west (Photograph 11-7). In San Mateo County, south of San Francisco, the San Andreas fault separates the San Francisco–Marin block on the east from the Point Reyes–Montara block on the west.

The Point Reyes–Montara block is a region of low, but locally rugged, relief underlain by Cretaceous granitic rocks, such as quartz diorite (80 million to 90 million years old), which include fragments of the older Sur Series gneiss and marble. Eroded remnants of Upper Cretaceous, Tertiary, and Quaternary sedimentary rock units remain on this western block, overlying or faulted against the crystalline rocks. Mount Wittenberg (elevation 1,403 feet) is the highest point on Point Reyes Peninsula; Montara Mountain has an elevation of 1,898 feet. This block includes the offshore continental borderland under the ocean, extending out to the continental slope and including the Farallons. Outcrops of quartz diorite appear prominently at Point Reyes, on Montara Mountain, and on the Farallon Islands.

The San Francisco–Marin block includes the region east of the San Andreas fault to the Hayward fault zone near the base of the Berkeley Hills. The San Francisco Peninsula is the northernmost extension of the Santa Cruz Mountains, while the Marin Peninsula forms the southern extension of the Mendocino Range. Mount Tamalpais, dominating the Marin skyline, has an elevation of 2,604 feet. The oldest, or "basement," rocks exposed in the San Francisco–Marin block are those of the Franciscan Formation; no granitic or Sur Series rocks have been found. The waters of San Francisco Bay proper are on this block.

**11-6** *Airborne image* (a) *of parts of San Francisco Peninsula and East Bay, with index to some of the features shown* (b). *This is not a photograph, but is an image obtained by radar on a high flight funded by the National Aeronautics and Space Administration and monitored by the U.S. Geological Survey. The image offers great promise for delineating certain geologic structural features far better than in the usual aerial photography. Note the bar scale and the geologic features named in* (b). *Comparison with the San Francisco map sheet of the* Geologic Map of California *will disclose other geologic features, such as the Hayward fault in the East Bay* (*NASA radar image.*)

**11-7** *Rounded hills of the Coast Ranges in Marin County. Aerial oblique photo looking west across Bodega Bay and Bodega Head. Point Reyes in the far distance. Both points are preserved because of resistant granitic rock. The broad San Andreas fault zone crosses the bay and headland from left to right in the middle ground, approximately coinciding with the bay, spit, sand dunes, and beach. (Aero Photographers, Sausalito.)*

The Berkeley Hills block is the elevated area between the Hayward fault on the west and the Calaveras fault zone on the east. Franciscan rocks are exposed in and west of the Hayward fault zone; thick Cretaceous marine sedimentary formations lie east of the fault and are overlain by Tertiary sedimentary and volcanic rocks. Franciscan rocks are widely exposed in the Bay block and also form the basement rock formation throughout the Diablo Range, of which the Berkeley Hills is a branch. Dominant structure of the Berkeley Hills is synclinal (Figure 11-9). The Berkeley Hills skyline ridge rises to a maximum elevation of 1,905 feet at Vollmer Peak.

Strictly speaking, these topographic and geologic "blocks" of the bay area are not precisely bounded by the San Andreas, Hayward, and Calaveras faults, for these three active, related faults of the San Andreas system strike at very low angles across some structural axes and cut strongly across topographic forms. The San Francisco Bay depression is irregular in outline; it is not simply bounded by these faults; the faults climb well into the present hills in some places on both sides of the bay.

The heterogeneous assemblage of marine sedimentary, volcanic, and metamorphic rocks of the Franciscan Formation and serpentine are typically developed in the bay area and are widely exposed in the low Berkeley Hills west of the Hayward fault, in the islands of the bay, and throughout the Marin and San Francisco Peninsulas. Graywacke—a dark-gray compact sandstone—predominates, but black shale, red

**11-8** *Duxbury Reef, near Bolinas, Marin County, at low tide. Rock exposed in this wave-cut terrace is steeply folded sandstone, shale, and chert of the middle Miocene Monterey Formation, about 1.5 miles west of the San Andreas fault. The several large white rocks scattered over the surface are not boulders but limy concretions that have been weathered out of the Monterey Formation. Concretions are formed in place by chemical deposition; note that the one in the foreground is thinly stratified. (Sarah Ann Davis photo, California Division of Mines and Geology Bull. 181.)*

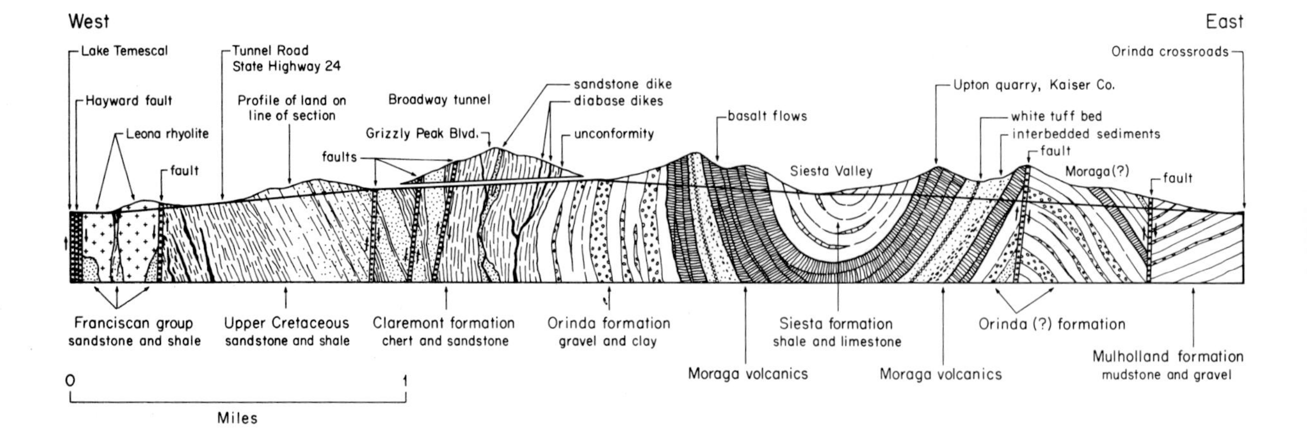


**FIGURE 11-9** *Geologic structure section across the Berkeley Hills.*

*Data from the boring of the Broadway tunnel through the Berkeley Hills and the deep cuts in approaches from either end along State Highway 24 afford an unusually detailed geologic picture of this branch of the Diablo Range east of San Francisco Bay. In 1969, a lower-level tunnel for Bay Area Rapid Transit trains passed through a similar section, plus the active Hayward fault zone where an average right-lateral (east block moving south) displacement of about 1/4 inch per year is taking place.*

*In the section, the Hayward fault zone is seen to involve shattered and broken rocks of the Franciscan Formation, which have been intruded by pluglike bodies of the Plio-Pleistocene Leona Rhyolite. East of the fault zone is a thick section of easterly dipping marine Upper Cretaceous rocks which include sandstone, shale, and conglomerate. East of the Cretaceous strata is a much-faulted section of thin-bedded chert and shale of the Miocene Monterey Formation (called Claremont Formation in the Berkeley Hills), which has been intruded by diabase dikes. So intensely folded is this part of the section at the crest of the Berkeley Hills that the beds are actually overturned —that is, they have been folded through more than 90 degrees.*

*The Franciscan, Cretaceous, and Claremont Formations are all of marine origin, but all later formations were deposited on the land by streams and in lakes. The early Pliocene Orinda Formation, lying unconformably on the Claremont shale, consists of stream-deposited gravel, sand, and clay, which came from a high-standing block where the bay now is. Orinda deposition was followed by outpourings of the Moraga volcanics—both lava flows and ash—in turn followed by late Pliocene lake-deposited shale and limestone of the Siesta Formation. On the eastern flanks of the Berkeley Hills the Orinda Formation (of which the Mulholland Formation is a part) appears again. Orinda, Moraga, and Siesta Formations are involved in a syncline. (After O. E. Bowen and B. M. Page, California Division of Mines and Geology, 1951.)*

radiolarian chert, greenstone (altered volcanic rock), and minor gray limestone are characteristic. Metamorphic rocks, consisting of blue glaucophane, green chlorite and actinolite, red garnet, and green jadeite are a small but distinctive part of the Franciscan Formation. The green serpentine, often much sheared and broken, is widespread. Fossils are extremely scarce; however, a few molluscs and cephalopods of Cretaceous types have been found, and the thin limestone members have yielded numerous foraminifera of Mid-to-Late Cretaceous age. Franciscan rocks of the Berkeley Hills, like those in other parts of the Diablo Range, appear to be Late Jurassic in age. A few radiometric ages obtained from the metamorphic rocks are Late Jurassic.

Within San Francisco, there are striking exposures of colored chert and other Franciscan rocks in Golden Gate Park, in the vicinity of Twin Peaks, and along the Alemany Freeway. Other chert beds and beautiful exposures of pillow basalt (greenstone) are exposed in great freeway cuts just north of the Golden Gate Bridge. The three highest peaks of the Bay area—Mount Tamalpais, Mount Diablo which dominates the Diablo Range east of the Berkeley Hills in the East Bay, and Mount Hamilton, site of Lick Observatory southeast of the bay near San Jose—are all composed of Franciscan rocks and serpentine.

Many rock quarries in the bay area have mined the hard red chert and graywacke. The islands in the bay have extensive exposures of Franciscan rocks. Yerba Buena—through which the Bay Bridge highway passes in a tunnel—is mostly graywacke; Red Rock, seen best from the Richmond-San Rafael Bridge, is red chert; Alcatraz is graywacke; and Angel Island, a new state park, exposes an exotic group of metamorphic rocks and minerals as well as the less spectacular sedimentary rocks. The rocks on Angel Island are graywacke, black shale, chert, greenstone, blue schists, and serpentine. More than thirty-five different minerals have been found in the schists, including actinolite, biotite, chlorite, feldspars, jadeite, epidote, garnet, glaucophane, hornblende, lawsonite, muscovite, pyrite, quartz, and tremolite.

Upper Cretaceous shelf-facies marine sedimentary rocks, unmetamorphosed and fossiliferous, crop out in scattered, folded, and faulted remnants on both sides of the bay and on both sides of the San Andreas and Hayward faults. Most common rock types are thin-bedded, alternating sandstone and shale. The fact that these rocks of shallow-water types are, at least in part, equivalent in age to the granitic rocks and to Franciscan rocks of the West Bay poses for geologists one of the greatest problems connected with the Coast Ranges. Precise correlations within the Cretaceous across the San Andreas and Hayward faults are not yet possible. The Cretaceous strata reach tremendous thickness—30,000 feet, plus—on the eastern flank of Mount Diablo and in the western side of the Great Valley. Some of the marine rocks in the Hayward fault zone are probably lower Cretaceous.

Marine Paleocene rocks, quite similar to the upper Cretaceous, also appear in folded and faulted remnants on both sides of the bay. Paleocene strata of the Martinez Formation are well exposed, for example, on the eastern flank of the Berkeley Hills, east of the Calaveras fault, and at San Pedro Point on the northern flank of Montara Mountain on the San Francisco Peninsula. At the last locality, Martinez conglomerate may be seen lying unconformably on Upper Cretaceous strata in rock cuts of State Highway 1. The Devil's Slide on the coast (see the San Francisco map sheet of the *Geologic Map*) at this point exposes a large area of Upper Cretaceous shale. Paleocene conglomerate of the Laird Formation crops out at Point Reyes in Marin County. Remnants of marine Eocene strata overlie the Paleocene, but the thickest and most complete Eocene section is on the flanks of Mount Diablo. Certainly, shelf seas covered the entire bay area in late Cretaceous, Paleocene, and Eocene times, with a possible brief withdrawal at the close of the Cretaceous Period.

In the Coast Ranges, the Oligocene Epoch was a time of restricted seas. Thin marine Oligocene rocks crop out a couple of miles east of the Calaveras fault near Walnut Creek; southwest of the Bay in the Santa Cruz Mountains marine Oligocene is well represented by the San Lorenzo Formation. No marine Oligocene rocks are known in California north of the bay area. Lower Miocene seas were also quite restricted and no lower Miocene is found immediately around the bay. However, shallow-marine coarse sandstone, shale, and conglomerate of the lower Miocene Vaqueros Formation crop out east of Half Moon Bay and in the Santa Cruz Mountains south of that latitude.

Advancing seas, following early Miocene time, probably covered most or all of the Bay area in middle and late Miocene time, including the Berkeley Hills as well as the San Francisco–Marin and Point Reyes–Montara blocks. These marine rocks are known as the Monterey Formation (Claremont shale, in the Berkeley Hills). They consist of thin-bedded siliceous and diatomaceous shale, chert, foraminiferal shale, a small amount of sandstone, and, in some places, a considerable thickness of a variety of extrusive and shallow intrusive volcanic rocks. On the road to Point Reyes and along Skyline Boulevard at the crest of the Berkeley Hills, thin-bedded contorted beds of Monterey siliceous shale are well exposed in the road cuts. In the Berkeley Hills, volcanics of this

**11-10** *Looking northwest along Big Sulfur Creek, Sonoma County, at turbine-generator unit No. 3 (27,500 kilowatts) of power plant at The Geysers. Natural steam, blowing from wells in the background, is conducted through the gathering line to the generator (right), barometric condenser (center) and cooling towers (left). Heterogeneous rocks of the Franciscan Formation underlie and crop out in these Coast Range hills. (Pacific Gas and Electric Company photo.)*

**11-11** *Foundry-sand operation at the old Tesla coal mines, Corral Hollow, in the northeastern part of the Diablo Range, Alameda County. The low-grade lignite coal is no longer mined here, but interbedded white sands are still mined from time to time. The steeply dipping strata are white sand, clay, and lignite of the middle Eocene Tesla Formation. These sediments were deposited in low-lying swampy areas near the margins of the Eocene seas. (Charles W. Jennings, photo.)*

age are represented by diabase dikes; but on the southern part of the San Francisco Peninsula, between Half Moon Bay and Palo Alto, as much as 2,000 feet of basaltic rocks appear in the middle Miocene section. Similar volcanic rocks appear both east and west of the San Andreas fault.

The fossiliferous upper Miocene-lower Pliocene Purisima Formation comprises a maximum thickness of 10,000 feet of marine sandstone, mudstone, and diatomaceous and foraminiferal shale in the Half Moon Bay area. This thins very rapidly northward and grades upward into the upper Pliocene-lower Pleistocene Merced Formation along the coast south of San Francisco. Elsewhere in the Point Reyes-Montara block and in the San Francisco-Marin block, the Purisima does not appear and was probably not deposited. In the Berkeley Hills, late Miocene-early Pliocene time is represented by the nonmarine conglomerate, sandstone, and mudstone of the Orinda Formation, which overlies the Claremont shale with slight angular unconformity. Detritus in the Orinda conglomerate consists of Franciscan rocks, and current directions indicated are from west to east. Thus, from latest Miocene to middle Pliocene time the bay block was a highland that was shedding sediments eastward to a subsiding trough on the site of the Berkeley Hills. Several miles farther east, the Orinda Formation grades

**11-12** *Looking north at Mount Saint Helena (elevation 4,344 feet) off State Highway 29, a few miles north of Calistoga. Often referred to as a "volcano," Saint Helena is a peak formed by erosion of the folded Pliocene Sonoma Volcanics, called here the Palisades, consisting of tuff-breccia. In this area, the Sonoma Volcanics lie unconformably on rocks of the Franciscan Formation. (Mary Hill photo.)*

into similar rocks of the Siesta Formation. Intercalated in the Pliocene part of the Orinda-Mulholland section are basaltic and andesitic flows, agglomerate, and tuff of the Moraga Formation. These volcanics are strikingly exposed in road cuts just east of the Broadway Tunnel, which cuts through the Berkeley Hills, and on the Berkeley skyline on Grizzly Peak Boulevard, where they are often called the Grizzly Peak volcanics. Potassium-argon dates of about 10 million years have been obtained in these volcanics.

In late Pliocene and early Pleistocene time, in the West Bay, a shallow, narrow seaway developed from the vicinity of Lake Merced southeastward through Merced Valley between the newly elevated San Bruno and Montara Mountains. More than 5,000 feet of sand, silt, gravel, sandy mudstone, and layers of volcanic ash accumulated as the Merced Formation. On the southwestern margin of this trough, along the rift valley of the San Andreas fault, stream and alluvial-fan gravels, sand, and mud accumulated as the Santa Clara Formation, still very little consolidated. Thinner, horizontal beds of the Merced Formation crop out over wide areas north of the bay in Sonoma County and are interbedded, in part, with the late Pliocene Sonoma volcanics. On the east side of the bay and in an uplifted block along the east side of the Hayward fault are exposures of deformed strata of gravel, sand, and silt, probably partly equivalent in age to the Santa Clara Formation—these strata are known as the Irvington Gravel. The beds have yielded an abundant vertebrate fauna, the type for the mid-Pleistocene Irvingtonian stage for land mammals. The fauna includes the ground sloth, dog, saber-tooth tiger, mammoth elephant, horse, and camel.

The major orogenic epoch in the Coast Ranges, accompanied by intense folding, faulting, uplift, and local depression, began very strongly in late Pliocene time and continued into the early Pleistocene. Formations of Purisima and Orinda age and older are most intensely folded; some are even overturned. Early Pleistocene strata of the Merced, Santa Clara, and Irvington Formations are gently dipping to strongly folded, especially adjacent to the active San Andreas and Hayward faults. Late Pliocene to mid-Pleistocene time was the greatest epoch of uplift and folding of the Berkeley Hills and also of the West Bay hills, and was the time of depression of a large part of the San Francisco block to form the bay trough. A series of late Pliocene or early Pleistocene rhyolite plugs—Leona Rhyolite—punctuates the Hayward fault zone along the western slopes of the Berkeley Hills (Figure 11-9).

Late Pleistocene to Holocene marine and nonmarine sediments accumulated in the San Francisco Bay-Santa Clara Valley trough and along the rift valley of the San Andreas fault. In and adjacent to the bay are, from bottom to top, the Alameda Formation, consisting of several hundred feet of continental and marine gravel, sand, silt, and clay; the nonmarine Temescal Gravel; and the marine Merritt Sand. Late Pleistocene sediments on the San Francisco Peninsula are principally the bedded beach and dune sands of the Colma Formation. A carbon 14 date shows the Colma Formation to be older than 30,000 years; it was probably deposited before the last advance of the ice late in the Pleistocene Epoch.

Tectonic uplift and depression and displacements along the active faults within the San Andreas system continue to the present day. One of the great difficulties in reconstructing bay area geologic history is in evaluating presently unknown cumulative displacements on the faults in this system. Just how this late Coast Range orogeny relates to plate tectonics we are not sure!

## THE CONTINENTAL BORDERLAND

Continental California does not stop at the shore line. Today's shore line is only a geologically temporary feature; it has not lasted for long but is everchanging. The Channel Islands off the southern coast—Santa Catalina, San Clemente, San Nicolas—are structurally parallel and geologically

**11-13** *Aerial oblique view southeastward along the San Andreas rift valley on the Marin Peninsula, from the head of Tomales Bay (foreground) to Bolinas Bay in the distance. (Robert E. Wallace photo, U.S. Geological Survey.)*

**11-14** *View northwestward along the Garcia River in the northern segment of the San Andreas rift valley to the ocean in the far distance near Point Arena. Dense forest and old logging activities obscure details of fault displacements, except in the open area just south of the point. (Robert E. Wallace photo, U.S. Geological Survey.)*

similar to the Peninsular Ranges. The east-west chain consisting of Santa Cruz, Santa Rosa, and San Miguel Islands is a direct western projection of the Transverse Ranges. Similarly, Coast Range geologic features trend obliquely out to sea in northwesterly directions (see the large geologic map).

Where then does the continent (continental platform) end? And what are the major features of that part of the California Coast Ranges below sea level?

The outer margin of the continent is properly taken at the junction of the high-silica granitic crust of the continent and the oceanic crust consisting of low-silica, high-magnesia basalt and serpentine. This junction is approximately marked by the 10,000-foot submarine contour on the large geologic map and lies at the foot of the

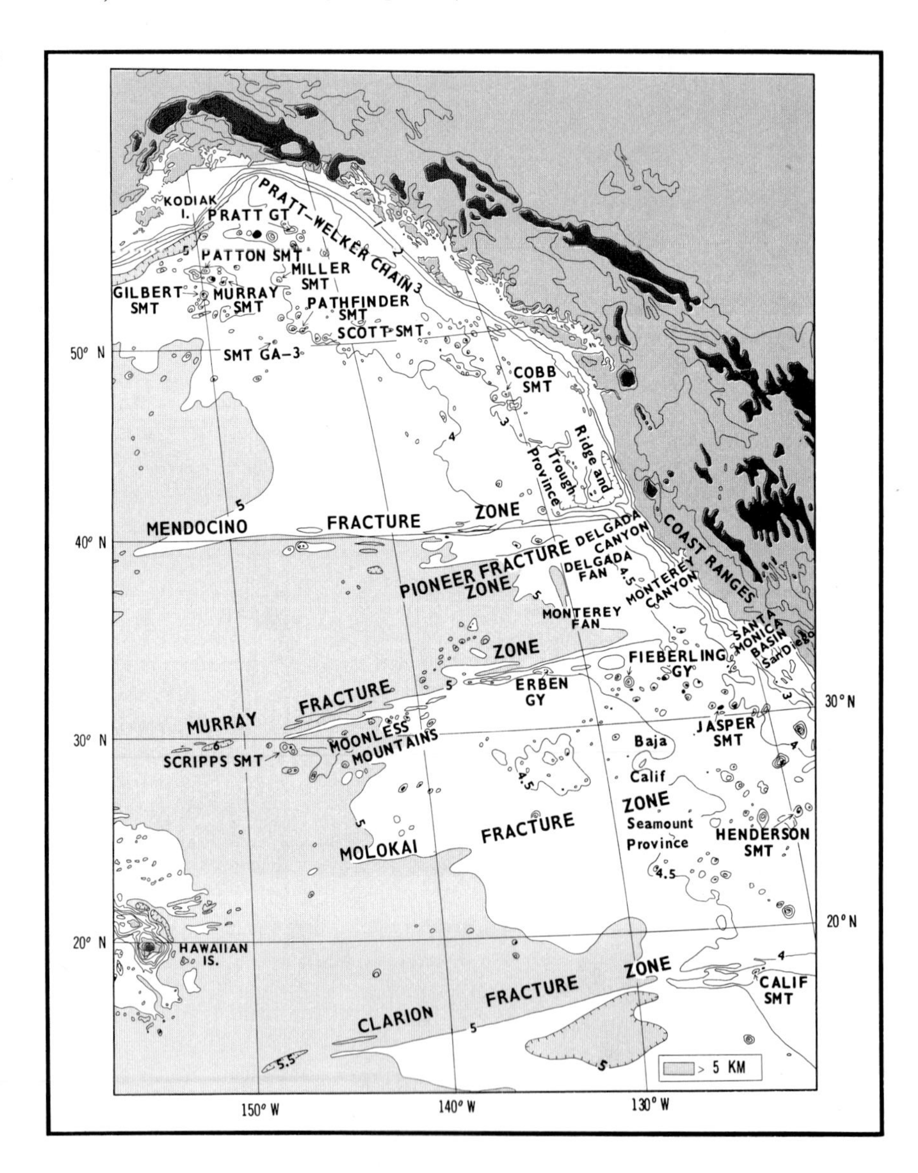


**FIGURE 11-15** *Features of the California borderland. From Menard,* Marine Geology of the Pacific, *McGraw-Hill Book Company, 1964, by permission.*

continental slope, from about 40 to 140 miles offshore. The continental slope rises shoreward relatively steeply at average angles from about 1 in 70 (1 foot vertically in 70 feet horizontally) to about 1 in 10 up to the outer edge of the continental shelf. The continental shelf is a gently sloping continuation of the land seaward; it is marked roughly by the 500-foot submarine contour on the geologic map. Slope of the shelf averages about 3 degrees, and it is very narrow compared to many continental shelves elsewhere around the world. The widest shelf area lies off San Francisco Bay.

Extending outward from the foot of the continental slope is an area called the continental rise, which grades imperceptibly down a slope of about 1 degree to the deep plain of the Pacific Ocean at an average depth of about 2.5 miles (Figure 11-15). The continental rise west of the Coast Ranges consists of two great deep-sea fans—the Delgada in the north and the Monterey in the south. These are made up of sediments and are associated with submarine canyons and deep channels, which probably transport the sediments by turbidity currents across the shelves and slopes and onto the rises. Exploration of the upper parts of submarine canyons has shown that they are very similar in all respects to normal on-shore canyons formed by erosion by running water. They are commonly named after on-shore geographic features. Indeed, many of the submarine canyons head into on-shore canyons and rivers. Their courses may be steep and narrow, their channels may expose bedrock, and they may meander on gentle slopes. Some of the notable submarine canyons offshore from the Coast Ranges, north to south, are Mendocino, Delgada, Noyo, Bodega, Pioneer (off San Francisco Peninsula), Monterey, and Sur Canyons.

Perhaps the greatest of all the offshore features, and also one of the least understood, is a series of three tremendous westerly striking fracture zones, which extend from the coast for hundreds of miles into the western Pacific Ocean (Figure 11-15). These are the Mendocino fracture zone, extending west of Cape Mendocino and Mendocino Submarine Canyon (Redding map sheet of *Geologic Map of California*); the Pioneer fracture zone, a lesser zone extending west of Pioneer Submarine Canyon (San Francisco map sheet of *Geologic Map*); and the Murray fracture zone, extending west of the Transverse Ranges. The fracture zones consist of mountain ranges, ridges, narrow troughs, escarpments, and sometimes volcanoes, forming very straight bands more than 100 miles wide.

The Murray fracture zone seems to continue eastward into the Transverse Ranges. The Mendocino fracture zone has less clear continuations onto land, but perhaps the east-west structural trend of the lower Eel River Basin, a few miles north, represents some reflection of this fracture-zone structure. The Pioneer fracture zone seems to have no landward extension.

The fracture zones are certainly great, complex, transform-fault systems of fundamental structural importance. Magnetic studies from oceanographic vessels indicate that large horizontal, strike-slip movements have taken place. The blocks of oceanic crust between fracture zones may be moving eastward under the continental borderland of California, rafted along on convection currents within the earth's upper mantle, although the Pacific plate as a whole appears to be drifting toward the northeast and subducting under the Aleutian Peninsula.

*Convict Lake lies in a glaciated basin on the eastern slope of the Sierra Nevada.*

# 12 yosemite valley and the range of light

■ California's greatest mountain range bends north-northwesterly in the eastern part of the state for 400 miles forming a formidable physical barrier to east-west communications—a tremendous attraction to vacationers, climbers, hunters, and fishermen, as well as to serious students of its plant and animal life, minerals, rocks, and glacial history.

Referring to Chapter 1, we are reminded that the Sierra Nevada is a great uplifted block of granitic rocks—including remnants of older metamorphic rocks—which has been tilted westward in late geologic time. While the average slope to the west is about 2 degrees, the eastern slope is many times that. This means that the high; saw-toothed, glaciated peaks of the crest of the range are 25 to 80 miles from the eastern margin of the Great Valley, while the crest is but a few miles west of the eastern valleys—Owens Valley, for example.

The northern and extreme southern end of the Sierra Nevada are lowest; peaks average 6,000 to 7,000 feet in elevation. Thus, the Sierra is distinctly lower as it merges northward and northwestward into the Cascade and Klamath Mountains, respectively; in fact, northern Sierran geologic structures and rock formations appear to "dive" below Upper Cretaceous to Holocene sedimentary formations. The southern end of the Sierra becomes lower and turns gradually more westward into the west-trending Tehachapi Mountains around the southern end of the San Joaquin Valley and merges into the Coast Ranges. These facts of geology and topography profoundly influenced routes of western migration and early development of California a century ago.

Because of its great height and area, its spectacular exposures of rock formations, and the obviously prominent position it holds in the state's structural framework, geologic history of the Sierra Nevada has been discussed more completely than that of any other natural province in Part Two of this book ("California through the Geologic Ages").

There are no known Precambrian rocks exposed in the Sierra but some judgment of what went on in Precambrian times comes from the adjoining White Mountains and from the Death Valley region to the southeast. Even the complete lack of Precambrian rock formations west of the Sierra allows some inferences to be made (Chapter 6). Remnants of Paleozoic rock sections show that the region of the present-day Sierra had a marine Paleozoic history (Chapter 7). First building of the Sierra Nevada in Late Jurassic time, the great Sierran orogeny, and the development of the granite batholith (Figure 8-8) may be reviewed in Chapter 8 for the major events in the history of this mountain range. Chapter 9 follows the past 65 million years of spasmodic renewed uplift, faulting, and volcanic activity, culminating in the Ice Age glaciation (Figures 9-12 and 9-13) and fashioning of the modern Sierra. A continuation of the dynamic processes of mountain building is summarized in Chapter 10.

Magnificent trails—of which the John Muir Trail along the crest of the High Sierra from Tuolumne Meadows to Mount Whitney is best known—provide access to hikers and climbers. Numerous trails extend out from the national parks into the wilderness areas and the national forests. Hardly a peak remains that has not been scaled by rock climbers; even the 3,000-foot vertical granite wall of El Capitan in Yosemite Valley has been climbed

**12-1** *Eastern face of Mount Whitney, a mass of granitic rock sculptured by alpine glaciation; photo taken due west from 11,000 feet elevation. (Ernest S. Carter photo.)*

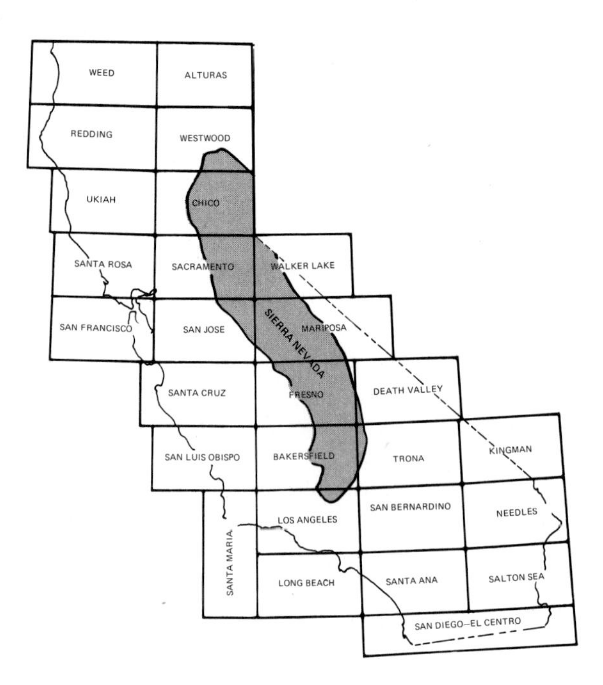


**FIGURE 12-2** *Index to map sheets* (Geologic Map of California) *covering the Sierra Nevada province. (California Division of Mines and Geology.)*

several times (Photograph 3-2). Fortunately for the less energetic visitors to the Sierra Nevada, practically all its principal geologic features can be viewed and the elements needed to unravel its history can be studied along Sierran highways.

The map sheets (*Geologic Map of California*) covering the Sierra Nevada (index, Figure 12-2) show all highways, topographic contours, and all principal geologic formations and structures on a scale of 1 inch equals 4 miles.

Where are the principal highways and what do they show?

## SIERRAN HIGHWAYS

No road within the Sierra Nevada follows its entire length. However, U.S. Highway 395 skirts the eastern side of the range, giving spectacular views of the steeply faulted eastern slope (see the drawing above this chapter title) and the Sierran crest and affording access to the eastern face by numerous short roads. State Highway 14 extends for 49 miles from Mojave around the base of the extreme southern Sierra Nevada to junction with U.S. 395. From the junction, Highway 395 extends northward in continual sight of the high summit peaks for about 410 miles to Susanville at the lower northern end of the range (see Photograph 3-13). From that point, State Highways 36 and 32 skirt westward around the Sierra Nevada but within the southern margin of the Cascade volcanic rocks.

In the western Sierra foothills is State Highway 49, the famous Mother Lode Highway. Its southern end is at the gold-mining town of Mariposa. From there, it winds northerly for 140 miles to Grass Valley and Nevada City, in the northern gold belt. From Nevada City, Highway 49 turns easterly for 80 miles across Yuba Pass (elevation 6,701 feet) to Sierraville on the eastern side of the Sierra. Through the Mother Lode gold country Highway 49 links the old, historic, gold-mining towns of the forty-niners and exposes the rock formations and fault zones of the western metamorphic belt, including the Late Paleozoic Calaveras Formation, Late Jurassic Mariposa (Photograph 5-11) and Jurassic Logtown Ridge metamorphosed volcanic

**12-3** *Temple Crag and first lake at entrance to Palisade Glacier, west of Big Pine, Inyo County, on U.S. Highway 395. (Frnest S. Carter photo.)*

rocks—all intruded by Late Jurassic granitic rocks. On top of these, of course, are Tertiary volcanic rocks and gold-bearing gravels. The most popular and widely distributed of all publications of the California Division of Mines and Geology is Bulletin 141, *Geologic Guidebook along Highway 49.*

**12-4** *The High Sierra due west of Bishop, Inyo County, on U.S. Highway 395. In spite of the extreme, sharp relief, the uniform level of the Sierran crest and eroded remnants of gentle slopes near the top elevations suggest the old (Miocene?) uplifted and tilted block from which the mountain peaks were carved. (Ernest S. Carter photo.)*

## DONNER SUMMIT AND ACROSS THE SIERRA VIA INTERSTATE 80

Interstate Highway 80 is the high-speed, all-year freeway linking the metropolitan San Francisco Bay area, Reno, and points east. At legal speeds, Sacramento—the state capital—is $1\frac{3}{4}$ hours from San Francisco (89 miles) and Reno is 3 hours (134 miles) from Sacramento. Even at freeway speeds, the traveler can hardly miss the spectacular scenery of the Sierra Nevada and some of its salient geologic features. By taking advantage of frontage and local bypass roads and the old highway U.S. 40 between Sacramento and Reno, Nevada, the geologically minded traveler can see in one full day the best and most

complete section exposing Sierran geologic history. Geology along Interstate 80 from Sacramento to Reno can best be followed on the Sacramento and Chico map sheets of the *Geologic Map*.

After crossing the Sacramento River by Interstate 80 on a high bridge, off-ramps lead into the capital city. In north Sacramento the highway proceeds northeast across the American River and its flat-lying alluvial sediments onto slightly higher floodplain, river-bench, and terrace deposits of Quaternary age. The topography is low and rolling, and buff-colored sand and gravel appear in shallow road cuts for about 17 miles (from Sacramento) to Roseville. Just beyond Roseville, the very gradual rise up the western foothills begins. Volcanic mudflows of the upper Miocene-lower Pliocene Mehrten Formation appear in low hills east of the highway. The Mehrten is not well exposed here but is distinguished from the valley gravels by the blocky fragments of volcanic rock exposed by weathering and removal of their matrix of ash. The highway continues for a few miles toward Rocklin through low ridges of the

**12-5** *Dana Peak at Tioga Pass. Note the beautiful rock glaciers, glacial moraines, and talus cones. None of these small glaciers flow; they have receded during ten years of observation by the photographer. (Ernest S. Carter photo.)*

**12-6** *Looking west up Hot Creek, Mono County, at the High Sierra Nevada; Mount Laurel top center and Mount Morrison to the left. Rocks of the plateau in the foreground are Late Tertiary rhyolites that have been altered and whitened by rising hot spring waters. A state geologist sits on a rock overlooking Hot Creek toward the lower left. (Mary Hill photo.)*

mudflows which are very gently dipping toward the valley. Just west of Rocklin the Mehrten Formation can be seen underlain by sands of the Eocene Ione Formation.

At Rocklin, the most westerly outcrops of Sierran granitic rock appear as good, fresh exposures of dark quartz diorite and granodiorite of probable Late Jurassic age. The large quarries on the south side of Rocklin have been worked since the 1860s and furnished stone for the state capitol building at Sacramento. Fourteen miles beyond Rocklin at Newcastle is a narrow belt of greenschist which is older than the granite in contact with it and has probably been metamorphosed from volcanic rock. The highway cuts through a 1½-mile-wide exposure of granitic rock and then a dense dark greenstone of volcanic origin in the approach to Auburn, 32 miles northeast of Sacramento. On the hills on either side of the highway are remnants of Mehrten andesitic mudflows. Just beyond Auburn is the interchange for State 49, the Mother Lode Highway.

About 0.75 mile beyond the State 49 junction is a narrow belt of black schist of the late Paleozoic Calaveras Formation in contact with greenstone; from this point on, the road cuts through dark brown Calaveras slate and complexly faulted and interfolded dark green metavolcanic rocks for several miles. At Weimar, 16 miles northeast of Auburn, is a narrow belt of green serpentine at the fault contact between slate and schist of the Calaveras Formation and the sheeted slate of the Late Jurassic Mariposa Formation. The well-rounded, pure-white quartz cobbles that are used for miles in the center separation strip of the highway are from gravels in the nearby Bear River which were left after being washed for gold as early as the 1850s.

North of nearby Colfax is Colfax Hill, clearly visible from the highway, which is capped by Mehrten andesite gravels underlain by Eocene auriferous (gold-bearing) gravel and sand. Just beyond Colfax, the deep canyon of the North Fork of the American River can be seen. The highway then crosses a fault that marks the eastern margin of the Mariposa Formation. There follows a thick section of Calaveras slate, schist, and greenstone.

At Gold Run, about 8 miles beyond Weimar, is a wonderful exposure of highly colored brown and reddish gravels and coarse sands in high banks left at the edge of an old hydraulic pit. The cuts were made in a Tertiary river channel that here trends north-northeast. Precise age of the gravels is not known but it is probably between Eocene and late Miocene. Before hydraulic mining was outlawed because of the huge amounts of gravel debris left, it was the cheapest means of recovering gold from gravel. The miners directed huge, high-speed jets of water at gravel slopes from monitors, similar to firehose nozzles, and washed the gold-bearing gravel into sluice boxes. The remaining gravels contain too little gold per cubic yard to be economically mined today.

Geology for the next few miles, between Gold Run and Baxter, is very complex. Granitic rocks, metavolcanic rocks, Calaveras schist, and serpentine are overlain by patches of Tertiary auriferous gravel, white rhyolite ash, and andesite flows and mudflows. East of Baxter, Mehrten mudflows overlie the rhyolite ash which may belong to the late Miocene Valley Springs Formation. Here these younger Tertiary formations overlie and obscure the great Melones fault zone, which is usually considered to mark the eastern boundary of the Mother Lode belt as it turns due north into the northern gold belt. Beyond Baxter the highway climbs to a gently rising high ridge at elevation about 5,000 feet, high above the little railroad towns of Blue Canyon and Emigrant Gap (26 miles east of Weimar). The ridge consists of Mehrten mudflows overlying granitic and metamorphic rocks, all in turn overlain by glacial till. Just beyond Emigrant Gap is the junction with State Highway 20. State 20 runs westward along Mehrten-capped Washington Ridge, parallel to the Yuba River, for 27 miles to the northern gold belt towns of Grass Valley and Nevada City.

Shortly beyond Emigrant Gap the highway turns due east through granitic rocks and then passes along close to the turbulent Yuba River through a thick section of schist and hornfels of the Triassic Sailor Canyon Formation, well exposed in canyon walls along the highway at Cisco, about 4 miles beyond the State 20 turnoff. Beyond Cisco to Soda Springs (9 miles) granitic rock, partly covered by glacial debris, is exposed. High ridges to the north are capped by andesite flows. The highway passes in view of a succession of glacial lakes.

Between Soda Springs and Donner Lake (6 miles) the freeway passes almost unnoticeably over Donner Summit (elevation 7,239 feet), through a succession of rock types—granite, rhyolite, andesite, basalt—and then winds down a spectacular, steep fault scarp to the level of glacial Donner Lake. From view-point turnoffs, the still more spectacular descent of old U.S. 40 can be seen about a mile south; on the high slopes beyond it are the snowsheds of the main-line Southern Pacific Railroad. Faults in this zone sharply cut the latest Tertiary volcanic rocks and have been very recently active. Extensive deposits left by glaciers surround Donner Lake.

Interstate 80 from Donner Lake to Truckee (about 2 miles) passes along a glaciated valley that is flanked on both sides by dark flows of andesite and basalt. At Truckee, State Highway 89 takes off south across Quaternary volcanics and glacial deposits to Lake Tahoe (14 miles) and north across late Tertiary volcanics to Sierraville (25 miles) to junction with the east end of State 49.

East of Truckee, Interstate 80 follows the Truckee River across the Quaternary lake beds of Martis Valley and flat-lying flows of Late Tertiary-Quaternary andesite and basalt farther east in the river gorge to the Nevada state line. Boca Dam, about 7 miles northeast of Truckee, was slightly damaged by a Richter magnitude 6.5 earthquake originating in the eastern Sierra Nevada fault zone about 4 miles west of the Boca Reservoir in 1966.

### THE HIGH SIERRA VIA YOSEMITE VALLEY AND TIOGA PASS—STATE HIGHWAY 120

From the little mid-San Joaquin Valley town of Manteca—readily accessible by freeways from west, north, and south—State Highway 120 goes due east for 20 miles to Oakdale on the Stanislaus River (San Jose map sheet of the *Geologic Map*). The highway first crosses for several miles over the flat alluvial deposits of the valley floor and then, approaching Oakdale, over very low rolling hills where road cuts reveal stream-laid sand and gravel with interbedded lake silts and clays of Plio-Pleistocene age.

Just east of Oakdale, flat-lying beds of the Pliocene Mehrten andesitic sands and gravels, tuffs, agglomerates, and mudflows appear prominently in the low, rounded foothills of the Sierra Nevada. A turnoff, about 10 miles east of Oakdale, across the Stanislaus River to Knight's Ferry gives a chance to see good exposures of Eocene Ione sands and clays once mined here.

Returning to State 120, latite lava flows cap the low hills just east of Knight's Ferry. The highway then winds across the foothill exposures of "tombstone" rocks (Photograph 8-6)—Jurassic greenstone—Mariposa slate, and serpentine. Western margin of the serpentine is the Bear Mountain fault zone that bounds the Mother Lode belt.

Fourteen miles east of Knight's Ferry is the "Yosemite Junction" with State Highway 49 (the Mother Lode Highway) and State 108 (the Sonora Pass Highway). Instead of turning at this junction, stay on Highways 49 and 108 for about 4 miles to Woods Crossing at Woods Creek. Here is prominently exposed the 3-foot-thick, nearly vertical white quartz vein that here represents the Mother Lode. Quartz has intruded a complex fault contact between serpentine and Mariposa slate. The old headframe of the Harvard gold mine can be seen on the Lode $\frac{1}{4}$ mile to the north. Turning south at Jamestown 2 miles beyond Woods Crossing, takes the traveler back onto State Highway 120 which follows the very complex Melones fault zone (eastern Mother Lode fault system) for about 18 miles. Turning eastward at Big Oak Flat, the highway winds upward across 35 miles of exposures of Calaveras schist, slate, dark sandstone, and metavolcanic rocks to the boundary of Yosemite National Park. Although nearly vertical and locally complexly folded, these late Paleozoic strata dip essentially toward the granite rocks of the park, which were intruded into the trough of the Sierra Nevada synclinorium.

From the western park boundary on east into Yosemite and out again across Tioga Pass to Mono Lake, the geology is best followed on the

Mariposa map sheet of the *Geologic Map*. About 22 miles of the winding Big Oak Flat road across granite, with minor inclusions of Calaveras schist, is traveled to reach the floor of Yosemite Valley.

## YOSEMITE VALLEY

Incomparable Yosemite Valley! Lying literally in the heart of the Sierra Nevada, this deeply entrenched valley has an unique and grand beauty that has made it one of the greatest attractions of all the western parks (see Photographs 5-6 and 8-3).

Long known to the Indians, who called it Ahwahnee ("deep, grassy valley"), Yosemite Valley was "discovered" by the white man in 1849. The name Yosemite probably comes from *Oo-soo-ma-te* ("grizzly bear"), an Indian name for the Yosemite Indians. As early as 1864, Yosemite and the Mariposa Grove of giant redwoods (*Sequoia gigantea*) were set aside as a grant by the United States government. The much larger Yosemite National Park was established in 1890.

Greatest impetus to government interest in Yosemite came from the early-day exploration, studies, and writings on the valley by California's great naturalist, John Muir. In 1868, Muir, then thirty years old, walked from Oakland to Yosemite Valley. Viewing the snow-capped, saw-toothed peaks of the Sierra Nevada (literally, "snowy mountains") from Pacheco Pass in the Coast Ranges, he called it "the most divinely beautiful of

**12-7** *Leavitt Peak (elevation 11,570 feet), Mono County, about 3 miles south of Sonora Pass on Highway 108. Gently folded, stratified Pliocene volcanic rocks of explosive origin. The volcanics lie unconformably on Cretaceous granitic rocks (not shown). (Charles W. Chesterman photo.)*

all the mountain chains I have ever seen." Muir thought that rather than Snowy Mountains, it should be called the Range of Light (see Photograph 9-9). Although not a trained geologist, Muir was a keen and accurate observer; he was the first to ascribe formation of the deep, steep-walled, flat-floored valley to sculpture by glaciation. Most geologists of his time—notably, California's state geologist in the 1860s, J. D. Whitney—believed that cataclysmic faulting had rent apart the rocks to form the valley (Illustration 8-3).

An inspiration to naturalists, writers, photographers, and millions of public visitors, and a challenge to rock climbers, hikers, and skiers, Yosemite Valley grandly demonstrates fundamental processes and facets of geologic history that have attracted geologists to its study. Geologists study the valley as a striking example of stream erosion during uplift and tilting of a mountain range (see the drawing at beginning of Chapter 3) modified by glacial erosion and deposition and for its clear and perfect exposures of a succession of granite intrusions (plutons) that formed the Late Cretaceous batholith.

Yosemite Valley is set in the highest, most rugged western central Sierra Nevada. Highway 120 crosses Tioga Pass, east of the Valley, at an elevation of 9,941 feet—only 8 air-line miles from the 6,400-foot level of Mono Lake. Westward, the tilted Sierran block drops off gradually for about 100 miles to the San Joaquin Valley near sea level. Mount Whitney (elevation 14,495) is about 100 air-line miles southeast of Yosemite.

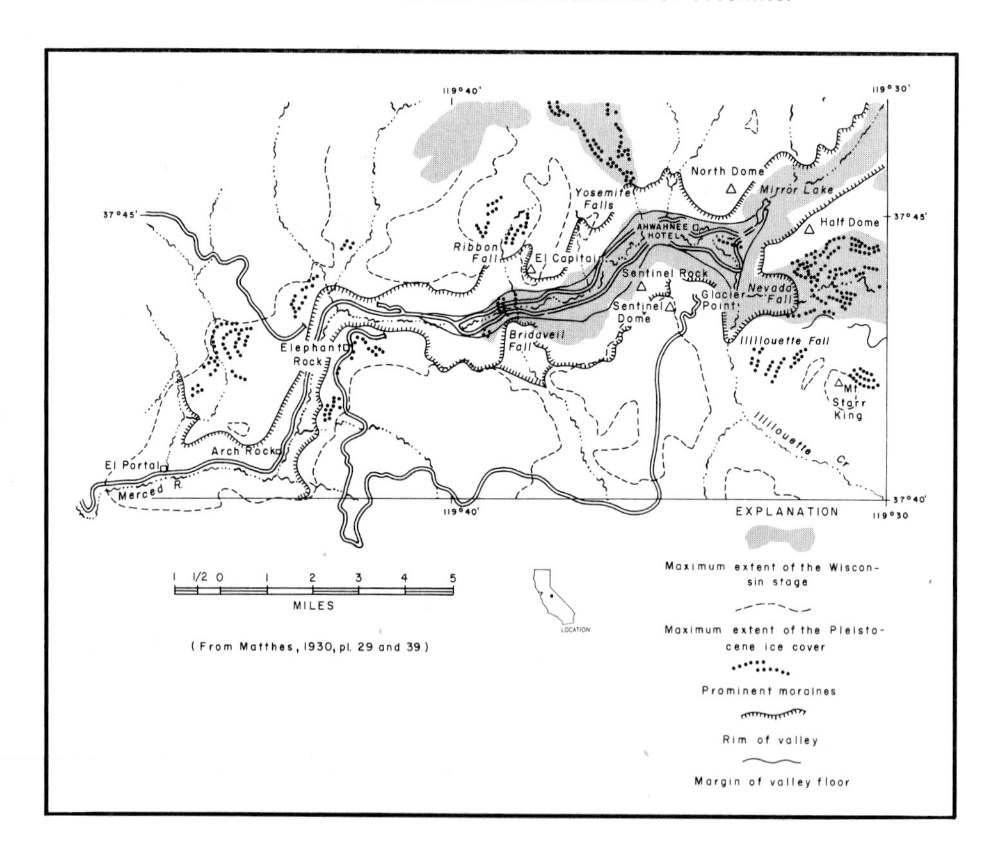


**FIGURE 12-8** *Generalized geologic map of the Yosemite Valley area. (California Division of Mines and Geology Bull. 182.)*

The valley may be entered by one of three highways: State 120 or State 140 from the west, or State 41 from the south by way of Fresno or Madera. The Yosemite All-Year Highway—State Highway 140—extends from Merced and follows the Merced River for many miles through its deep canyon onto the floor of Yosemite Valley.

**12-9** *Looking east from the entrance to Yosemite Valley. El Capitan on left, Half Dome in the center distance, and Bridalveil Fall dropping from its hanging valley on the right. (National Park Service photo.)*

### the rocks and the batholith

Oldest rocks in the Yosemite region are slate, chert, quartzite, and schist of the Calaveras Formation, derived from mud, clay, silica, and sand deposited in extensive Late Paleozoic seas, perhaps from Devonian to Permian time. These rocks are best exposed and most easily seen below elevation 3,000 feet in the lower part of the Merced River canyon, east of the Mother Lode and several miles west of the entrance to Yosemite Valley. Similar rocks, as well as Jurassic-Triassic metavolcanic rocks, are exposed in a narrow belt in the vicinity of Tioga Pass. On the east side of the pass, near Mono Lake, are metamorphosed sedimentary rocks of Ordovician and Silurian ages.

These intensely folded and faulted pregranite rock formations dip generally eastward from the west side of the range and dip westward from the east side of the range forming the gigantic, complex syncline into which the magma to form the Sierra Nevada batholith was intruded (Photograph 7-5 and Figure 8-8). A width of about 35 miles between the western and eastern blocks of the older rocks exposes only granite—in several

**12-10** *Yosemite Falls. Several different joint systems in the massive granite are accented by shadows and by the distribution of trees. Many of the vertical streaks, however, are marks left by intermittent dripping water. (National Park Service photo.)*

varieties (Photograph 2-11). For more than 60 miles along Highway 120 from the floor of the valley up the Big Oak Flat Road past Tenaya Lake and through Tuolomne Meadows to Tioga Pass, granite is the only rock seen! Geologists estimate that a thickness from 9 to 17 miles of the older overlying rocks was removed by erosion to expose the granite in Eocene time—about 60 million years ago.

Magma rose into the trough of the pre-Sierra Nevada synclinorium in a series of pulses beginning perhaps as long ago as Late Permian and Early Triassic time in the eastern Sierran region. Renewed intrusive activity took place, with mountain building, in the western foothills region in the Late Jurassic Period. Granitic rocks exposed in the Yosemite National Park are all of Late Cretaceous age, dated radiometrically from 94 million to 85 million years old. Seven major pulses of granitic magma, represented now by seven different plutons ranging in composition from quartz diorite through granodiorite and quartz monzonite to true granite are exposed in the sheer walls of the valley. Some of the granitic monoliths—such as El Capitan, Sentinel Rock, Half Dome, and Glacier Point—can be studied through 3,000 vertical feet of exposures! The principal plutons seem to have been emplaced oldest to youngest, from west to east.

### sculpturing of the landforms of the yosemite

Through investigations of the rock formations, observations of remnants of the ancient Sierran land surfaces that are still preserved in some places in the mountains, and the assistance of radiometric dates, geologists can look back to Eocene time, 40 million to 60 million years ago. The Sierra Nevada of that time was vastly different from what we see today! The shore line of shallow seas of the Pacific Ocean lay at the margin of the western Sierra foothills, and the Sierra Nevada had a gently rolling upland surface. Mount Whitney was perhaps 4,000 feet above sea level. Westward-running rivers—like the Merced—flowed with gentle gradients in broad, shallow valleys. The best reconstruction of the Eocene long profile of the Merced River defines a broad valley cut 800 to 1,500 feet below such mountains as Half Dome. The floor of Yosemite Valley, now at elevation 3,000 feet, was then about 800 feet above sea level.

For millions of years, the Sierra Nevada remained a range of rolling hills no more than a few thousand feet high at their crests. Then, about the close of the Miocene Epoch—about 10 million to 11 million years ago—uplift and westward tilting began. This has continued spasmodically and irregularly to the present day. By the close of Pliocene time, about 3 million years ago, the combination of uplift and erosion had left the floor of Yosemite Valley at an elevation of about 1,800 feet. Effect of uplift and westward tilting was, of course, to greatly increase the velocity of runoff down the western slopes, and westward-flowing streams like the Merced River, given added downcutting power, rapidly incised themselves to form deep, narrow valleys of V-shaped cross section. Streams tributary to the Merced and parallel to the trend of the Sierra, were not so greatly affected and did not so entrench themselves.

Today, snowfall on the highest Sierran peaks in Yosemite National Park is enough to feed and maintain only very small glaciers in protected northside valley heads at elevations generally greater than 12,000 feet. However, closely following rapidly increasing uplift about 3 million years ago and also related to the advances of continental glaciers over interior North America, climates became cooler and precipitation increased. Snow accumulated on 9,000-foot slopes and above, and alpine glaciers began to form. These mountain-valley glaciers came 25 miles down the Merced Canyon to elevations as low as 2,000 feet. Probably at least four advances and retreats of the glaciers took place, with minor fluctuations in between. At its maximum extent, the Merced glacier completely filled Yosemite Valley and extended over onto the uplands.

**12-11** *Half Dome at the east end of Yosemite Valley; Tenaya Canyon on the left, upper Merced River in the foreground. The top of Half Dome is about 4,850 feet above the river. (National Park Service photo.)*

**12-12** *The Merced River drops 594 feet over Nevada Fall as it enters Yosemite Valley over a giant, glaciated, granite staircase. (National Park Service photo.)*

The deep, solid, moving ice profoundly affected the mountain and valley landscapes. Along the high Highway 120, across the granite surface, particularly near Tenaya Lake and Tuolomne Meadows, glacial polishing and striations and patches of glacially deposited gravels are beautifully visible (Photograph 3-8). The sharp V-shaped valleys were eroded to U-shapes; granitic rocks were rounded, polished, and striated or scratched (Photograph 1-2). Projecting spurs, cliffs, and walls of the valley were eroded to form plane, vertical cliffs, and the valley, especially where the combined Merced and Tenaya glaciers merged, was gouged to a deep rock basin. Present depth of sediment in this part of the Yosemite Valley is 2,000 feet. Not only did glacial erosion modify the surface, but also deposits of glacial rock debris left lateral and terminal moraines, as well as ground morainal material on the valley floor. Glacial erosion was closely controlled and affected by jointing in the granite (Photograph 3-14). The vertical faces of Half-Dome, El Capitan, and other walls of the valley are joint-plane surfaces. The rounded domes, however—like Half Dome, Basket Dome, and Sentinel Dome—are spherical joint surfaces left by expansion as erosion removed heavy overlying rock; they continue to spall off in onionskin forms.

Since the latest major retreat of the ice about 9,000 years ago, sculpturing of the cliffs has occurred by prying out of huge blocks bounded by planar joints and by spalling of curved sheets in the spherically jointed rocks. Erosion has gone on most rapidly and extensively where the fractures are closest. The sixty-five living glaciers of the High Sierra Nevada have been retreating since John Muir's day, but we have no means of knowing whether this is a "final" phase, or whether the ice will again advance during the next few thousand years.

Huge talus cones now extend downward from the granite walls to the valley floor, representing postglacial rock slides and falls. Expanding water, as it freezes, is an important instrument in prying off rock slabs. During the Owens Valley earthquake of 1872, John Muir, living in Yosemite at the base of Sentinel Rock at the time, witnessed extensive rock falls triggered by earthquake shaking! This great earthquake was due to extensive faulting in the east-side Sierra Nevada fault zone. Quite evidently, the geologic forces that have shaped the high Sierra and Yosemite Valley continue today, undiminished, as they have operated for the past several millions of years.

**12-13** *Liberty Cap, over Nevada Fall. Although glaciers swept around its base and undermined the jointed granite, its spherically jointed crest was never quite topped by ice. Note the low-dipping joint and shear zone that cut across the base of the dome. (Ernest S. Carter photo.)*

**12-14** *Nevada Fall and the glaciated valley above it through which the Merced River flows. On the left is the base of Liberty Cap (compare photograph 12-13). (Ernest S. Carter photo.)*

# 13 beneath the great valley

*A slough of the Mokelumne River winds across the flatlands of the wine country near Lodi at the south end of the Sacramento Valley.*

■ The Great Valley of California—more than 400 miles long and an average of 50 miles wide—includes an area of about 20,000 square miles, or more than one-tenth the area of the state. Introduction to the Great Valley (also called Central Valley or Sacramento-San Joaquin Valley) is found in Chapter 1 of this book. For the most part it is monotonously level, rising gradually to the Sierra Nevada on the east and much more abruptly to the Klamath Mountains, Coast Ranges, and Tehachapi Mountains on the north, west, and south, respectively. The only significant break in the topography of the Great Valley is the 2,000-foot high Sutter Buttes in the center of the Sacramento Valley. Sacramento Valley occupies one-third of the Valley area, while the San Joaquin Valley includes two-thirds. Surface geology of the Great Vâlley and its margins is shown on the *Geologic Map of California* (index, Figure 13-1).

## VALLEY HIGHWAYS

Although its economy is primarily agricultural, the valley is dotted with towns and small cities, as shown in the drawing that opens this chapter. Sacramento, the state capital, is the metropolitan area (population 700,000 in 1977) in the southern Sacramento Valley; Fresno—population about 200,000—is the largest metropolitan area in the San Joaquin Valley. Gas and oil wells in the valley have produced petroleum products valued at $13 billion.

Interstate Highway 5 is a north-south freeway that runs the length of the state, from Oregon to Mexico, as well as the length of the Great Valley. In the valley, it generally clings to the west side and so affords a close view of the Coast Ranges. State Highway 99—long the only major north-south route—now merges into Interstate 5 at both the north and south ends of the Central Valley. A close network of all classes of state and local roads completes ready access to all parts of the valley.

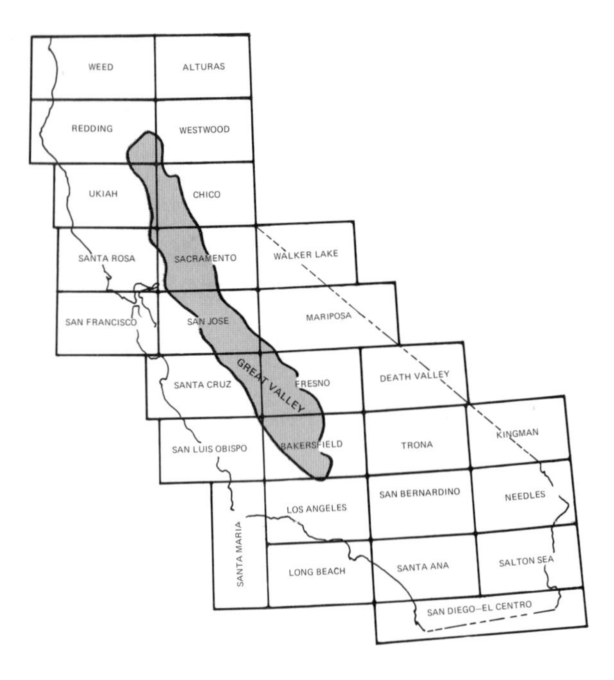


**FIGURE 13-1** *Index to map sheets* (Geologic Map of California) *covering the Great Valley and its margins.* (*California Division of Mines and Geology.*)

## THE VALLEY LOWLANDS

For the most part, the Great Valley extends for many miles as a broad, flat plain; so wide that the Coast Ranges and Sierra Nevada can scarcely be seen from its center through the usually hazy atmosphere. On clear winter days, visibility sometimes increases to as much as 100 miles and the flat-floored valley becomes beautiful by contrast with the snow-capped, serrated crest of the Sierra and the high, rounded hills of the Coast Ranges. Looking out the windows of Sacramento's tall buildings, one can see Mount Diablo at the north end of the Diablo Range, the Mendocino Range to the northwest, the Sierran crest to the east and even the tip of Lassen Peak, 130 miles due north at the southern end of the Cascade Range. Or one can see the distant Sierra Nevada from the top of Mount Diablo.

From all parts of the Sacramento Valley, the sharp jagged peaks of the Sutter Buttes (Photograph 13-3) stand out in sharp relief as a unique feature of the Great Valley (Chico map sheet of *Geologic Map*). They form a circular, serrated skyline of peaks 10 miles in diameter, clearly reflecting their recent volcanic origin in their topography. They are made up of a high, central core of andesite and tuff surrounded by a ring of sedimentary rocks, and in turn by an outer ring of andesite breccia and tuff, which merges into the older sediments of the valley floor. Rhyolite porphyry has intruded the sedimentary strata and volcanics of the central core. Apparently, slow, quiet intrusion was the earliest activity in late Pliocene time, but closing volcanism in the Pleistocene Epoch was explosive and a volcano 1 mile in diameter spread volcanic fragments widely around the central vent. About this time, volcanism was widespread around the northern end of the Sacramento Valley in the Coast Ranges and northwestern foothills of the Sierra Nevada, adjacent to the valley. In the Coast Ranges volcanic flows and sediments were deposited with stream-laid gravel, sand, and lake sediments to make up the Plio-Pleistocene Tehama Formation. In the foothills on the Sierra Nevada side the same formation is called the Tuscan.

These extend for many miles northeastward into the volcanics of the Cascade Range and Modoc Plateau.

Of great interest and also of economic importance is the structural relationship between the volcanics and the older sedimentary formations at Sutter Buttes. Marine sandstone, shale, and minor conglomerate of Late Cretaceous to Eocene age and the gravels, sands, and volcanic sediments of the Plio-Pleistocene Tehama Formation were turned upward into sharp folds surrounding the volcanic plugs. The resulting traps were ideal for petroleum accumulation and account for the prolific production of natural gas from the Sutter Buttes field. (A large-scale map of

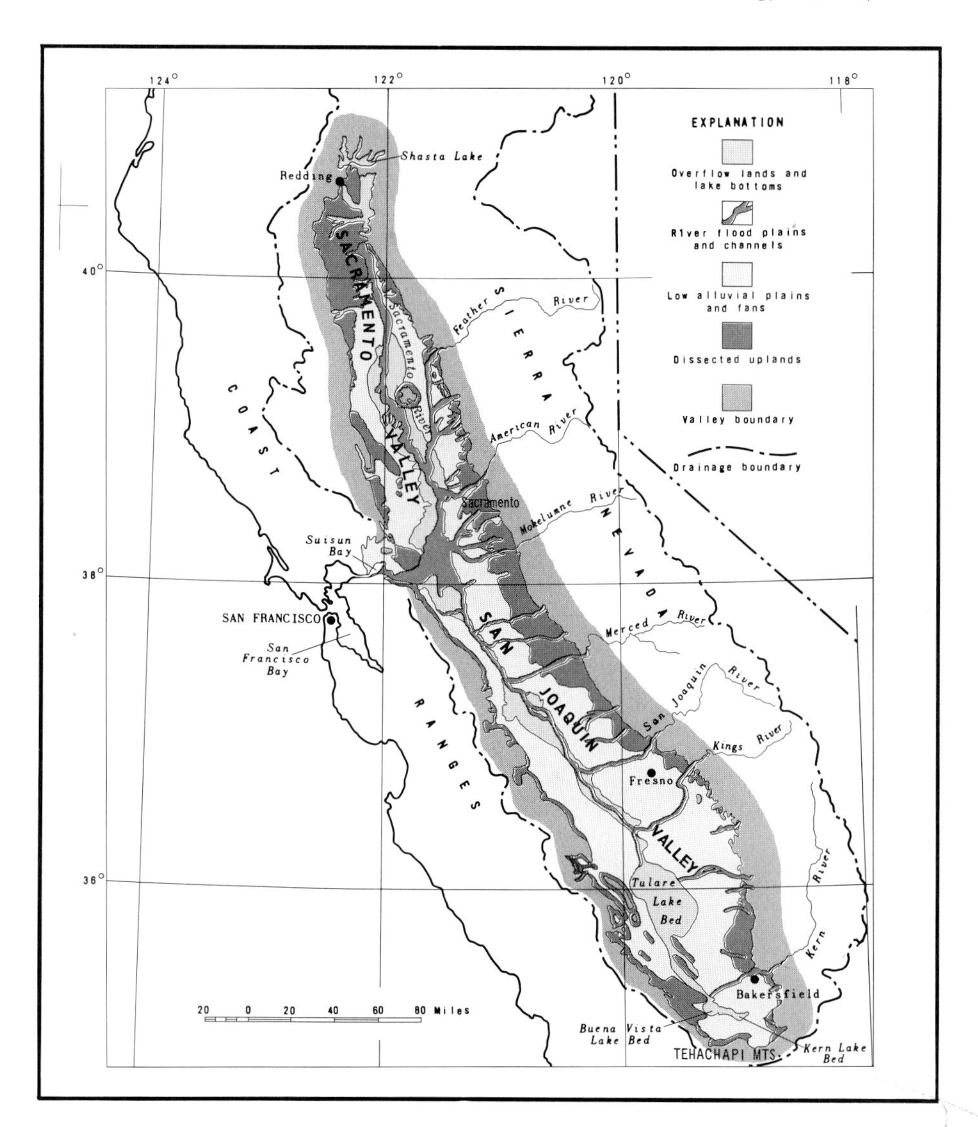


**FIGURE 13-2** *Map of the Great Valley showing drainage and principal landforms.* (*California Division of Mines and Geology Bull 190.*)

**13-3** *Looking northwest across the Sacramento Valley at Sutter Buttes. (John S. Shelton photo.)*

"Marysville" Buttes has been published as Plate 4, Bulletin 181, California Division of Mines and Geology.)

In the southwestern San Joaquin Valley, the Elk Hills, Lost Hills, and Kettleman Hills are low parts of the Coast Ranges that expose Pliocene and Pleistocene sediments. They are separated enough from the closely adjacent Temblor Range so that they appear as "islands" surrounded by valley alluvial fan deposits (Illustration 9-30).

An overall view of the valley floor is concisely given by Figure 13-2. Here are shown the overflow lands and lake bottoms, the principal river floodplains, channels, and deltas, the low alluvial plains and fans, and the slightly higher, marginal uplands. Before man's activities, there were extensive lakes formed by flood-season overflow in both the Sacramento and San Joaquin Valleys. The largest of these is the combined Tulare, Buena Vista, and Kern Lakes. In June, 1969, after a winter of heavy rain and snow, and during melting of the thick Sierra Nevada snowpack, runoff from the Kern and Kings Rivers caused the man-built levees to be overtopped and for a short time, Tulare Lake was the largest freshwater lake in California (130 square miles in area).

A great freshwater lake, known as Lake Corcoran, spread over much of the western part of the San Joaquin Valley about 600,000 years ago in late Pleistocene time. Existence of the lake is known from the fine lake clays, volcanic ash, and diatomite which make up the Corcoran Clay Member of the Plio-Pleistocene Tulare Formation. It is also interesting to note that the very last seas to occupy any part of the Great Valley left this southwestern part of the San Joaquin Valley at the close of the Pliocene Epoch about 3 million years ago.

What lies beneath this vast mat of sediments that has been accumulating in the Great Valley for the past 3 million years? How and to what extent has this basin been involved in the geologic history of the marginal Coast Ranges, Sierra Nevada, Klamath Mountains, Cascade Range, and Transverse Ranges provinces? Finally, how do we get at Great Valley geologic history?

**13-4** *The neat piles of gravel in rows along the American River east of Sacramento near Folsom, off U.S. Highway 50, are tailings left by the floating dredges after removing the gold. (Mary Hill photo.)*

## STRUCTURE AND GEOLOGIC HISTORY OF THE GREAT VALLEY

To answer the last question first: Obviously, the field geologist studies outcrops of the older formations all around the margin of the valley (plus the few areas of outcrop in its interior—like Sutter Buttes and Kettleman Hills) and projects the marginal geology out under the valley sediments. For example, the gentle western slope of the Sierran granitic and metamorphic rocks can be projected far out under the valley floor to the Sacramento River delta west of Sacramento; also, the steeply eastward-dipping Mesozoic strata of the east side of the northern Coast Ranges extend out under the western floor of the Sacramento Valley. The other prime means of attack has been by geophysical methods—mapping gravity changes and differences in magnetic effects, and studying the characteristics and velocities of earthquake waves, both natural and those induced by artificial explosions. Since the early 1930s, these two general approaches have guided the drilling of thousands of oil and gas wells. Great Valley sediments of Cretaceous to Pliocene age have become prolific producers of oil and gas (gas in the Sacramento Valley and oil and gas together in the San Joaquin Valley). Wells are the best sources of information, for they give specific data on the rock formations in slim columns down to depths of as much as 4 miles.

All these sources of information have been used to draw the two sections that show structure and imply geologic history of the valley: a section across the Sacramento Valley from west to east (Figure 13-5) and a section along the length of the Great Valley from north to south (Figure 13-6).

## MESOZOIC FORMATIONS AND HISTORY

The oldest formation that properly belongs to the Great Valley sequence is the late Jurassic Knoxville Formation. This is mostly dark shale and mudstone that was derived from the rising Sierra Nevada and Klamath Mountains and was deposited in cool, seas on the continental shelf and slope at a time when formation of the valley basin was just beginning. Knoxville shale is thickest on the northwestern side of the Sacramento Valley (20,000 feet thick west of the Orland-Willows area). Marine sediments, from the same general sources,

**FIGURE 13-5** *A geologic section across the Sacramento Valley from west to east.*

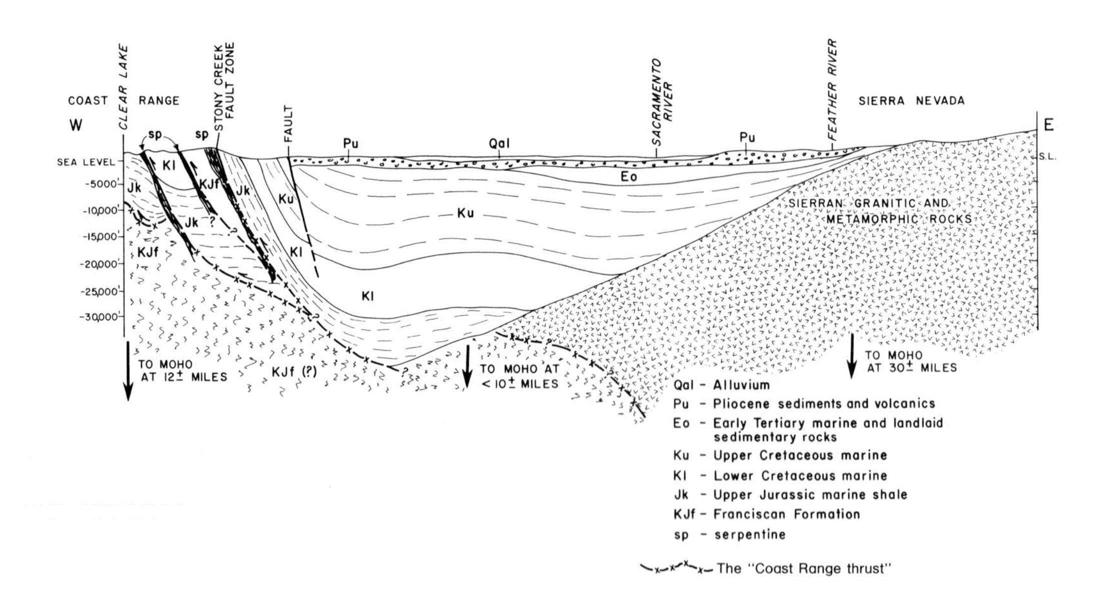

continued to be deposited through the 60 million years of Cretaceous time. Up to 20,000 feet of Lower Cretaceous shale and sandstone and another 15,000 feet of Upper Cretaceous shale, sandstone, and conglomerate form the very steep, eastward-dipping section along the western side of the Sacramento Valley. In middle Late Cretaceous time much more conglomerate and coarse sandstone began to be deposited, probably reflecting new uplift and mountain building in the Sierra Nevada and Klamath Mountains. It can be readily seen in the two sections that these shelf-and-slope sedimentary units thin rapidly southward and to the east where they approach the old shore lines along the western base of the Sierra. Note, in Figure 13-5, that the latest Cretaceous sediments extended farthest east across the Sierra Nevada "basement" rocks as the mountains were eroded to lower elevations during Late Cretaceous time.

While the marine (miogeosynclinal) sediments of the Great Valley sequence were being deposited—mostly as turbidites—in the valley area on Sierra Nevada basement rocks, eugeosynclinal sediments and volcanics of the Late Jurassic to early Late Cretaceous Franciscan Formation were being deposited in a trench on materials of the upper mantle at great depths at the foot of the continental slope farther west on the site of the present Coast Ranges. Tremendous faults—intruded by serpentine, which may have come from the upper mantle—on the western side of the valley (Stony Creek fault zone, Figure 13-5) separate rock formations of the miogeosynclinal type from those of the Franciscan eugeosynclinal type. There is a strong possibility that rocks of the Great Valley sequence have been thrust far westward over Franciscan rocks in the Coast Ranges. Somewhere under the western side of the Valley such a fault zone probably separates Franciscan basement from Sierran basement, perhaps as shown in Figure 13-5 (is there any clue to the ages of these faults?).

Far below all the Mesozoic formations is the Moho—bottom of the earth's crust on the upper mantle. This is about 12 miles deep under the Coast Ranges but plunges to 30 miles or more in the Sierran root. Along the center of the Great Valley, geophysicists have traced a gravity "high," suggesting that mantle material may rise—as intrusives—to much shallower depths along a central valley "ridge."

Any of the many roads that extend from the Great Valley westward into the Coast Ranges expose the Mesozoic rocks described, but the thickest and best sections are found west of the Sacramento Valley, along, for example, State Highway 20 from Williams to Clear Lake and State 261 from Willows to Elk Creek. Most interesting, but narrow and winding, are the county roads from

**FIGURE 13-6** *A geologic section along the length of the Great Valley from north to south.*

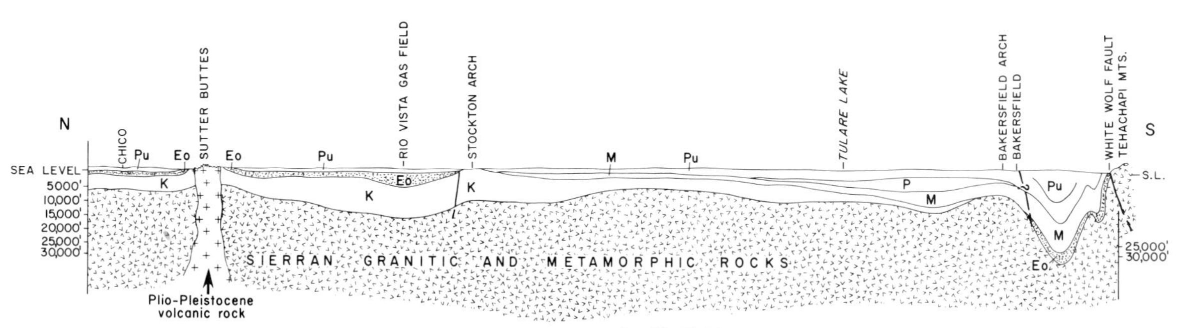


Pu - Plio-Pleistocene stream and lake sediments
P - Pliocene marine sediments
M - Miocene marine sedimentary rocks
Eo - Early Tertiary marine sedimentary rocks
K - Cretaceous (mostly upper) marine sandstone and shale

Corning and Red Bluff through Paskenta and across the 6,000-foot Yolla Bolly Mountains through Covelo and down the Middle Fork of the Eel River to U.S. Highway 101. Many good highways cross the Diablo Range west of the San Joaquin Valley. One of the best to show the Mesozoic and Tertiary section is State Highway 152 from Los Banos by San Luis Reservoir to Gilroy by way of 1,368-foot Pacheco Pass.

## CENOZOIC STRUCTURES AND HISTORY

Paleocene and early Eocene seas spread widely through the Sacramento Valley and San Joaquin Valley (Figure 9-1); but these two basins were almost separated by the rising Stockton Arch, a faulted ridge extending from the Sierra Nevada to the northern part of the Diablo Range in the southern Coast Ranges. The San Joaquin basin extended only as far south as a similar ridge called the Bakersfield Arch. Active northeast-trending faults in East Bakersfield and in the foothills at the southern end of the San Joaquin Valley (White Wolf fault) may be related to the Bakersfield Arch.

The Sierra Nevada was a low hilly area during early Tertiary time. Margins of the Great Valley were uplifted at the end of early Eocene time and the seas receded. Then in late Eocene time (Figure 9-2) the valley broadly subsided and the seas became more extensive again; not deep enough to cover the Stockton Arch, the seas did overwhelm the Bakersfield Arch and extend the San Joaquin basin south of Bakersfield into a deep trough that persisted throughout the rest of the era (Figure 13-6).

Along the eastern side of the southern Sacramento Valley and extending continuously into the eastern side of the northern San Joaquin Valley, upper Eocene rocks consist of the unusual white quartz sand and sand-clay mixtures of the Ione Formation. These sediments were derived from long, deep weathering of the lowlands of the Sierra Nevada and were deposited on floodplains and in shallow lakes and lagoons. They merge westward under the Great Valley into marine sandstone and shale. During much of the Tertiary period, a long, narrow trough—called the Vallecitos Syncline—made a principal sea connection from the southern San Joaquin Valley across the Diablo Range, which was already forming by Cretaceous time. The major southern Sacramento Valley connection to the sea was across the present Sacramento delta–San Francisco Bay area.

Oligocene seas occupied narrower and smaller parts of the same basins than in the late Eocene, but by early Miocene time the only remaining seas in the Great Valley were in the southern San Joaquin Valley (Figure 13-6). These early Miocene seas probably had an open-sea connection by way of a trough across the southern Coast Ranges (Figure 9-3). Late Miocene seas were rather more extensive again (Figure 9-4) and finally the last seas in the Great Valley disappeared at the close of the Pliocene Epoch, about 3 million years ago, after lingering longest in the southern San Joaquin Valley between Coalinga and Bakersfield (Figure 9-5).

The Great Valley—most strongly along its western side—participated in the mid-Pleistocene Coast Range orogeny. Many folds and faults were developed in the western and central valley at this time, known in detail from extensive exploration for petroleum. The numerous faulted, anticlinal trends have yielded large amounts of oil and gas. Some of the anticlinal folds—Kettleman Hills is an example—are so young that the anticlines appear as topographic hills. A great many faults accompanied the extensive folding on the valley floor and around its margins.

Thus, the geologist driving the long, "uninteresting," flat miles across the floor of California's Great Valley looks at the mountains beyond, recalls the records from thousands of oil wells and from countless measurements by seismic and gravity crews; and he pieces together the buried history of more than 100 million years.

# 14 klamath mountains highways

*The Trinity Alps, in the heart of the Klamath Mountains province.*

■ The 12,000-square-mile area of the Klamath Mountains province is not one mountain range but many, including the Siskiyou, Trinity, Trinity Alps, Scott, Scott Bar, Marble, South Fork, and Salmon Mountains. The broad view is that of plateau surfaces, about 5,000 to 7,000 feet in elevation, which represent old erosion surfaces that may be of Miocene age. These surfaces have been deeply dissected, principally by the Klamath and Trinity Rivers and their tributaries, to form steep and rugged ranges that are not readily accessible. Scott Valley is the only broad, alluviated valley within the province!

High average precipitation has produced a dense forest cover, which tends to increase the difficulties of geologic exploration. The large Klamath and Trinity Rivers carry most of the runoff. Population is scattered and low; Weaverville, the largest town, has 2,000 people. Historically, the mining of metals has been the principal occupation—gold starting with James Marshall's discovery in 1848, and later silver, copper, zinc, chrome, and iron sulfides. Lumbering and the summer tourist trade now support most of the population.

## GENERAL GEOLOGY

In its rocks and geologic history, the Klamath Mountains province is a northwesterly continuation of the Sierra Nevada, but the direct connection is obscured by overlying upper Cretaceous and younger rocks of the northern end of the Sacramento Valley and the Cascade Range-Modoc Plateau volcanic province.

Although geologically related, the Klamaths differ in a number of broad respects from the Sierra: (1) The Klamath Mountains are much lower (Thompson Peak, in the Trinity Alps in the heart of the Klamath Mountains province, is the highest at 8,936 feet elevation); (2) individual ranges in the Klamaths trend in different directions but generally more northerly than the Sierra; (3) structure of the Klamath Mountains is that of a

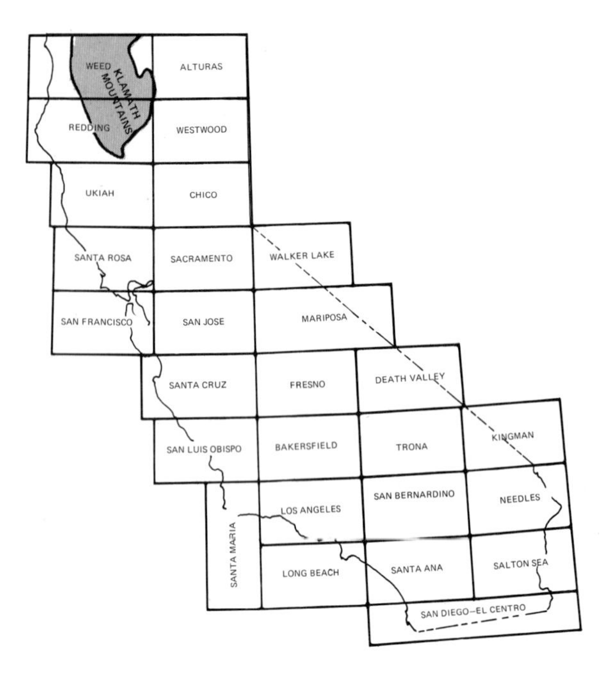


**FIGURE 14-1** *Index showing map sheets* (Geologic Map of California) *covering the Klamath Mountains province. (California Division of Mines and Geology.)*

broad arc open to the east; (4) the Klamath Mountains expose more of the pre-Nevadan Paleozoic and Triassic to Mid-Jurassic rocks and less granite; and (5) most of the exposed granitic rocks that have been dated are very Late Jurassic in age—probably intruded at the time of the Nevadan orogeny. None are as young as the granitic rocks of Late Cretaceous age that form the central Sierran batholith. The Klamath Mountains have not been elevated so greatly or eroded so deeply as the Sierra Nevada.

Geology of the Klamath Mountains province stands out sharply on the geologic map (Figure 14-2), but is shown in greater detail on the Redding and Weed map sheets of the *Geologic Map of California* (index, Figure 14-1).

## ROCKS AND GEOLOGIC HISTORY

In spite of extremely complex and difficult geology, the Klamath Mountains province lends itself to two broad types of simplification and organization. In terms of *geologic history* (what happened in *time*), the rock formations divide themselves into the pre-Nevadan or older rocks and the post-Nevadan or younger rocks (Figure 14-2). Nevadan granitic rocks, ultramafic rocks, and the older series make up the bulk of the Klamaths; these rocks are intensely folded in some places and in the central and western parts of the province, many are metamorphosed. The younger strata are gently folded to flat marine Cretaceous strata and land-laid Cenozoic beds. The younger rocks are found overlapping the older series along the southeastern flanks of the province.

In terms of *structure,* a series of arcuate belts is very apparent, consisting of the eastern Klamath Mountains belt, central metamorphic belt, western Paleozoic and Triassic belt, and the western Jurassic belt (Figure 14-3). Each belt is characterized by rocks of certain ages and certain types. Geologists who qualify as experts on the province postulate low-angle faults—thrust faults—between the plates, with the more easterly blocks thrust upward and westward for many miles. Many of these fault zones appear to have been intruded by sheets of serpentine and peridotite, perhaps originating in the upper mantle. Subsequently, the thrust-fault contacts and intruded ultramafic rocks have been folded and faulted.

## the older-rock belts

The eastern Klamath Mountains belt consists essentially of eastward-dipping stratified rock formations of all ages from Ordovician to Middle Jurassic, making a column more than 40,000 feet thick (Figure 14-3 and Table 14-1).

The central metamorphic belt is made up principally of hornblende and mica schists, called the Salmon hornblende schist and the Abrams mica schist. They are separated from the eastern Klamath Mountains belt by sheets of ultramafic rock in fault zones. Radiometric dating of the metamorphic minerals shows that regional metamorphism took place in Late Pennsylvanian to Early Permian time. This was probably an orogenic period. The Salmon and Abrams Formations are

**FIGURE 11-2** *Geologic map of northwestern California and southwestern Oregon. (California Division of Mines and Geology Bull. 190.)*

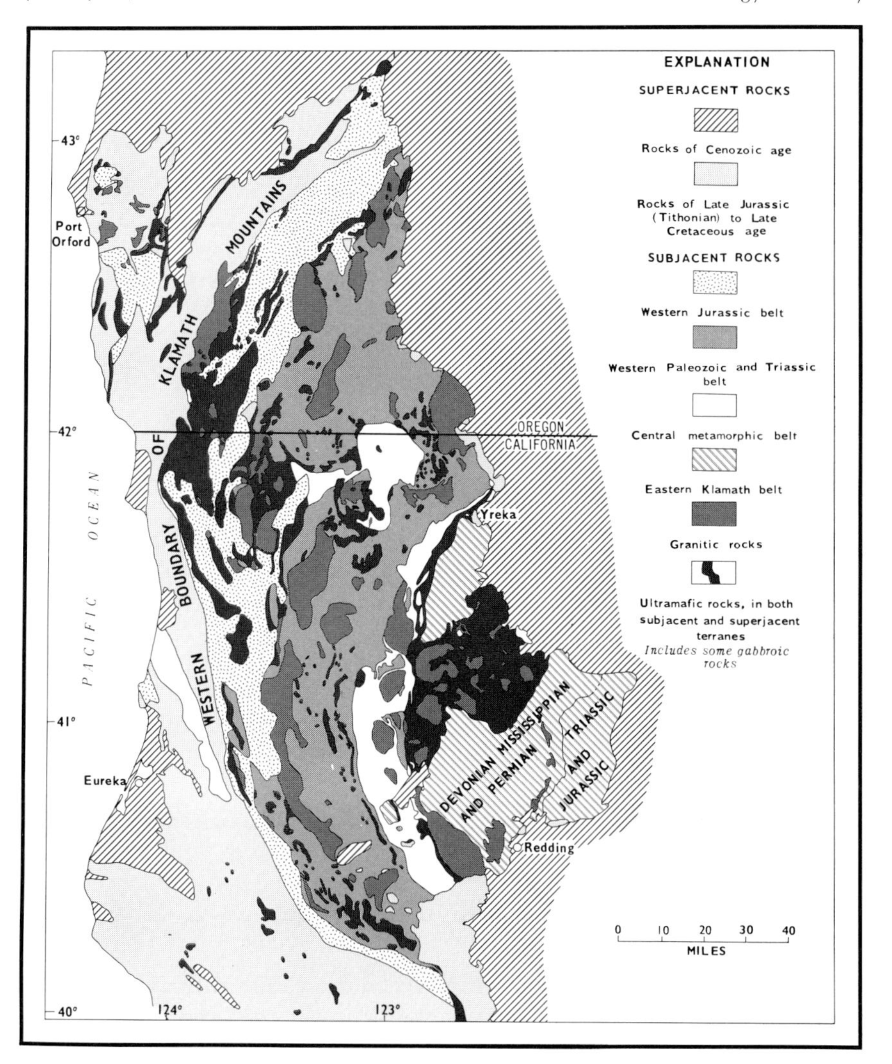

**TABLE 14-1 Geologic column in eastern Klamath Mountains**

| AGE | FORMATION | MAXIMUM THICKNESS (feet) | DESCRIPTION |
|---|---|---|---|
| Holocene and Pleistocene | River terraces | 200 | Gravel, sand, mud. Gold-bearing. |
| Miocene | Marine terraces | 150 | Shale, sand, conglomerate—flat beds. Only on ridge tops near western boundary. |
| Oligocene | Weaverville | 2,000 | Folded and faulted strata consisting of shale, sandstone, conglomerate, tuff, and lignite. Deposited on swampy floodplains. |
| Late Cretaceous and early Cretaceous | Local names | 2,000 | Well-bedded mudstone, shale, and conglomerate. All shallow-water marine and contain many fossils such as pelecypods and ammonites. Beds are usually dipping 10°–30°; they lie unconformably on granite and all older formations. |
| Late Jurassic | Granitic plutons | | Mostly quartz diorite intruded during the late Jurassic Nevadan orogeny; minimum ages around 130 million years. |
| Late Jurassic | Ultramafic rocks | | Peridotites and serpentine in thick sheetlike bodies. May be of different ages but in the west they intrude late Jurassic Galice Formation and are intruded by latest Jurassic granite. |
| Middle and early Jurassic | Potem, Bagley, and Arvison | 6,700 | Fossiliferous tuffaceous sandstone, shale, limestone; interbedded volcanic breccia and tuff. Bagley consists of andesite flows and breccias. |
| Triassic | Several formations including Bully Hill rhyolite, Pit shale, and Hosselkus limestone | 13,000 | Fossiliferous marine limestone, shale, conglomerate, volcanic tuff and breccia, rhyolite lava flows, and siltstone. |
| Permian | Dekkas andesite | 3,500 | Lava and fragmental volcanics. Marine fossils in tuff beds. |
| | Nosoni | 2,000 | Fossiliferous mudstone and tuff. |
| | McCloud limestone | 2,500 | Gray limestone of Pennsylvanian to early Permian age, containing corals and fusulina. |
| Pennsylvanian | Baird | 5,000 | Mudstone, limestone, chert, and volcanics. Abundant shallow-water marine fossils. |

**TABLE 14-1** ***(continued)***

| AGE | FORMATION | MAXIMUM THICKNESS (feet) | DESCRIPTION |
|---|---|---|---|
| Mississippian | Bragdon | 6,000 | Interbedded marine shale and sandstone. |
| Devonian | Kennett | 400 | Mudstone and tuff; corals and brachiopods in limestone beds. |
| | Balaklala rhyolite | 3,500 | Flows and explosive fragments. |
| | Copley greenstone | 3,700 | Andesitic lava flows and breccia. Pillow lavas show extrusion under water. |
| Silurian | Gazelle | 2,400 | Graywacke, mudstone, conglomerate; limestone containing corals, brachiopods, and trilobites. Graptolites in shale. |
| Ordovician | Duzel | 1,250 | Graywacke, chert, limestone with corals and brachiopods. May be, in part, Silurian in age. |

extensively intruded by late Jurassic granitic and ultramafic rocks.

The western Paleozoic and Triassic belt is a very complex belt of slightly metamorphosed sedimentary and volcanic rocks that were probably deposited in a eugeosynclinal basin. Limestones present have generally been recrystallized to marble and the fine-grained sediments and volcanic rocks to greenschist. Scarcity of fossils has made age determinations difficult. These rocks of low metamorphic grades undoubtedly include equivalents of the more systematic, unmetamorphosed, fossiliferous formations of the eastern Klamath Mountains belt, but only in a few cases can such correlations be made. Rock formations of this belt have been more extensively intruded by ultramafic rocks (late Jurassic?) and latest Jurassic granitic rocks than have the other belts.

Principally, four groups of rocks make up the western Jurassic belt. Along the northern margin of the Klamath Mountains lies the Galice Formation, which consists of slightly metamorphosed dark slaty mudstone with lesser graywacke sandstone and conglomerate and a lower unit of metamorphosed volcanic rocks. The sedimentary unit is at least 3,000 feet thick, and the volcanic unit is about 7,000 feet thick. Late Jurassic fossils date the Galice Formation. It is believed to be the equivalent of the Mariposa Slate in the western Sierra Nevada.

The South Fork Mountain schist forms a narrow zone that extends for 150 miles in California along the western and southern boundary of the Klamath Mountains. It occupies the eastern slope and crest of an extremely long, narrow ridge, called South Fork Mountain, which may well be one of the major Klamath Mountain thrust-fault zones. The formation consists of quartz-mica schist of sedimentary origin, and chlorite-epidote-feldspar schist of volcanic origin. Age and relationships of the South Fork Mountain schist are unknown, but it closely resembles some of the schists of the Franciscan Formation and it appears to grade southeastward into rocks that contain sparse Early Cretaceous fossils. The Galice Formation and South Fork Mountain schist are intruded by ultramafic rocks and all, in turn, by latest Jurassic granitic rocks.

### the younger formations

Along the northwestern side of the Sacramento Valley, a section of shallow-water marine sedimentary strata of Late Jurassic to Late Cretaceous age—the Great Valley sequence—forms an east-dipping section more than 50,000 feet in maximum thickness. The section seems to have been thrust upward and over Franciscan rocks along the Stony Creek fault. As this section is followed northward it thins rapidly and is cut off by complex southeastern extensions of the South Fork Mountain fault zone. In fact, the South Fork Mountain and Stony Creek faults may be parts of the same great fault system (Coast Range thrust). On the eastern flanks of the Klamath Mountains just west of Red Bluff, the entire Great Valley sequence is represented by only 2,000 feet of gently-east-dipping mudstone, sandstone, and conglomerate beds of early and late Cretaceous age!

North of the southeastern part of the South Fork Mountain fault and west of the northern end of the valley, highly fossiliferous Lower Cretaceous strata lie directly on eroded surfaces of latest Jurassic granitic rocks and on the pregranitic rocks of the eastern Klamath belt. Fossiliferous Upper Cretaceous strata extend all around the northern end of the Sacramento Valley where they are overlapped by the land-laid, coarse sediments of the Plio-Pleistocene Tehama, Red Bluff, and Tuscan Formations.

Few and only thin rock formations of Tertiary age are found within the Klamath Mountain province, although continental formations of this

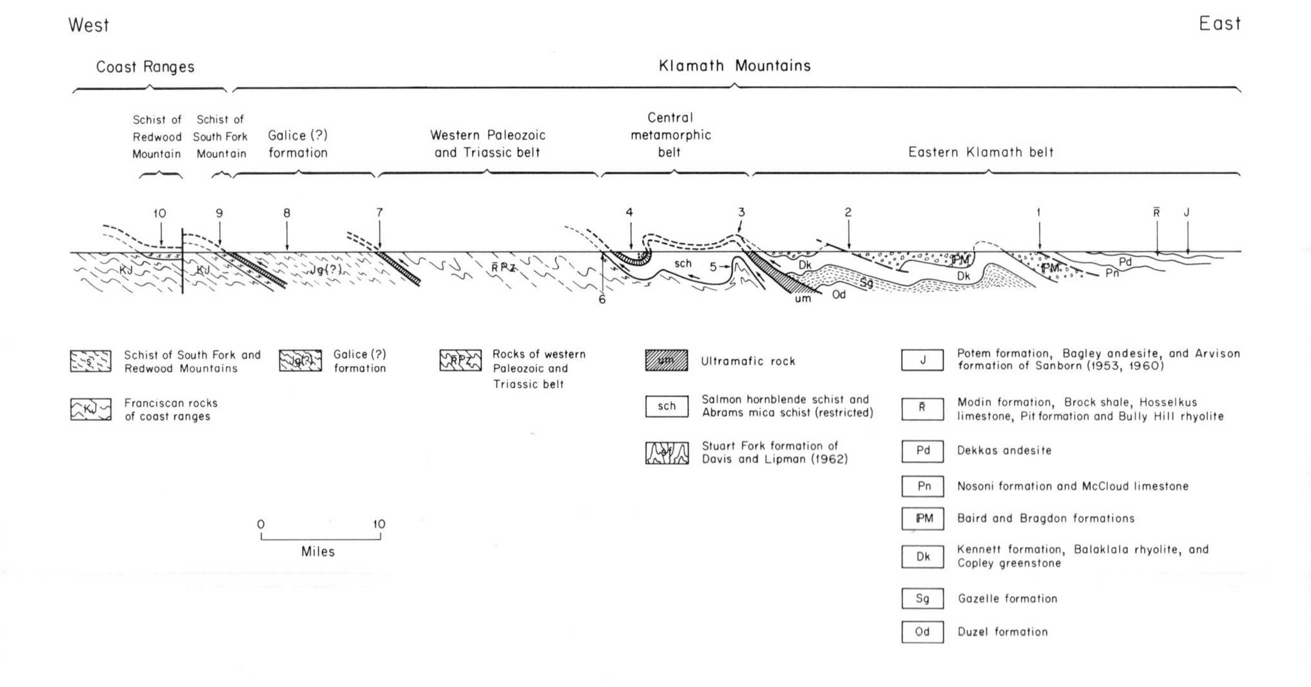


**FIGURE 14-3** *Schematic section across Klamath Mountains and part of northern Coast Ranges. Granitic plutons are not shown. (1) Structural break between Pennsylvanian and Permian strata, which suggests possible orogeny. (2) Low-angle thrust fault. (3 and 4) Thrust fault and intruded ultramafic sheet along which rocks of the eastern Klamath belt have been thrust westward over rocks of the central metamorphic belt in a subduction zone (see also points 7 and 9). (5 and 6) A folded thrust fault—central metamorphic belt formations thrust over western Paleozoic and Triassic belt rocks. (7) Serpentine in thrust fault with western Paleozoic and Triassic belt rocks thrust over Galice Formation in the western Jurassic belt. (8) East-dipping slaty cleavage in the Galice Formation. (9) South Fork Mountain schist and South Fork Mountain thrust fault. (10) "Redwood Mountain" schist—probably displaced South Fork Mountain schist. (Courtesy W. Porter Irwin, California Division of Mines and Geology Bull. 190.)*

period are thick along the northern and eastern borders of the province. Principal unit is the Oligocene Weaverville Formation, well exposed near Weaverville about 45 miles west of Redding on State Highway 299. The Weaverville Formation consists of sandstone, shale, conglomerate, and lignite, which were probably deposited in shallow lakes and swamps. The beds have yielded many fossil plants. Weaverville beds are folded and faulted and lie unconformably on Cretaceous and older rocks. Evidently, uplift and extensive erosion of some importance took place in the interval of time from the latest Cretaceous to early Oligocene. Then, post-Oligocene deformation again disturbed the area.

Thin, flat-lying fossiliferous marine beds of Miocene age occur as ridge-top patches in the northwestern Klamaths in Del Norte County. Extensive high terraces, sometimes covered by thick, red, oxidized soils containing nickel minerals, may be of Miocene age in the western Klamath Mountains province.

Quaternary sands and gravels, some of Pleistocene age as shown by vertebrate fossils, occur as alluvium and terrace deposits.

Like the Sierra Nevada, higher parts of the Klamath Mountains were glaciated. The Trinity Alps had at least thirty alpine valley glaciers, up to 13 miles long, during the times of latest glaciation in the Pleistocene Epoch. Four episodes of glaciation have been recognized in the Trinity Alps—the last three perhaps correlating with the Tioga, Tenaya, and Tahoe glacial advances in the Sierra Nevada. All the classical evidences of alpine glaciation—polished and striated rock surfaces, serrated ridges, pyramidal peaks, cirques, moraines, and small glacial lakes—have contributed to the beauty and interest of the Trinity Alps (see drawing at the beginning of this chapter).

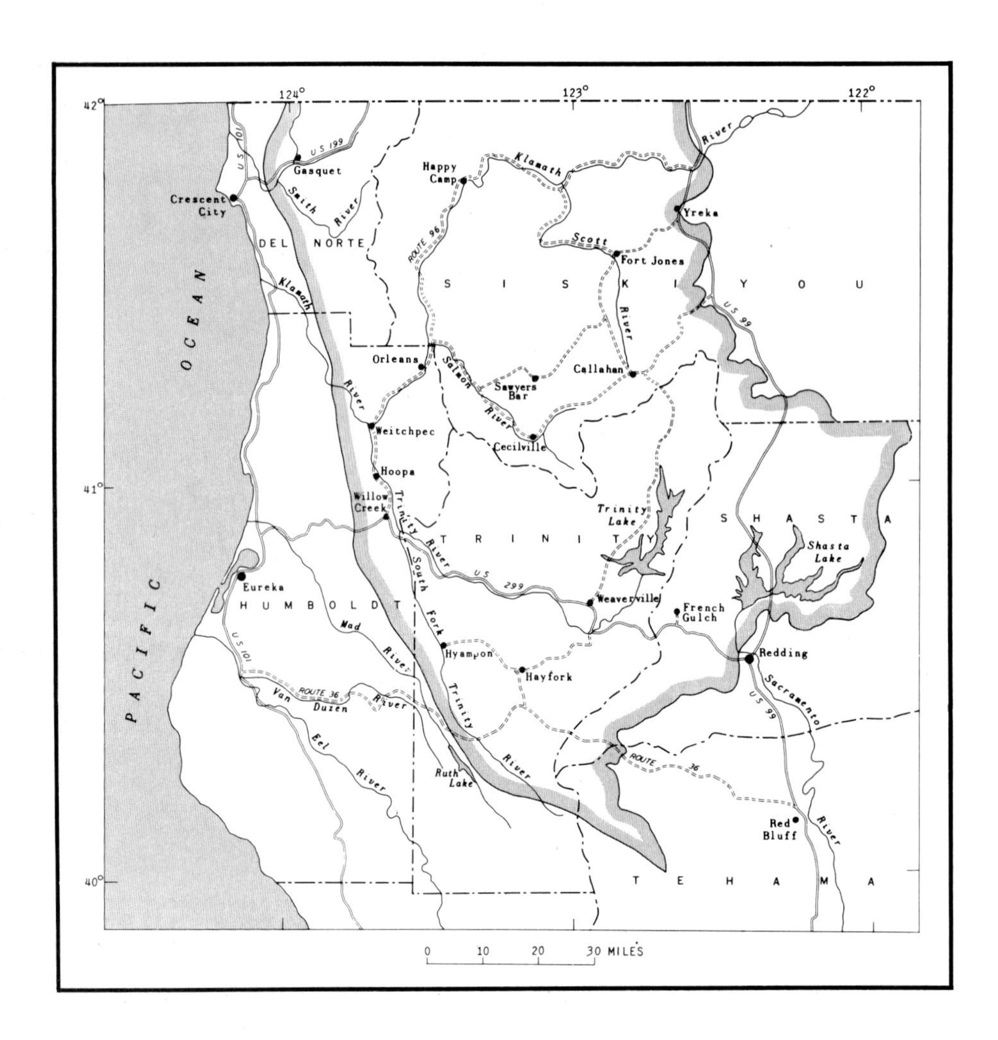


**FIGURE 14-4** *Map of northwestern California showing outline of Klamath Mountains province, drainage, towns, and routes of travel. (California Division of Mines and Geology Bull. 190.)*

## KLAMATH HIGHWAYS

The Klamath Mountains lie mainly between the Coast Highway, U.S. 101, and the Great Valley Highway, Interstate 5, State 99 (Figure 14-4). Highway 101 extends, entirely in the Coast Ranges, through Eureka and Crescent City. Interstate 5 extends northward from Redding, at the north end of the Sacramento Valley, for about 45 miles across the Klamath Mountains by way of Shasta Lake and the upper Sacramento River and then continues northward into Oregon, skirting through the western margin of the Cascade Range. Eureka, Redding, and Red Bluff are the takeoff points for interior-Klamath trips.

State Highway 299 is the only completely paved road crossing from east to west. It extends westward from Redding to junction with U.S. 101 at Arcata (145 miles)—a trip that takes most of a day—and northeastward from Redding to the Nevada border (176 miles). U.S. Highway 199 cuts across the northwestern corner of the state for 45 miles from Crescent City to the Oregon border. State Highway 36 winds westward across the southern Klamaths from Red Bluff on Interstate 5 to Fortuna on U.S. 101. State Highway 96 is an interesting road, through the heart of the Klamaths, from Willow Creek on State 299 down the Trinity River to junction with the Klamath River at Weitchpec, then up the Klamath River to join Interstate 5 a few miles north of Yreka. From Weaverville on State 299, State 3 winds northward along Lewiston Lake and the upper Trinity River to junction with Interstate 5 at Yreka. This highway gives access, by short side roads to the west, to the Trinity Alps vacation country. All roads (except 299) that cross the Klamaths are principally summer routes and not to be counted on at all times!

**14-5** *Dripstone in Shasta Caverns, Shasta Lake in Pennsylvanian-Permian McCloud Limestone. (Mary Hill photo.)*

**14-6** *"Gray Rocks" on east bank of the McCloud River (see Photograph 14-5) photographed by L. C. Graton, U.S. Geological Survey, before flooding by Shasta Lake.*

**14-7** *Pre-Silurian schist (dark rock, right center) intruded by light-colored granitic rock. Head of Stuart Fork in the Trinity Alps. (U.S. Geological Survey photo by H. G. Ferguson.)*

*Natural entrance to a lava tube, Lava Beds National Monument.*

# 15 the cascade volcanoes and modoc highlands

■ California has an abundance of volcanic rock formations, of all ages and in all the natural provinces; but the Cascade Range and adjoining Modoc Plateau are exclusively volcanic over an area of 13,000 square miles. Because of their youthful age—from Oligocene volcanic rocks to historically active volcanoes (Photograph 3-9 and drawing that opens Chapter 10)—conical peaks, domes, lava flows, volcanic sediments, and many other types of volcanic features can be seen in various stages of preservation. East of the partly forested Cascades, the semiarid climate supports only scanty vegetation cover, which does little to obscure rock outcrops.

## GENERAL GEOLOGY

Although dominated by geologically young volcanic rocks, northeastern California lends itself to division into three quite distinctive areas. On the west is the high chain of volcanic peaks that make up the Cascade Range. Volcanic rocks of the Cascades merge eastward into a broad volcanic highland or upland called the Modoc Plateau. At its eastern margin this "plateau" rises to the mountainous volcanic highland of the Warner Range. The eastern side of the Warner Range is marked near its base by the spectacular Surprise Valley fault—a fault much like the faults that farther south separate the westward-tilted Sierra Nevada from the Great Basin (Basin Range) province to the east. East of the Surprise Valley fault the Great Basin province spreads widely across Nevada and into Utah.

### cascade range

The Cascade Range in California is exclusively of volcanic origin and is characterized by beautiful, spectacular volcanic cones like Lassen Peak and Mount Shasta. Cones of this type cap the Cascades northward across Oregon and Washington into British Columbia.

In Oregon, the oldest volcanics were intruded by small bodies of granite-like rocks; in Washington, a series of granitic and older metamorphic rocks—much like those of the Sierra Nevada and Klamath Mountains in California—is well exposed as the basement rocks of the Cascade Range. Doubtless, deep under the Cascade volcanics in California—and probably underlying the Modoc Plateau as well—are similar rocks. This is suggested, for example, by windows exposing granite and metamorphic rock at Eagle Lake in the southeastern Modoc Plateau, at least 30 miles north of the nearest similar exposures in the northern Sierra Nevada, and by similar windows of Klamath Mountains province rocks, 30 miles east of the Klamath Mountains province, appearing through volcanics at the northern base of Mount Shasta.

The great variety of Cascade volcanic rocks reflects an extremely complex history of volcanism. They have been broken down by volcanologists into the Western Cascade series and the High Cascade series.

Rocks of the Western Cascade series are Miocene in age and are composed of basalt, andesite, and dacite flows, and inter-layered rocks

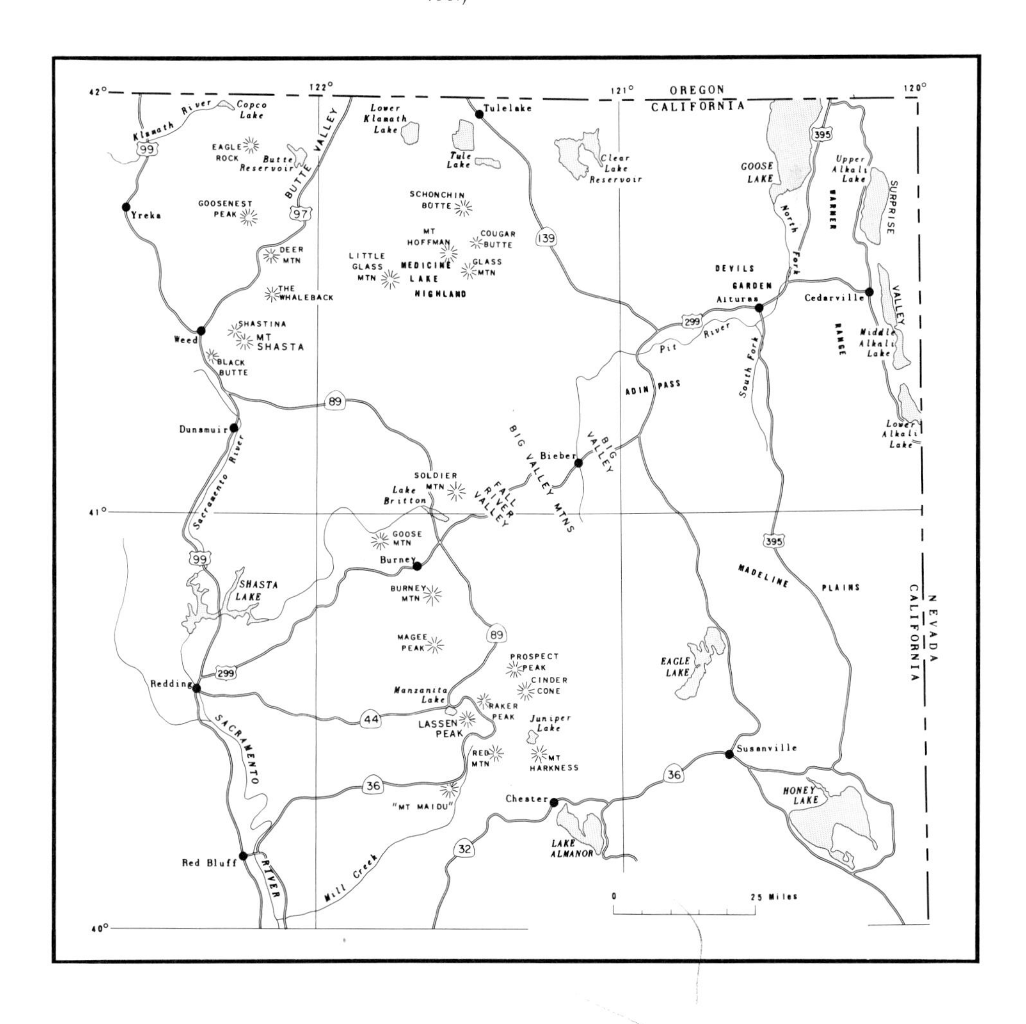


**FIGURE 15-1** *Index map of northeastern California showing highways and some of the features referred to in the text. (California Division of Mines and Geology Bull. 190.)*

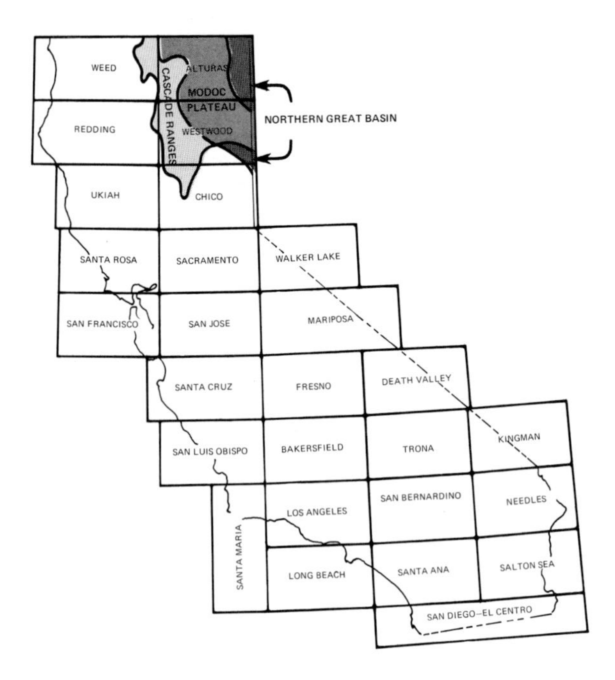


**FIGURE 15-2** *Index to map sheets* (Geologic Map of California) *covering the Cascade Ranges and Modoc Plateau. (California Division of Mines and Geology.)*

of explosive origin, including rhyolite tuff, volcanic breccia, and agglomerate (see Table 2-2, Principal Types of Igneous Rocks). They may often be recognized as belonging to the Western Cascade series because of a pervasive slight greenish cast, caused by general permeation of warm waters near the end of the Miocene Epoch. Uplift and erosion probably also occurred at that time. The series is exposed along the western side of the Cascade Range in a belt up to 15 miles wide and extending 50 miles south of the Oregon border to the town of Mount Shasta (Weed map sheet of *Geologic Map of California*).

After a short interval of uplift and erosion in late Miocene to early Pliocene time, volcanism became active again; and the High Cascade volcanic series was erupted slightly east of—but overlapping—the Western Cascade rocks to form a belt up to 40 miles wide and 150 miles long within the state. The oldest of these rocks is near the southern end of the Cascade Range. The oldest High Cascade rocks are overlain by the late Pliocene Tuscan Formation southwest of Lassen Volcanic National Park. The Tuscan Formation rocks are made up of mudflow breccias, agglomerates, and volcanic sandstone. Near the base of the Tuscan is the white rhyolitic Nomlaki Tuff Member, which was formed by incandescent "flows" of volcanic ash.

The early High Cascade rocks were very fluid basalt and andesite that were erupted from fissures to form low shield-type volcanoes and lava flows. The eruptions later became higher in silica and were more explosive. In Pleistocene time, some of the great composite cones like Mount Shasta, Burney Mountain, and Brokeoff volcano (near Lassen Peak) had their beginnings. Rhyolite lava and ash, and dacite were erupted to form volcanic peaks and domes. Eruptions of some of these materials have continued spasmodically into modern times.

The Medicine Lake Highland, about 35 miles east of Mount Shasta, is an eastward projection of the Cascade Range (Photograph 1-8). It is underlain by the Oligocene-to-Miocene Cedarville Series and the Pliocene Warner basalt, both more typical of the Modoc Plateau than of the Cascade Range. In the Medicine Lake area, a shield volcano

**15-3** *Looking north from 18,000 feet along the chain of Cascade volcanoes in California—Lassen Peak near center and Mount Shasta in the background. (Ernest S. Carter photo.)*

**15-1** *Subway Cave, a lava tube in the Hat Creek Valley area of Shasta County, just off State Highway 89 a few miles north of Lassen Peak. The entrance has been formed by collapse of the roof of the tube (see also Illustration 5-4). (Mary Hill photo.)*

20 miles across was built up by fluid pyroxene andesite in late Pliocene–early Pleistocene time. After reaching a height of about 2,500 feet, the top of the shield collapsed to form a basin, or caldera, 4 miles wide, 6 miles long, and 500 feet deep. A whole series of cones was then built up in and around the rim of the caldera. Eruptions of andesite, dacite, and rhyolite later built up masses like Mount Hoffman and the obsidian mass of Glass Mountain (Photographs 2-6 and 3-4).

## modoc plateau

The Modoc Plateau (Illustration 1-7) is not the flat-topped upland that is usually associated with the term "plateau"; it is rather a broad highland area built up of irregular masses of volcanic materials of great variety, but predominantly basalt. No sharp boundary, either in ages of volcanism or rock types, exists between the plateau and the Cascade Range. Numerous cones and shield volcanoes scattered over the plateau and extensive block-fault systems with largely vertical movement are responsible for a surface with considerable relief.

The Cedarville Series (Photograph 3-9) is the oldest group of volcanics in the Modoc Plateau. It consists of interbedded andesitic lava flows, rocks of explosive origin, and lake deposits dated by plant fossils from Oligocene to Miocene age. This older rock series has been block-faulted; and individual blocks have been tilted more than the later rocks, although some of the faults may be active even today. The Cedarville Series is best exposed in the Warner Range at the eastern margin of the Modoc Plateau; for example, it is readily seen on Highway 299 between Cedar Pass and Surprise Valley.

The Alturas Formation, of Pliocene age, which overlies the Cedarville Series, is also composed largely of lake beds and volcanic flows and tuff. Lake-bed tuff and diatomite are extensive in the vicinity of the town of Alturas. The Pliocene rocks also include basalt and andesite flows, mudflows, and dacite-to-rhyolite pyroclastic (explosive fragmental) rocks.

Very fluid flows of basalt were erupted widely over the plateau area and in the Warner Range to form the Warner basalt, probably ranging from late Pliocene to Pleistocene time. Many younger Pleistocene to Holocene basalt flows, cinder cones, small shield volcanoes, and interbedded lake deposits are difficult to separate from the older Warner basalt.

Many faults, mostly with vertical movements, extend across the plateau in northwesterly and northerly directions. Longest of these is the Likely fault (Alturas map sheet of the *Geologic Map*). Occasional small earthquakes and a host of very recent fault scarps testify to continuing activity.

### northern great basin

East of the Modoc Plateau, there is a series of north- and northeast-trending fault-bounded basins and ranges characteristic of the Great Basin province.

The Warner Range rises to an elevation of 9,983 feet in Eagle Peak near its southern end. No definite boundary exists between the Modoc Plateau and the Great Basin. The average elevation of the Modoc Plateau and also of Surprise Valley, east of the Warner Range, is about 4,500 feet. The Warner Range has been uplifted at least 5,000 feet and tilted toward the west along the tremendous Surprise Valley fault. A small area of glacial till lies on top of the Warner Range in a rather difficult spot to see about 6 miles north-northwest of Eagle Peak. The range consists of the Oligocene-Miocene Cedarville Series, topped by Warner basalt. Rhyolite intrusions of post-Cedarville age are found in its northern parts. Hot springs and mud volcanoes mark the face of Surprise Valley fault. A paved secondary road running the entire length of the valley follows the fault.

The valley is a long narrow graben, now occupied by a series of large saline lakes—remnants of Pleistocene Lake Surprise which must have filled the valley to a depth of a couple of hundred feet during the Ice Age.

Honey Lake Valley is a 4,000-foot high, wedge-shaped fault valley between the northern end of the Sierra Nevada and the southeastern Modoc Plateau. A great northwest-trending fault marks the base of the granitic Sierran escarpment, with less conspicuous faults on the Modoc side of the valley (Westwood map sheet of *Geologic Map of California*). The granitic Fort Sage Mountains, at the southeastern end of Honey Lake Valley, are bounded by faults; the one at its western base developed a 6-inch vertical scarp at the time of the Herlong earthquake in 1951.

## "VOLCANIC" HIGHWAYS

Any one of the many federal or state highways that lead into, or across, the Cascades, the Modoc Plateau, or the northern Great Basin reveals spectacular features of recent volcanism in all its many forms! Specific geologic features are easily checked on the Redding, Alturas, and Westwood map sheets of the *Geologic Map of California*. The numerous county and local roads must be checked locally; they are not always open.

Interstate 5 generally lies just west of the Cascade Range. It offers access to Lassen Volcanic National Park by way of State 36 from Red Bluff, State 44 from Redding, and State 89 from Mount Shasta via McArthur-Burney Falls Memorial State Park. State Highway 299 from Redding to Alturas and the Warner Range intersects State 89 about 35 miles north of Lassen Park. From Weed, U.S. Highway 97 crosses the Cascades to Klamath Falls, Oregon; and U.S. 395, the eastern Sierra Nevada highway, extends northward along the Honey Lake fault zone and through the eastern part of the Modoc Plateau by way of Alturas into Oregon. This latter highway offers access to interior State Highway 44 to Lassen Volcanic National Park (Photograph 2-5) and Redding, and State Highway 139 which runs north past Eagle Lake and Lava Beds National Monument to Klamath Falls. North of Lassen Park, Highway 44 circles to the east

**15-5** *View southeast along semiarid, steep eastern face of the Warner Range. Light-colored playa surface of Surprise Valley at left center; desert Hays Canyon Range, Nevada, in background. (Emerson Studios photo, courtesy Shasta-Cascade Wonderland Association.)*

around Sugarloaf Volcano (elevation 6,555 feet) and into Hat Creek Valley. Subway Cave, located in this area, is a collapsed Quaternary lava flow (Photographs 15-4 and 15-6), which is easily visited. Spectacular, very recent fault scarps mark the steep eastern slopes of Hat Creek Valley.

It is not practical to detail the many fascinating features along northeastern California's "volcanic" highways. Failing that, we shall look a little more closely at a few key volcanic areas that offer particular interest for the visitor and represent typical examples of the geology of the Cascade-Modoc region.

## THREE CLASSIC VOLCANIC AREAS

Three classic volcanic areas—Mount Shasta, Lassen Peak, and Lava Beds National Monument—serve as fine examples of volcanic geology in the Cascade Range-Modoc Plateau province. The geology of each is perhaps better known than most of the province and certainly these three are the areas most frequently visited.

### mount shasta

The great composite cone of Mount Shasta is typical of the Cascade volcanoes. Its beautiful double cone (Shasta and Shastina) is visible on a clear day for 100 miles in every direction. Its summit elevation of 14,162 feet puts the top of the mountain 10,000 feet above the little town of Mount Shasta, which is located on its low western flanks. Its diameter, scaled from the Redding map

**15-6** *Interior of Subway lava tube, Hat Creek Valley area, Shasta County. (Mary Hill photo.)*

**15-7** *Mount Shasta and its lower satellite cone, Shastina, looking north. The small town of Mount Shasta on Interstate Highway 5 appears in the dark lower left. An excellent highway from the town takes the visitor well up onto the base of the cone. (Ernest S. Carter photo.)*

sheet of the *Geologic Map of California* is 16 miles. Just this one volcano has a volume of 80 cubic miles! A fine scenic paved highway, built primarily for skiers, winds up the mountain from the small town of Mount Shasta on Interstate 5. It crosses the andesite flows and breccias of Cascade Gulch, circles a plug dome, then crosses late Pleistocene glacial deposits, and ends at an elevation just under 8,000 feet.

At the bottom of this vast accumulation lie the rock formations of the eastern Klamath Mountains belt and upper Cretaceous marine sedimentary sandstone and shale, both exposed in windows through the volcanics. Curving around the western base of the volcano are the Miocene altered volcanics of the Western Cascade Series—well exposed for 30 miles along Interstate Highway 5. On top of these are the late Pliocene andesites that represent the older rocks of the High Cascade Series.

The main cone of the mountain has been built up entirely in the Quaternary Period. The earliest Pleistocene lavas were basaltic andesite; the later (upper) lavas were pyroxene andesite and some dacite; pyroclastic rocks are interbedded. Low on the southern margin of Mount Shasta is Everitt Shield Volcano, which was formed by eruptions of very fluid basaltic andesite that flowed down the Sacramento River canyon below Shasta Springs for about 40 miles. The lava flow can be easily recognized because the basaltic andesite of which it is composed exhibits prominent columnar jointing. From a point near Everitt Shield Volcano a long fissure, shown on the Redding map sheet, extends due north through the summit cone of Mount Shasta. Shasta Springs was named for the large volume of spring water that issues from the Sacramento River gravels underlying the Everitt flows at this location.

Late in the history of Mount Shasta, a new vent developed high on the slopes of the volcano to form the summit cone of Shastina (Photograph 15-8), about 1.5 miles west of the crater of Shasta itself. This was located on a late east-west fissure. The last eruptions of Shastina, which occurred after the last glaciation, built two small domes within its crater. The very prominent cone of Black Butte, whose western slope Interstate 5 passes over just north of the town of Mount Shasta, is a hornblende andesite dome.

During the later part of the Ice Age, ice extended as small alpine glaciers down to the base of Mount Shasta on the west and out onto the margin of the Modoc Plateau on the east. The tiny glaciers still left now cover a total area of about 2 square miles and extend no lower than 9,000 feet.

The latest eruptions of Mount Shasta have come from the summit vent of the main cone, now occupied by a snow field about 600 feet across. The materials erupted were pumice, cinders, lapilli, blocks, and volcanic bombs of hypersthene andesite. An explorer in the area reported an explosive eruption of the mountain in 1786. Features of the summit of Shasta are best seen in the aerial photo, Illustration 9-11.

### lassen peak

Lassen Peak is the most southerly of the High Cascade volcanoes, a 10,453-foot peak located about 80 miles south-southeast of Mount Shasta (Illustration 2-5). This beautiful landmark is visible from the whole northern end of the Sacramento Valley and on clear days can be seen from as far south as the capital city of Sacramento (drawing at the beginning of Chapter 10 and Photograph 10-1). The geology of the area is clearly represented on the Westwood (Susanville) map sheet of the *Geologic Map of California*.

Lassen Volcanic National Park includes a classic area of about 170 square miles of composite volcanoes, cinder cones, and hot springs, which reveal a varied volcanic history from Pliocene time to the present day. The park is easily reached by fine highways from all directions. State Highway 44 enters the northwestern corner of the park at Manzanita Lake, 49 miles from Redding. State 36 enters the southwestern corner of the park from

**15-8** *Looking northeast over Shastina along the chain of Cascade volcanoes. The latest activity of Shastina formed two small plug domes within its crater. (Ernest S. Carter photo.)*

**15-9** *Lassen Peak from the air. (Ernest S. Carter photo.)*

Red Bluff in precisely the same distance. These two highways are connected across Lassen Park by a fine road that winds up around the slopes of the peak to an elevation of more than 8,500 feet. From its highest point, just above Lake Helen, an easy mountain trail leads to the summit of Lassen Peak (Photograph 10-1).

There can be little doubt that basement rocks of the Sierra Nevada extend below Lassen Peak; but gravity studies show that the lighter volcanic rocks lie in a deep basin, with the denser basement rocks far below. Only 15 miles southeast of Lassen Park, northwest-striking Paleozoic and Mesozoic rock formations, intruded by granite, are exposed. They are overlapped unconformably by Pliocene volcanics of the Tuscan Formation. The northwest-trending structures continue across the national park in the form of faults and chains of volcanic cones. About 16 air-line miles west of the park, windows through the volcanics have exposed underlying Eocene land-laid gravels of the Montgomery Creek Formation and marine Upper Cretaceous beds underlying the gravels.

A detailed account of the geologic history of Lassen Volcanic National Park by Prof. Howel Williams of the University of California has been published in Bulletin 190 of the California Division of Mines and Geology, with a number of striking photos of Lassen in eruption during the period 1914–1917. Summit features of the Lassen plug dome and surrounding volcanic features may be seen in the vertical aerial photo, Illustration 2-5.

Williams concluded that the earliest volcanic activity occurred in Pliocene time in the southern part of the park. The oldest rocks were basaltic flows, which were followed by a series of very fluid pyroxene andesite lavas. At the time these flows were deposited, a flood of lavas poured from vents and fissures over 30 square miles. During the buildup of a thick plateau of fluid lavas over the park area, a huge volcano had gradually been rising in the southwestern corner of the park. This volcano, now identified as

**15-10** *Petroglyph Point in Lava Beds National Monument. Remnants of shoreline sediments of prehistoric Tule Lake may be seen at the bottom of the lava cliff. (Mary Hill photo.)*

Brokeoff Cone, probably reached a height of 11,000 feet and a diameter of 15 miles. Remnants of the long-collapsed Brokeoff Cone are seen in Brokeoff Mountain, just outside the southwestern corner of the park. On the northeastern slope of Brokeoff Cone, a vent later opened just about under the present site of Lassen Peak. From this vent there erupted streams of fluid dacite, in Pleistocene time, to form the black, glassy columnar lavas that can now be seen encircling the peak. Shortly thereafter, more viscous, gas-poor, dacite lava welled up sluggishly to form the plug dome of Lassen Peak. A number of lesser dacite domes are found in the park area and a line of dacite cones appears at the northwestern base of Lassen Peak. These erupted ash and pumice about 200 years ago. Bodies of viscous dacite form the Chaos Crags (Photograph 2-5). Cinder Cone, in the northeastern part of the park, is a beautiful symmetrical dark cone that was formed by explosive eruptions about A.D. 500.

There is no recorded history of eruption of Lassen Peak until May, 1914, when Lassen burst into activity and erupted explosively. A year later, lava rose in the small summit crater and spilled over the northwestern and northeastern rim causing extensive mudflows by the sudden melting of snow. On May 22, 1915, a great blast came from the northeastern side of Lassen's crater, flattening and burning all trees and other plants to form what is called the Devastated Area. Activity thereafter declined and ceased in 1917.

Hot springs and fumaroles are numerous in the Lassen Park area and can hardly be missed along the highway across Lassen Peak. The springs contain sulfuric acid, hydrogen sulfide, sulfur, silica, water, clay, and mud; little mud volcanoes are common.

There is no sound geological reason to believe that volcanic eruptions have permanently ended in California's High Cascades.

### lava beds national monument

Lava Beds National Monument is an area of about 80 square miles, just south of Tule Lake National Wildlife Refuge and just north of the Medicine Lake Highlands (Alturas map sheet of *Geologic Map of California*). The general landscape is seen in Illustration 1-8, taken from Little Mount Hoffman, a few miles south of the monument, and

**15-11** *Spatter cone developed on a lava tube, Lava Beds National Monument, Siskiyou County. This was formed when a fountain of highly fluid basaltic lava broke through along fractures to the surface from a lava tube. (Charles W. Chesterman photo.)*

Illustration 1-7, taken from Schonchin Butte in the monument area. The monument is reached by short access roads from State Highway 139 (Oregon 39) and is about 35 miles south of Klamath Falls, Oregon.

Geologically, the monument is a part of the Modoc Plateau and is made up solely of the geologically recent Modoc basalt. Major feature of Lava Beds National Monument is a succession of pahoehoe ("ropy"), Hawaiian-type lava flows containing more than 300 known lava tubes. Although very recent, none are judged by volcanologists to be less than 5,000 years old. The tubes range from a few feet to about 75 feet in diameter (Photograph 5-4). Some have several levels, and many have collapsed to form winding trenches across the plateau. Lava stalactites from the roofs of the tubes and lava columns built up on the floors as stalagmites are common. The lava tubes formed as the surfaces of the highly fluid basalt chilled and solidified while the still molten lava inside continued to flow. Pressure domes appear on the surfaces of the pahoehoe lavas, and lava spurting out on the surface has formed spatter cones a few feet high (Photographs 15-11 and 15-12).

The interest of historians in the lava beds is that the tubes were used as hiding places and bases by the Modoc Indians during the Modoc War of 1872–1873. The intricacies of the complex passageways, known only to the natives, enabled Captain Jack and his Indians to hold off U.S. troops for the better part of two years!

The sharp-edged, highly irregular, clinkery aa flows are not as widespread as the pahoehoe but do occur; an example is the flow from Schonchin Butte, a small, but prominent, explosive volcanic cone adjacent to the road and near the center of the monument. At least a dozen such cones, a few feet to 700 feet high, are within the borders of the monument. Less violent explosive activity formed spatter cones instead of cinder cones.

Northeast of the main part of the monument, the Prisoners Rock is a cone of tuff that has been cut by erosion and the wave action of prehistoric Tule Lake. The extensive deposits left by this ancient lake cover many tens of square miles as shown on the Alturas map sheet. The western margin of Tule Lake basin is marked by Gillem Bluff, a very straight fault scarp 15 miles long that extends due north. It is typical of many such scarps in the Modoc Plateau province.

**15-12** *Spatter cone, Lava Beds National Monument, Siskiyou County. (Mary Hill photo.)*

Gibraltar Dam on the upper Santa Ynez River, Santa Ynez fault zone.

# 16 rocks and structures in the transverse ranges

■ The Transverse Ranges province stands out among all the other natural provinces; it extends for 325 miles directly across the structural grain of California (see Chapter 1 and Figure 16-1). In width, it pinches and swells from less than 15 miles to more than 60 miles. Narrowest parts are the western Santa Ynez Mountains and the Cajon Pass between the San Gabriel and San Bernardino Mountains; it is broadest from the Santa Monica coast across the Santa Monica, Topatopa, and Pine Mountains to the Tehachapi Mountains. The bounds of the province are well marked in its western and central parts but become vague and poorly defined in the east. Geologists differ, for example, about whether to include the Orocopia Mountains—just southeast of the Eagle Mountains—in this province. Figure 16-1 shows the position of the Transverse Ranges in relation to the other major provinces in southern California and also the names of the principal individual ranges within the province. Geologic details are best shown on the relevant map sheet of the *Geologic Map of California,* which are indexed in Figure 16-2.

How are the Transverse Ranges separated from the adjoining provinces? Margins of the Transverse Ranges province are extensively faulted; boundaries of the dozen or more individual ranges within the province are also approximated by major fault zones. The province boundary along the northern side of the western ranges (Santa Ynez, Topatopa, and Pine Mountains) is marked, from west to east, by the Santa Ynez River, Santa Ynez fault, and Big Pine fault. (The drawing at the beginning of this chapter shows Gilbraltar Dam on the Santa Ynez River in the Santa Ynez fault zone.) The Big Pine fault abuts the San Andreas fault where it strikes due east for a few miles. At this junction, on the south side, is 8,000-foot Frazier Mountain consisting of old crystalline rocks thrust over Pliocene sediments. North of Frazier Mountain the southern Sierra Nevada turns southwest and west into the Tehachapi Mountains. In the southern Coast Ranges, the Temblor Range turns southeast and east into the San Emigdio Mountains and Pleito Hills along the southern end of the San Joaquin Valley to meet the Tehachapi; the Sierra Madre and San Rafael

Mountains also bend to a more easterly trend as they extend southward toward the Pine and Topatopa Mountains. On the southern side of the Santa Ynez Mountains the trend is east along the Santa Barbara coastline (Photograph 10-3).

In the central Transverse Ranges, the northern boundary of the Sierra Pelona and San Gabriel Mountains is the San Andreas fault. Southern boundary is a series of reverse or thrust faults of the Sierra Madre–Cucamonga fault zone along the southern side of the San Gabriel Mountains. To the west, the bounding fault complex is the major Raymond-Malibu fault zone; the latter marks the southern margin of the Santa Monica Mountains. The Santa Monica Mountains extend due west across the continental borderland almost 100 miles into the Channel Islands—Santa Cruz, Santa Rosa, and San Miguel. Sea-floor topography strongly suggests that the Malibu fault zone also continues to the west. In fact, the Transverse Ranges, as a whole, are almost certainly a landward extension of the great Murray fracture zone, which extends

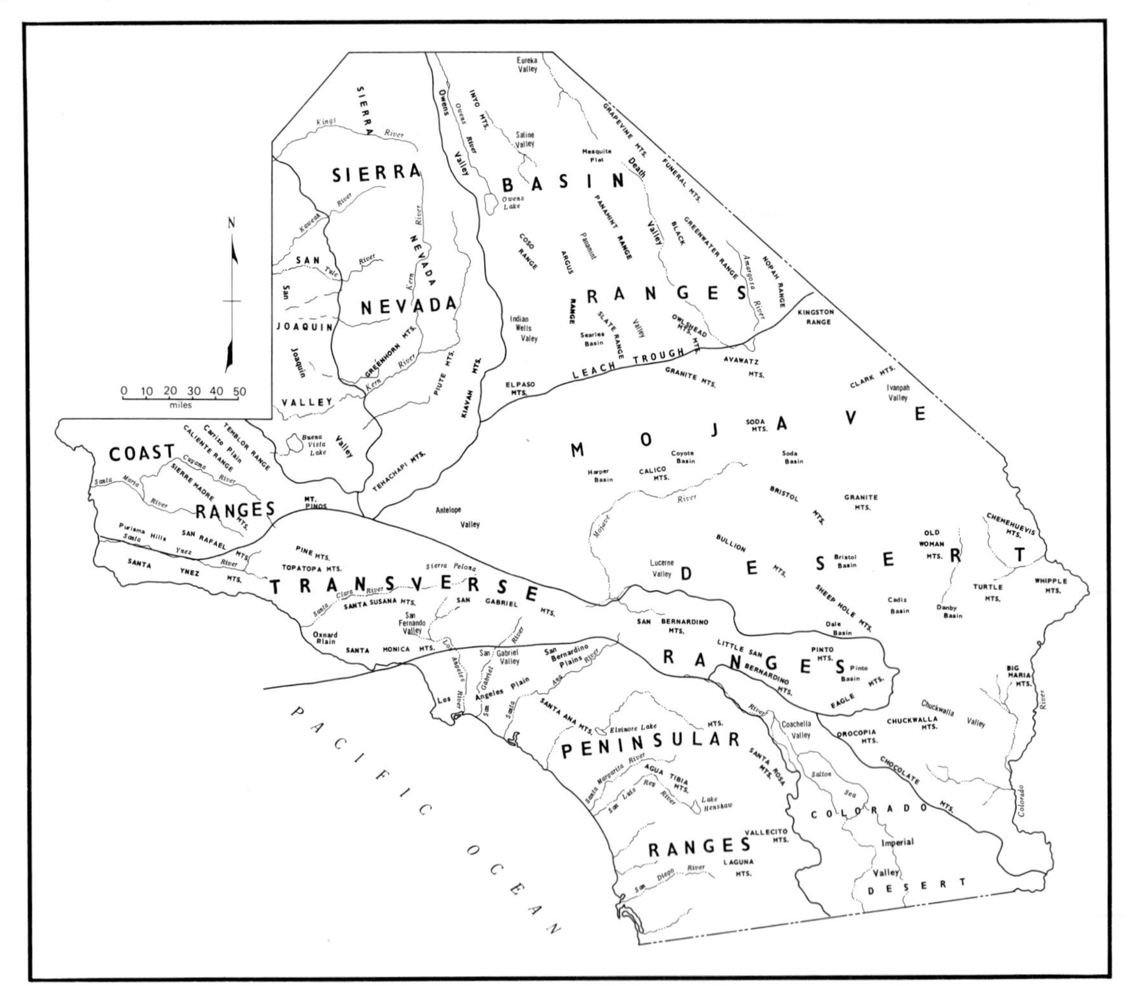


**FIGURE 16-1** *Outlines of the natural provinces in southern California showing names of the principal ranges, basins, and rivers. (California Division of Mines and Geology Bull. 170.)*

hundreds of miles due west across the deep-sea floor.

The San Andreas fault cuts obliquely across the eastern end of the San Gabriel Mountains and continues southeastward essentially as the southern boundary of the San Bernardino, Little San Bernardino, and Orocopia Mountains. The northern margin of much of the province is a series of discontinuous faults of large displacements along the northern flanks of all these individual ranges.

## GEOLOGIC HIGHLIGHTS

Rock formations in the Transverse Ranges cover a wider age range than in any other natural province in California. They range from Precambrian gneisses 1.7 billion years old to modern alluvium but with large gaps in the record, particularly in the Paleozoic Era. The rock types also show great diversity. All manner and types of structures are found and, because of important mid-Pleistocene mountain building, any formations older than late Pleistocene are likely to be strongly folded and faulted. Geology of each of the ranges within the province is distinctive and, in certain respects, unique!

What are the principal rock types and structures? What geologic history do they show? Where can some of the most interesting geologic features be seen?

## OLDER ROCKS

In general, the Transverse Ranges show a major break, or unconformity, in the sedimentary record in the Late Mesozoic Era. Similar periods of mountain building, igneous intrusion, and deep erosion have been recognized in the Sierra Nevada, Klamath Mountains, Coast Ranges, and elsewhere in California. The rocks above this important break in the Transverse Ranges are clastic sedimentary rocks of the shelf-and-slope, or miogeosynclinal, type, no older than latest Jurassic. The rock formations below the great unconformity are crystalline rocks ("basement"), including

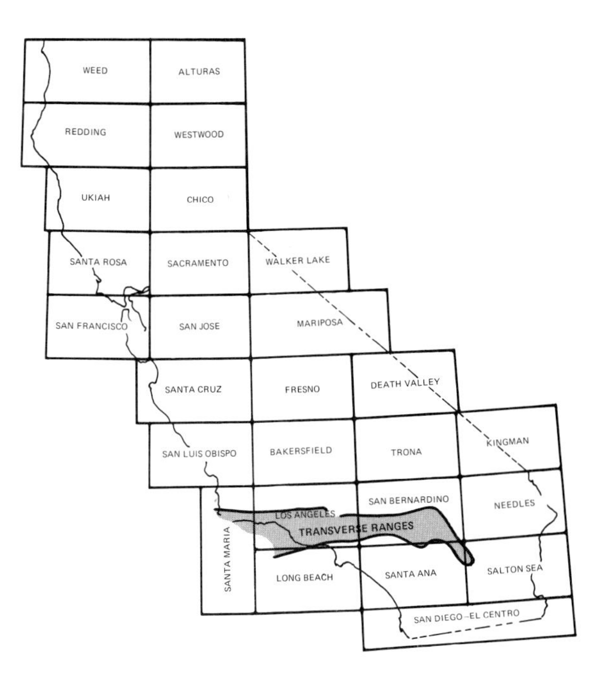


**FIGURE 16-2** *Index to map sheets* (Geologic Map of California) *covering the Transverse Ranges province.* (*California Division of Mines and Geology.*)

granite and other igneous rocks and a great variety of metamorphic rocks of all ages from late Cretaceous granite to Precambrian gneiss.

In the far western Santa Ynez Mountains and small areas along its northern margin, the basement rock formations exposed are typical Franciscan and serpentine. In this respect, the western Santa Ynez range is similar to the adjoining San Rafael Mountains in the southern Coast Ranges. The Franciscan here may be Late Jurassic in age as it appears to be overlain by dark Late Jurassic shale, which grades upward into Lower Cretaceous black shale. Elsewhere in the Santa Ynez Mountains, no rocks older than Late Cretaceous are exposed and no granitic rock appears. Whether the Franciscan Formation underlies the entire Santa Ynez Mountains is one of the unsolved geological problems.

Basement rock throughout the rest of the Transverse Ranges, wherever exposed, has been found to be granitic rocks and remnants of older schists and gneisses. This is the case, for example, in the adjacent Topatopa Mountains, Pine Mountain, and Frazier Mountain areas. Some of these gneisses are probably as old as Precambrian, as they are similar to those in the San Gabriel Mountains. These rocks, exposed in the Frazier Park area, are reached by a short side road west of Lebec on Interstate 5.

In the Sierra Pelona, and widely distributed along the San Andreas fault system through the Transverse Ranges, the oldest rock is Pelona schist, a thick sequence of low-grade metamorphic rocks of sedimentary and volcanic origin including quartz-albite-mica, actinolite, and chlorite schists, with lesser amounts of quartzite, marble, and hornblende schist. Age of the Pelona schist is unknown; experienced field geologists have mapped it as anything from Precambrian to Tertiary! However, it has been intruded by Late Cretaceous granite on the northern side of the San Gabriel Mountains. The Pelona schist is well exposed along the Bouquet Canyon Road just northwest of State Highway 14.

In the Santa Monica Mountains and their Channel Island projections, basement is Mesozoic granitic rock and a slate and schist formation known as the Santa Monica slate, which has been intruded by the granite. The Santa Monica slate is similar to parts of the Bedford Canyon Formation in the Santa Ana Mountains (Peninsular Ranges), which has been dated by marine fossils as early Late Jurassic. The Santa Monica slate is beautifully exposed by Interstate Highway 405 (San Diego Freeway) where it crosses the Santa Monica Mountains.

The San Gabriel Mountains are geologically complex, but it should be pointed out here that gneisses, schists, and granitic rocks from 1.7 billion years old to granite whose minimum age is 70 million years are found in that range! Earliest-known plutonic intrusion is the anorthosite body emplaced 1.22 billion years ago. Of great interest is the recent dating by radioactivity of granitic rocks in the central eastern San Gabriel Mountains at 245 million years, strong evidence for Permian mountain building. Similar basement-rock types are exposed in the Little San Bernardino-Eagle-Orocopia Mountains areas in the far eastern Transverse Ranges. Those who want to see the Orocopia schist, Precambrian anorthosite, and related rocks in the Orocopia Mountains will have to walk, for no roads directly cross the range! However, Interstate 10, State 195, State 111, and secondary roads outline the mountain block. The huge area of Joshua Tree National Monument—about 60 by 20 miles—encompasses much of the Little San Bernardino and Eagle Mountains and extends out into the Mojave Desert (Illustration 3-6).

In the large and high San Bernardino Mountains (Illustration 10-21 is a view of the range looking south)—55 miles long and maximum elevation 11,485 feet at San Gorgonio Mountain—are granitic rocks of various ages and gneisses of probable Precambrian age; but most interesting are about 10,000 feet of well-preserved quartzites and recrystallized carbonate rocks including limestone and dolomite of Paleozoic age. The Furnace limestone, in the upper part of the Paleozoic

**16-3** *Banded light and dark anorthosite, with dark bands of gabbro, in Soledad Canyon, northern side western San Gabriel Mountains. (G. B. Oakeshott photo.)*

**16-4** *Close-up of iron-ore beneficiation plant, Eagle Mountain. (Kaiser Steel Corporation photo.)*

section, contains Mississippian fossils. The Paleozoic section—including the Furnace limestone—and the Mesozoic granitic rocks are best seen north of Big Bear along State Highway 18, which runs from San Bernardino past Big Bear Lake resorts in the high mountains and down the northern side of the range by way of Cushenbury Canyon into the Mojave Desert at Lucerne Valley. On the Cushenbury grade, a huge cement plant that utilizes Furnace limestone can be seen.

## YOUNGER ROCKS

The "younger rocks" are those that lie stratigraphically above the major late Mesozoic unconformity. In the sequence of younger strata, the oldest are Lower Cretaceous marine black shale (possibly as old as Late Jurassic at the bottom) which lies on Franciscan basement in the far western Santa Ynez Mountains (Santa Maria map sheet of *Geologic Map*). Elsewhere throughout the Transverse Ranges rocks of Late Jurassic and Early Cretaceous age are missing, except as that geologic time may be represented by granitic intrusions.

### late cretaceous and early tertiary formations

Upper Cretaceous hard, gray marine shale and sandstone and conglomerate beds are very widespread in the western mountains of the Transverse Ranges, but none are known in the San Gabriel Mountains and the eastern ranges; they probably were never deposited (Figure 8-2). A section of several thousand feet of south-dipping Lower and Upper Cretaceous dark shale and sandstone is easily examined at the base of the northern side of the Santa Ynez Mountains on U.S. Highway 101 just south of the Santa Ynez River and Buellton. Here the highway cuts due south across the western Santa Ynez Mountains through a remarkable section of Cretaceous and Tertiary strata.

About 7 miles south of Buellton the highway crosses the Santa Ynez fault and from there south for 5 miles to the coast a steep, south-dipping section is exposed from Cretaceous through a succession of marine middle and upper Eocene sandstones and shales, Oligocene marine sandstone topped by Oligocene continental red and green beds, and marine lower and middle Miocene shale (upper Miocene is exposed a few miles to the west), siliceous shale, and chert. At the coast, Highway 101 turns due east on top of a late Pleistocene marine terrace. The highway continues for 30 miles into Santa Barbara in full view of great south-dipping beds of Eocene sandstone formations that form the crest and steep southern slopes of the Santa Ynez Mountains. Paleocene and lower Eocene strata are not found in the Santa Ynez Mountains or in the Transverse Ranges east of the San Gabriel Mountains. Paleocene marine conglomerate and sandstone occur as fault slivers in the high western San Gabriel Mountains and on their north flank.

No Eocene rock formations are known in the Transverse Ranges for about 130 miles east of the western end of the San Gabriel Mountains. Then, strangely enough, they are found in an isolated area of less than 2 square miles in the Orocopia Mountains. Since the occurrence of marine Eocene rocks is south of the San Andreas fault in the western San Gabriel area and north of that fault in the Orocopia area, many geologists have linked these two occurrences by postulating right-lateral separation of about 130 miles on the San Andreas fault.

### late eocene to early miocene rocks and history

Latest Eocene and Oligocene rocks consist principally of red beds of sandstone, mudstone, and conglomerate of the Sespe Formation. These were land-laid, as shown by vertebrate fossils during a time span of late Eocene to early Miocene time. Such rocks are found in the western Transverse Ranges, on the northern side

**16-5** *Looking north at the Santa Ynez Mountains a few miles west of Santa Barbara. Miocene rocks in the foreground; Eocene formations crop out in the high mountains in the middle ground. (Mary Hill photo.)*

of the San Gabriel Mountains (where they contain thick volcanics and are called the Vasquez Formation, Photograph 16-6), on the Channel Islands, and as far east as the northern end of the Santa Ana Mountains in the Peninsular Ranges province. Westward in the western Santa Ynez Mountains, Oligocene sandstone and shale become increasingly marine and in the most westerly outcrops near Point Conception those beds are entirely marine.

Shallow seas spread more widely in late Oligocene to early Miocene time (Figure 9-3); in them were deposited up to several hundred feet of near-shore, fossiliferous, coarse sands and gravels of the Vaqueros Formation. This formation is present along the southern side of the Santa Ynez Mountains, in the Ventura Basin (but not shown on Figure 16-1), Santa Monica Mountains, Channel Islands, northern Los Angeles Plain, and in small, isolated patches in the northeastern margin of the San Gabriel Mountains at Cajon Pass. The main San Gabriel Mountains and all of the Transverse Ranges to the east (to, but not including, the Orocopia) were above sea level through Eocene time and later.

### miocene rocks and history

Middle-to-late Miocene seas spread more and more widely over the western Transverse Ranges; and a deep trough developed on the site of the Ventura Basin, Santa Monica Mountains, Los Angeles Basin, and Channel Islands. Many thousands of feet of marine mudstone, sandstone, thin-bedded organic shales (including diatomaceous and foraminiferal types), cherty shale, brownish bituminous shales and sandstones, and diatomite (Photograph 16-7) make up the Miocene section. Formation names, like Modelo, Monterey, and Puente are used for these rocks. Particularly are these rocks thick in the southern Santa Ynez Mountains, the Santa Susana Mountains, and the Los Angeles Basin. The shore

**16-6** *Looking east at the characteristic spectacular outcrops of conglomerate and coarse sandstone of the Vasquez Formation, northwestern San Gabriel Mountains. (G. B. Oakeshott photo.)*

line of these seas extended northward into the foothills of the southwestern San Gabriel Mountains. At the maximum extent of late Miocene seas (Figure 9-4), all the areas named were covered; and seas of the eastern Ventura Basin extended up the Santa Clara River valley (Figure 16-1) along the northwestern end of the San Gabriel Mountains. In the last area and extending far east to Cajon Pass, the Miocene Epoch is represented by thousands of feet of conglomerate, coarse sandstone, and mudstone which were deposited by streams, on alluvial fans, floodplains, and in local lakes. Such beds make up the Tick Canyon, Mint Canyon, and Cajon Formations—the last spectacularly exposed at Cajon Pass.

Middle Miocene was a time of great volcanism in the Transverse Ranges. Thickest volcanic rocks of this age are in the western Santa Monica Mountains and are exposed in deep road cuts along U.S. Highway 101; but they are also farther east in the Santa Monica Mountains (Topanga Formation) and occur along the southern front of the western San Gabriel Mountains and further east along this mountain front where they are known as the Glendora Volcanics.

The central and eastern Transverse Ranges were undergoing rapid erosion during most of Miocene time. The Mount Pinos-Frazier Mountain area further west formed a barrier south of the San Joaquin Valley basin. Intermittent uplift had been occurring in all these areas since late Eocene time and was intensified and renewed in the late Miocene.

**16-7** *White diatomaceous shale of the late Miocene Modelo Formation exposed in a road cut on Coast Highway U.S. 101, near Point Dumé, southern side Santa Monica Mountains. (G. B. Oakeshott photo.)*

## pliocene and pleistocene history

With this rather widespread crustal warping and uplift in the mountain blocks, marine Pliocene sediments are largely confined to the deeper basins (Figure 9-5). The Ventura and Los Angeles Basins became as much as 5,000 feet deep and many thousand feet of sandstone, mudstone, silty shale, and conglomerate lenses occur in these basins. Along the Santa Barbara and Santa Monica coasts, there are much thinner Pliocene sands and mudstones. In the narrow and deep synclinal trough known as the Ridge Basin, along the San Gabriel fault zone, are more than 20,000 feet of stream and lake deposits. These are beautifully and spectacularly exposed in steep folds along Interstate 5 in the Ridge Route segment between the San Fernando Valley and the Tehachapi Mountains (Figure 16-1).

Nonmarine late Pliocene and Pleistocene beds are widespread along the margins of the Transverse Ranges. The western ranges were gently folded and elevated during Pliocene time and very marked crustal movements and uplift took place in the eastern ranges.

Seas persisted into the Pleistocene Epoch in the central Ventura Basin, Santa Barbara coast, and in the southwestern Los Angeles Basin at the northern end of the Peninsular Ranges. In the eastern Ventura Basin, western San Gabriel Mountains, and northern Los Angeles Basin, as much as 5,000 feet of stream and fan gravels of the lower Pleistocene Saugus Formation are found. In these areas, thin upper Pleistocene strata, from shallow-marine to land-laid beds, lie with a great angular unconformity on lower Pleistocene strata; lower Pleistocene beds are even vertical or overturned in some places (opening drawing of Chapter 5, Figure 5-17c).

## STRUCTURE

Structurally, the Transverse Ranges are so complexly folded and faulted, and there are such great differences in structural history between individual ranges that it is difficult to discuss the province without getting into a maze of detail on individual faults and folds. Faulting of almost every type occurs: (1) horizontal strike-slip; (2) normal, or vertical, movements; and (3) reverse and thrust movements. All types accompany folding. Both folding and faulting are of many different ages. Most structures are aligned east-west.

Folding seems generally to predominate in the western ranges that are made up primarily of sedimentary rocks. The Santa Ynez Mountains, for example, in their central part, consist of a south-dipping block of Cretaceous and Tertiary strata like one-half a huge anticline, chopped off on the north by the Santa Ynez fault. Highway 101 cuts due south across this block in the western Santa Ynez Mountains by way of Gaviota Pass and then follows due east along the foot of the mountains along a narrow coastal terrace, exposing this great structure to view. The Santa Monica Mountains are an anticline, but a much-faulted one, with a core of older Jurassic or Triassic rocks intruded by granite and flanked by younger Cretaceous and Tertiary sedimentary rocks.

To a large extent, the central and eastern ranges—San Gabriel, San Bernardino, Little San Bernardino, Eagle, and Orocopia Mountains (Figure 16-1)—which consist principally of great masses of pre-Tertiary crystalline rocks, seem to have responded to severe north-south–directed, compressive, mountain-building forces by major faulting, both compressional and strike-slip.

The generalized south-to-north structure section across the Ventura Basin (Figure 16-8) demonstrates that this thick sequence of Tertiary and Quaternary strata yielded principally by formation of a steep-sided syncline with older strata anticlinally folded and elevated along reverse faults on both flanks. Youthful age of the structures is shown by the fact that lower Pleistocene marine strata (San Pedro Formation) have been affected by both folding and faulting.

Individual folds in the Transverse Ranges have not been traced from one range to another; but a number of the major faults may be followed along the margins and even across two or more ranges, several even extend beyond the Transverse Ranges province.

The geologic map of California emphasizes the great through-going faults and drops out some of the lesser detail. Of all these, the San Andreas fault is the greatest! It extends along the northeastern side of the Frazier Mountain block, Sierra Pelona, and San Gabriel Mountains and then crosses Cajon Pass, separating the San Gabriel from the San Bernardino Mountain ranges and on southeast in a complex

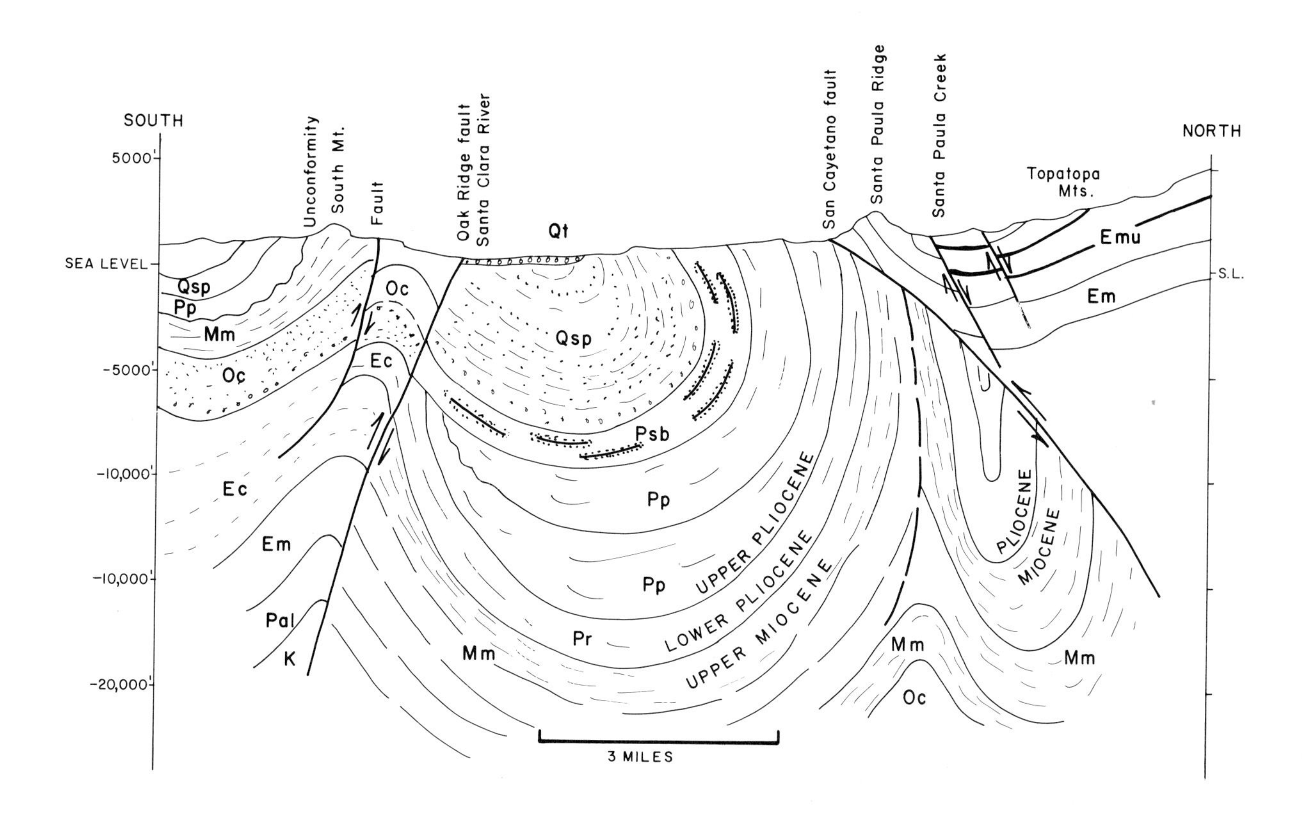


**FIGURE 16-8** *Structure section across Ventura Basin.* Qt, *upper Pleistocene terrace gravel;* Qsp, *marine lower Pleistocene clay, sand, and gravel of the San Pedro Formation;* Psb, *upper Pliocene to lower Pleistocene marine mudstone of the Santa Barbara Formation;* Pp, *marine sandstone, shale, and conglomerate of the upper Pliocene Pico Formation;* Pr, *marine lower Pliocene sandstone, shale, and conglomerate of the Repetto Formation;* Mm, *mudstone, shale, chert, and diatomite of marine Miocene formations;* Oc, *Oligocene continental red beds—sandstone, conglomerate, and mudstone of the Sespe Formation;* Ec, *upper Eocene continental beds in lower part of Sespe Formation;* Emu, *upper Eocene marine sandstone and shale;* Em, *marine Eocene sandstones;* Pal, *Paleocene marine sandstone, shale, and conglomerate;* K, *hard, dark sandstone and shale of marine upper Cretaceous formations.*

**16-9** *Looking south across the bed of the Santa Clara River, from near Santa Paula in the Ventura Basin, at South Mountain (Figure 16-8 shows the structure). Oakridge fault is followed by the river bed; beds of the late Eocene to early Miocene Sespe Formation appear on the mountain slopes. South Mountain oil field produces from Sespe sandstone at depth. (Mary Hill photo.)*

zone along the southwestern flanks of the San Bernardino, Little San Bernardino, and Orocopia Mountains. From Frazier Park to Cajon Pass, repeated movements on the fault have formed beautiful rift valleys with abundant evidence of recency of movement. This segment was displaced in a right-lateral manner—perhaps as much as 30 feet—during the great Fort Tejon earthquake of 1857. It has been inactive since. The fault trace can be closely and conveniently followed by secondary roads (see local road maps), which almost precisely follow the fault valleys, first, west to east through the little town of Frazier Park (Illustration 10-10), very shortly along Interstate 5 to Gorman, and then east-southeast along a straight trace through the tiny towns of Lake Hughes and Elizabeth Lake (named after the local sag ponds in the fault zone), and Leona Valley, Palmdale (Photograph 17-7), Little Rock, Valyermo, Wrightwood, and into Cajon Pass.

At Wrightwood is a good view of the great, active rock and earth slide on the north side of 8,000-foot-high Wright Mountain, the debris from which forms mudflows that continually threaten the small resort community (Photograph 16-10). In Cajon Pass, the fault zone, crossed by Interstate 15, is marked by the famous Blue Cut, where rock formations of several ages have been so shattered and mashed that little is left except unstable, constantly sliding, blue gouge. The blue cast is probably due mainly to Pelona schist. Southeast of the Blue Cut, Interstate 15 is followed for about 8 or 10 miles in one of the greatest fault zones to be found in the state—a zone involving the San Andreas fault proper and the very active San

**16-10** *Wright Mountain rockslide in early 1969, showing the vertical scarp in granitic rocks at the head of the slide near the top of the 8,000-foot mountain. In the foreground is some of the rubble which moved in surges down slope toward the small community of Wrightwood. (From a kodachrome by B. W. Troxel.)*

Jacinto fault, which merges into the San Andreas fault a few miles to the northwest.

The San Gabriel fault is an inactive fault that extends southeasterly, internally, within the San Gabriel Mountains; its geological relationships show that it has had both vertical and right-lateral displacement. The Oakridge reverse fault may be seen in the section, Figure 16-8. The Malibu-Raymond, Santa Ynez, and Sierra Madre fault zones must all be classed as major faults.

**16-11** *View northeast from upper Limerock Canyon toward Mendenhall Peak in the western San Gabriel Mountains (with Lookout tower right of center on skyline). Light-colored strata of the Saugus Formation dip gently to the right in the foreground, beneath the terrace. A prominent branch of the San Gabriel fault extends from the saddle, upper left, across the photo and behind the terrace. The main San Gabriel fault can be traced from the same saddle across the higher slopes just back of steep slopes in the middle ground, which expose light-colored granitic rocks. The skyline ridge, right of the saddle, consists of Mendenhall Gneiss.* (From a kodachrome by G. B. Oakeshott.)

Continuing activity of the Sierra Madre thrust faults was dramatically demonstrated by a maximum of about 7 feet of fault displacement (mountain block up) which caused the San Fernando earthquake of February 9, 1971, magnitude 6.4.

## folding and faulting

Geologic columns readily show some structural features—unconformities, for example—but structure is better shown on geologic structure sections and geologic maps. Figure 16-12 is a geologic structure section from south to north of a 20-mile-long segment across the western San Gabriel Mountains from San Fernando Valley to the Soledad Basin. Along with this section, a map showing the pattern of faults and folds (Figure 16-15) does much to give a clearer picture of structures in this part of the Transverse Ranges.

The San Gabriel fault dominates the internal structural features of the range. It strikes about southeast from Frazier Mountain, runs 65 degrees east of south across the western San Gabriels and turns due east into the core of the eastern San

Gabriels; total length is about 50 miles. It dips toward the north from about 70 degrees to vertical. There is evidence of considerable uplift on the north but there has also been 4 miles, or more, of right-lateral horizontal displacement. Latest movements on some segments of the San Gabriel fault have affected lower Pleistocene Saugus gravels but have not displaced late Pleistocene terrace deposits.

In the northern block, a series of northeast-striking, left-lateral faults makes up the pattern. In the block south of the San Gabriel fault, the prominent Little Tujunga syncline (Figure 16-15) parallels the San Gabriel fault; its northern limb is overturned at the eastern end. A series of low north-dipping reverse faults makes up the Sierra Madre fault zone along the southern margin of the San Gabriel Mountains (see the drawing at the beginning of Chapter 5 and Figure 5-17c). Late Pleistocene terrace gravels have remained flat and unfaulted. The age of much faulting and folding is clearly mid-Pleistocene, as noted farther west in the Transverse Ranges, but activity obviously continues to the present day. Uplift on the northern side of each fault has been as much as 2,000 feet.

The entire fault pattern of this area—right-lateral, strike-slip San Gabriel fault; left-lateral, strike-slip faults in the northern block; and reverse faults in the southern block—is consistent with great north-south compressive forces that no doubt formed the San Gabriel Mountains as well as the rest of the Transverse Ranges. The great mid-Pleistocene orogeny indicated was very widespread, involving uplift, tight folding, major faulting, and general destruction of the Tertiary basins by mountain-building processes. The domes and faulted anticlines that have yielded such enormous amounts of oil from late Miocene and Pliocene rocks in the Ventura and Los Angeles Basins were formed at this time.

**FIGURE 16-12** *Geologic structure section across the western San Gabriel Mountains, a central Transverse Range, Los Angeles County, from a point* B *on the south side in San Fernando Valley to a point* B′ *on the north side of the mountains. Top section continues from the right end into the bottom section, left to right.* B *to* B′ *is 20 miles. (California Division of Mines and Geology Bull. 172.)*

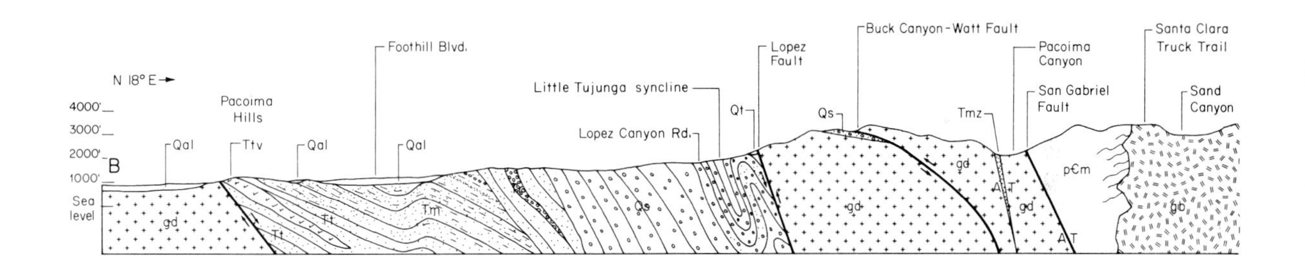


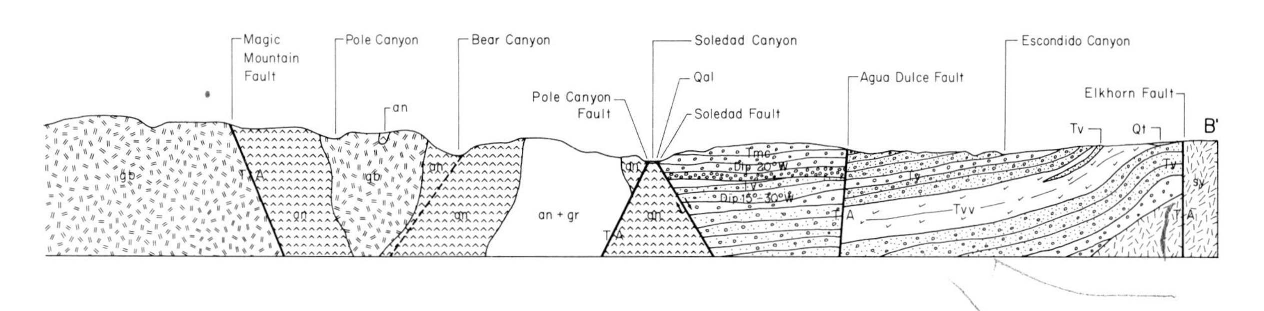

Crustal movements have continued to the present day, as shown by many elevated and warped terraces, by the active San Andreas fault along the northern side of the range, the San Jacinto fault which has been a frequent source of earthquakes, and the Cucamonga fault which has cut alluvial fans along the southern side of the eastern San Gabriel Mountains.

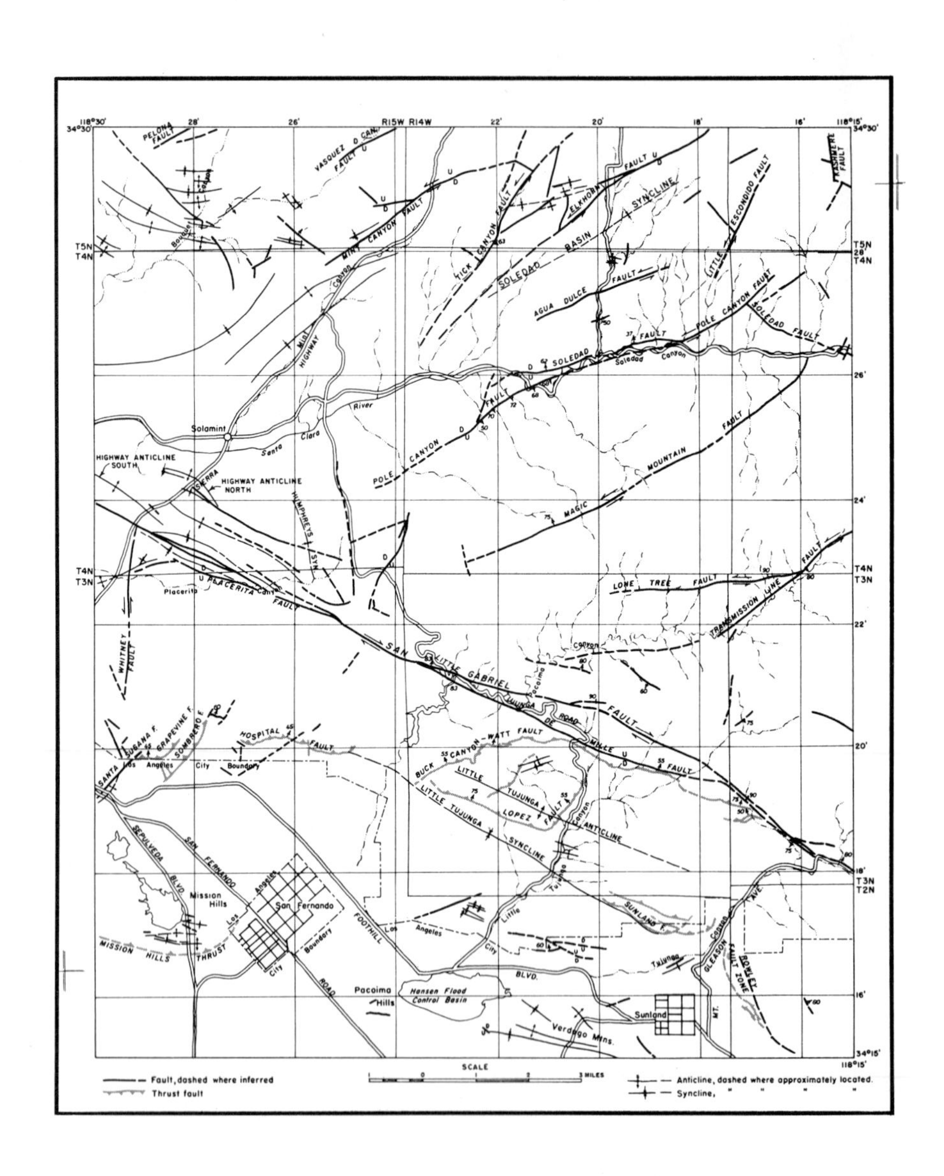


**FIGURE 16-13** *Pattern of faults and folds in an area of about 225 square miles at the western end of the San Gabriel Mountains. (California Division of Mines and Geology Bull. 172.)*

**16-14** *Looking southwest from Soledad Canyon at Pole Canyon fault. Light-colored anorthosite on right against dark gabbroic rocks. (From a kodachrome by T. E. Gay, Jr.)*

**16-15** *Steeply dipping red beds of the Vasquez Formation (left) in contact with Cretaceous gneissoid granite along a branch of the Pelona fault, which crosses Bouquet Canyon just north of San Fernando quadrangle. (From a transparency by G. B. Oakeshott.)*

# 17 death valley and the mojave desert

*View from the Racetrack road west across Saline Valley to the Inyo Mountains.*

■ The Mojave Desert and southern Great Basin occupy more than one-fifth of the area of the state (typical desert landscape, Figure 1-12). They form essentially a single natural province extending north from the Mexico-Arizona-California border for nearly 500 miles to the Nevada border north of Mono Lake. An introduction to the geography and geology of this vast area of mountain ranges and desert basins was given in Chapter 1. Rocks of the area have provided important information that has extended our knowledge of the Precambrian and Paleozoic eras of California. (History of these eras was summarized in Chapters 6 and 7.) Marine Triassic rocks and their history, Jurassic-Cretaceous orogenic movements, and granitic batholiths were discussed in Chapter 8. The Cenozoic history of the province is reflected in great accumulations of stream and lake sediments (often containing salt, gypsum, and borates), volcanic rocks and volcanoes, and extensive systems of folds and block faults that formed the basins and ranges seen today (Chapter 9). The principal ranges, basins, and rivers of the area and their relationships to the other natural provinces in southern California are shown in Figure 16-1. Details of the geology, on the scale of 1 inch equals 4 miles, are shown on appropriate map sheets of the *Geologic Map of California* (Index, Figure 17-1). Figure 4-9 shows a typical geologic column in the heart of the Mojave Desert.

## DESERT HIGHWAYS

In rapidly increasing numbers in the past few years, people have been turning to the desert as a place to live and for recreation. Death Valley National Monument, formerly closed for the summer, is now manned by rangers all year round. Excellent highways cross the deserts and ranges in all directions. Yet, the unwary traveler can get into serious trouble, especially in the extreme summer heat and dryness, if he lacks an adequate water supply or has an automobile breakdown; extra

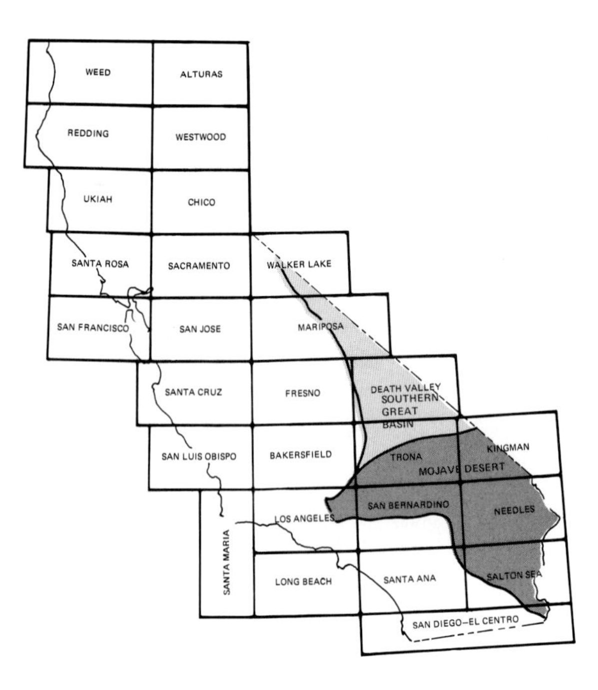


**FIGURE 17-1** *Index to map sheets* (Geologic Map of California) *covering the Mojave Desert and southern Great Basin provinces.* (*California Division of Mines and Geology.*)

water, two spare tires, and a shovel should be carried. Unpaved roads should be checked locally at any season. The best time to visit the Mojave Desert and southern Great Basin is from November 1 to May 1.

The major north-south route across the desert and Great Basin is U.S. Highway 395. Northward from San Diego, Highway 395 enters the desert at Cajon Pass and extends north from the pass for 110 miles to junction with State Highway 14 at the foot of the southern Sierra Nevada. For the next 250 miles, Highway 395 runs northward within the high, arid western marginal valleys of the Great Basin (Mono Lake and Craters are shown in Photographs 1-10 and 4-3). State Highway 14 crosses the western triangle of the Mojave Desert for 82 miles from Palmdale on the San Andreas fault to the junction with U.S. 395.

Five good east-west state highways cross the short distances east from Highway 395 across the ranges and basins to the Nevada border in the area north of Death Valley National Monument. State Highways 136 and 190 extend from Lone Pine at the north end of saline Owens Lake for 84 miles to the floor of Death Valley and for another 50 miles southeast across the Black Mountains and Amargosa Range to Death Valley Junction.

Several major highways take the long west-east route across the Mojave Desert. State Highway 58 crosses the desert from Mojave to the flourishing central-desert town of Barstow, passing the huge open-pit borate mine near Kramer. From Barstow, Interstate 15 crosses the desert en route to Las Vegas and points east, and U.S. Highway 66 extends southeastward to cross the Colorado River at Needles and continue into Arizona. Nine miles east of Barstow is the old silver-mining ghost town of Calico, now restored for tourists. Interstate 15 is joined at Baker by State Highway 127, which is the main access road to Death Valley from the south. Near the Nevada border, Interstate 15 passes the world's greatest rare-earth minerals mine at Mountain Pass. Seventy-nine miles southeast of Barstow on U.S. 66 are the very recent, perfectly preserved Amboy Crater and nearby Bristol Lake (dry) where calcium chloride brines are mined. Interstate 10, from the Los Angeles metropolitan area, crosses the

(a)

(b)

**17-2** (a) *Dry canyon cut in Pleistocene Little Lake lava flows during the last wet cycle associated with the Ice Age (see Figure 17-13). (Mary Hill photo.)* (b) *Fossil falls along the dry canyon shown in* (a). *(Mary Hill photo.)*

southeastern end of the Transverse Ranges between the Eagle and Orocopia Mountains, then traverses a series of low ranges in the southeastern Mojave Desert and crosses the Colorado River near Blythe into Arizona. State Highway 62 branches off Interstate 10 to Joshua Tree and Twentynine Palms, from which local highways extend south into Joshua Tree National Monument (Illustration 3-6).

## DEATH VALLEY AND THE MOJAVE DESERT

The name Death Valley is synonymous with extreme desert. Indeed, it is a land of extremes—the hottest, the driest, and the lowest land in the state! Geologically, it exposes some of the oldest rocks in California and some of the youngest structures in its still-active faults and warped lava flows of late Pliocene age. Death Valley and adjacent ranges provide a classic and characteristic example of Great Basin geology and landscapes.

Any route to Death Valley from the Los Angeles metropolitan area crosses at least 150 miles of desert. To experience, as completely as possible, desert and Great Basin landscapes, rock formations, and geologic history, we shall follow a circuit tour to and from Death Valley. The route of this geologic tour (Figure 17-5, index map) is from the San Andreas fault at Palmdale, at the northern base of the San Gabriel Mountains, across the narrow western wedge of the Mojave Desert, northeasterly across a series of basins and ranges, southward on the floor of Death Valley, back southwestward through Barstow in the heart of the desert, finally reaching the San Andreas fault again in Cajon Pass—about 40 miles southeast of Palmdale. This is a distance of about 450 miles (depending on what side trips the traveler takes); it encompasses a geologic time span from Precambrian to Holocene!

The circuit to and from Los Angeles can be made and the things we mention can be seen in about three days. Overnight accomodations are good at Mojave, Randsburg, Trona, Stovepipe Wells, Furnace Ranch, Furnace Creek Inn,

**17-3** *Sugarloaf Mountain, a perlite dome in the southwestern Coso Range, Inyo County. This is typical of these rhyolitic domes or plugs. Note the extremely steep flanks of the dome, formed as the viscous, glassy lava pushed up to the surface and solidified in place. To the right of the dome is a small, narrow valley which represents part of the moat that sometimes is left around the plug. The moat is probably the remnant of the crater into which the plug of rhyolitic glass was intruded. (Charles W. Chesterman photo.)*

**17-1** *Looking east across the northern end of dry Owens Lake, from the Sierra Nevada, in the foreground, across the Inyo Mountains to the distant Panamint Mountains. (Ernest S. Carter photo.)*

Shoshone, Baker, and Barstow, but cannot be counted on in season without reservations.

Maps 1 to 13 (given as figures in this chapter) show the geology in strips along the highways. The index map (Figure 17-5) and maps 1 to 13 are from Lauren A. Wright and Bennie W. Troxel, *Geologic Guide No. 1,* in California Division of Mines and Geology Bulletin 170 (out of print in 1969), as are also many of the points made in the text of this chapter. Detailed regional geology is shown, in color, on the Los Angeles, Bakersfield, Trona, Death Valley, and San Bernardino map sheets of the *Geologic Map of California.*

Map 1 (Figure 17-6) shows the geology of an interesting route across the San Gabriel Mountains in the Transverse Ranges. The route shown branches off State Highway 2 as the Angeles Forest Highway and crosses the mountains to Vincent and Palmdale.

## SAN ANDREAS FAULT TO SOLEDAD MOUNTAIN

Entering State Highway 14 at Vincent, the road crosses low Soledad Pass between the San Gabriel and Sierra Pelona Mountains and into the San

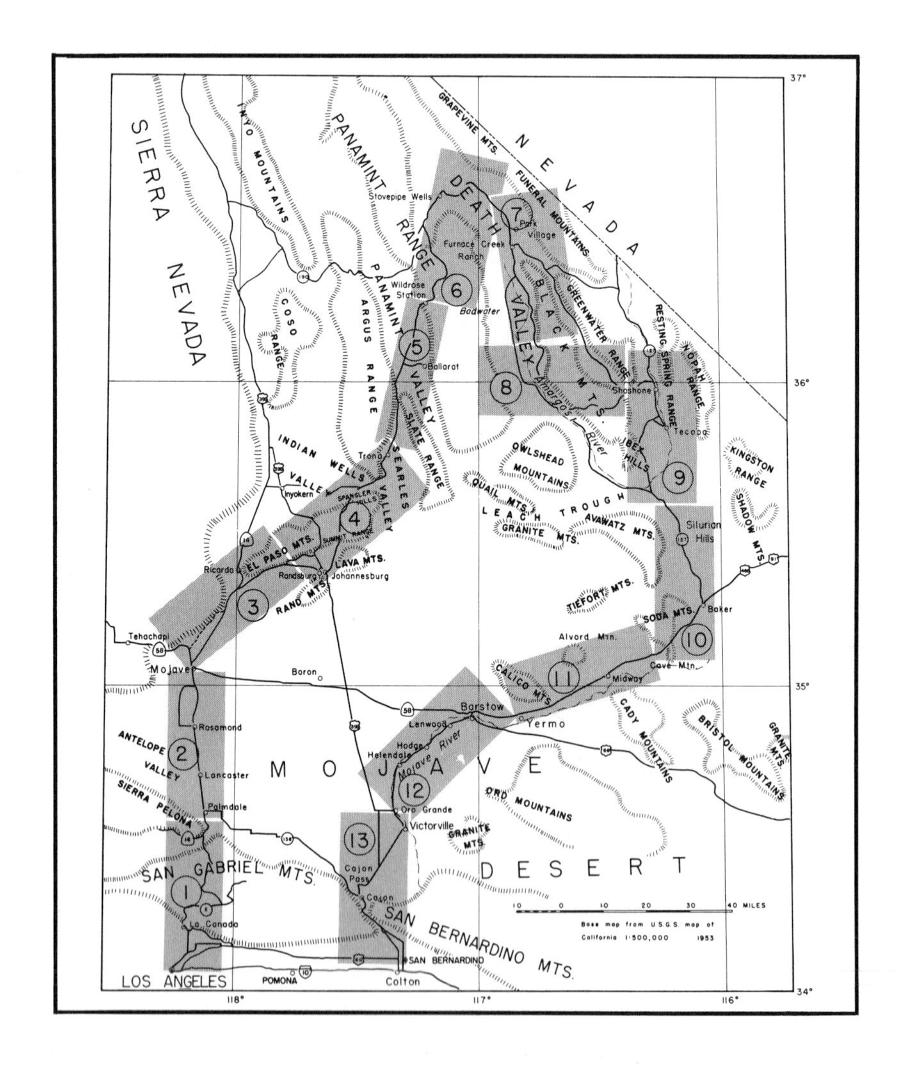


**FIGURE 17-5** *Index map showing highways and strip maps along the route through the Mojave Desert and southern Great Basin described in this chapter.*

Andreas fault zone and beyond it into the desert past Soledad Mountain (Map 2, Figure 17-8). Looking northwest or southeast, one can see many miles of strikingly straight rift valleys, which mark the 3-mile-wide fault zone (Photograph 17-7). Secondary roads follow the fault zone in both directions. The *p€* on map 2 is Pelona Schist of uncertain age, which is sliced into the fault zone in many places. Cretaceous granitic rocks (*KJgr*) and coarse late Tertiary land-laid sand and gravel (*Tp*) appear in the fault zone also. Northwest of the highway near Soledad Pass are dark volcanic rocks of the Vasquez Formation. Far to the southeast, the San Andreas fault is marked by a notch where the fault crosses into the northern flank of the San Gabriel Mountains. A second notch to the right is on the trace of the San Jacinto fault, an active source of earthquakes from San Bernardino southeast. A few hundred feet southeast of Highway 14, from a point 2 miles south of Palmdale, the Palmdale Reservoir has been built on the fault.

For 20 miles north to the Rosamond Hills the highway runs very straight across the flat, level, alluvial Antelope Valley (Figure 17-8). In Rosamond Dry Lake, an exploratory well was drilled years ago into lake beds and coarse alluvium and bottomed at 5,500 feet in the same material! The Rosamond Hills consist of granitic rocks (*Jgr*) flanked by red beds and volcanics of the Miocene Rosamond Formation (*Tm*).

Soledad Mountain appears prominently out of the desert floor a few miles north of the Rosamond Hills. It consists of rhyolite tuff and breccia, intruded by rhyolite volcanic plugs. The famous old Golden Queen Mine, which produced gold from the rhyolite that makes up the mountain, has been idle in recent years.

## THE GARLOCK FAULT ZONE TO CANTIL VALLEY

A couple of miles north of Mojave, State Highway 58 merges from the left and State 14 continues northeast and follows close to the Garlock fault zone. State 58 follows southwestward in that fault zone (Bakersfield map sheet of *Geologic Map*).

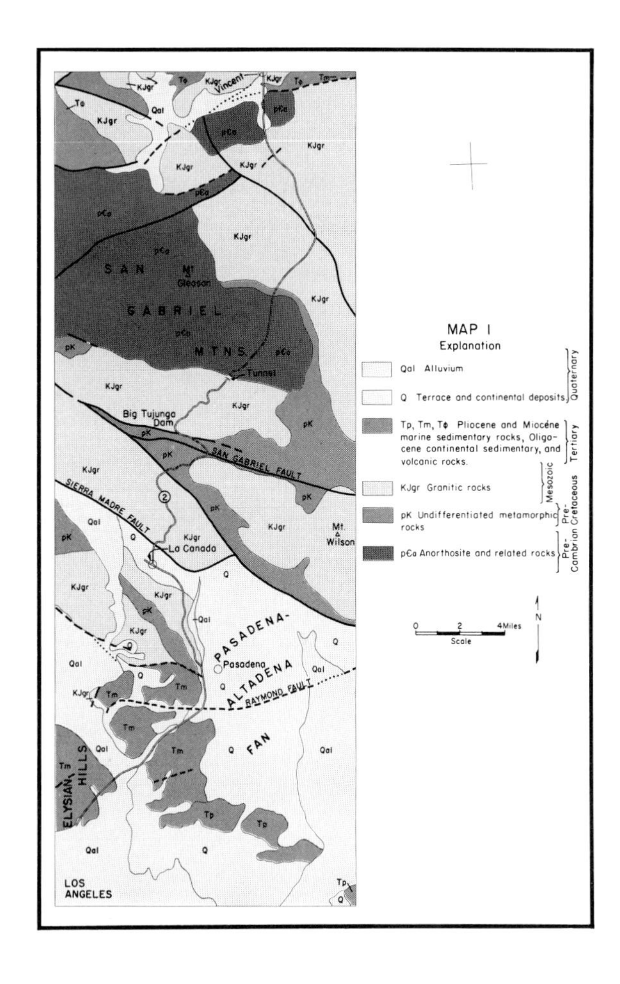


**FIGURE 17-6** *Map 1 showing geology along State Highway 2 from Los Angeles to Vincent, across the San Gabriel Mountains.*

**17-7** *Looking west across the San Andreas fault zone near Palmdale in southern California. The northwesterly striking trace of the fault lies in the rift valley extending across the picture from lower left to upper right corner. Note the grove of trees where an extra water supply comes to the surface in the fault zone. (John S. Shelton photo.)*

The Garlock fault is a major northeast-striking fault that intersects the San Andreas fault at Frazier Mountain. It generally forms the boundary between the Mojave Desert and Great Basin provinces. The segment of the trip from the Garlock fault to Cantil Valley can be followed on Map 3 (Figure 17-9).

From Highway 58, a short trip along the Los Angeles Aqueduct Road reveals some very interesting features of the Garlock fault zone. A 200-foot-high scarp, facing southeast, offsets Quaternary fanglomerate, showing extremely recent vertical movement. Characteristically, the Garlock fault is up on the northwest and the northwestern block moves westerly; that is, movement is *left lateral* in contrast to the right-lateral San Andreas fault. Here, the fault zone marks the front of the Sierra Nevada. About 18 miles northeast of the junction of Highways 58 and 14, Highway 14 crosses the Garlock fault, which cuts off the southern end of the Sierra Nevada frontal fault.

At Cantil Valley junction, Highway 14 turns more directly north and crosses two main branches of the Garlock fault: one in the alluvium at the base of El Paso Mountains and the other (El Paso fault) forming the contact between Quaternary sediments and exposures of granite. Highway 14 extends north about 6 miles into the beautiful, highly colored beds of the Pliocene Ricardo Formation in Red Rock Canyon. This consists of red conglomerate, siltstone, sandstone, tuff beds, and basalt flows (Photograph 17-10 and Trona map sheet of the *Geologic Map*). Red Rock Canyon has been the site of many western movies made over the years.

Return to the Cantil Valley junction and continue northeastward on a paved secondary road in the Garlock fault zone. Cantil Valley is a downdropped block (graben) between El Paso Mountains on the northwest and the Rand Mountains on the southeast. Koehn Lake is the undrained bottom of the graben. The Rand schist in the Rand Mountains is probably Precambrian in age. The Garlock fault has offset alluvium near Koehn Lake and recent surveys suggest that left-lateral slippage (creep) is now going on.

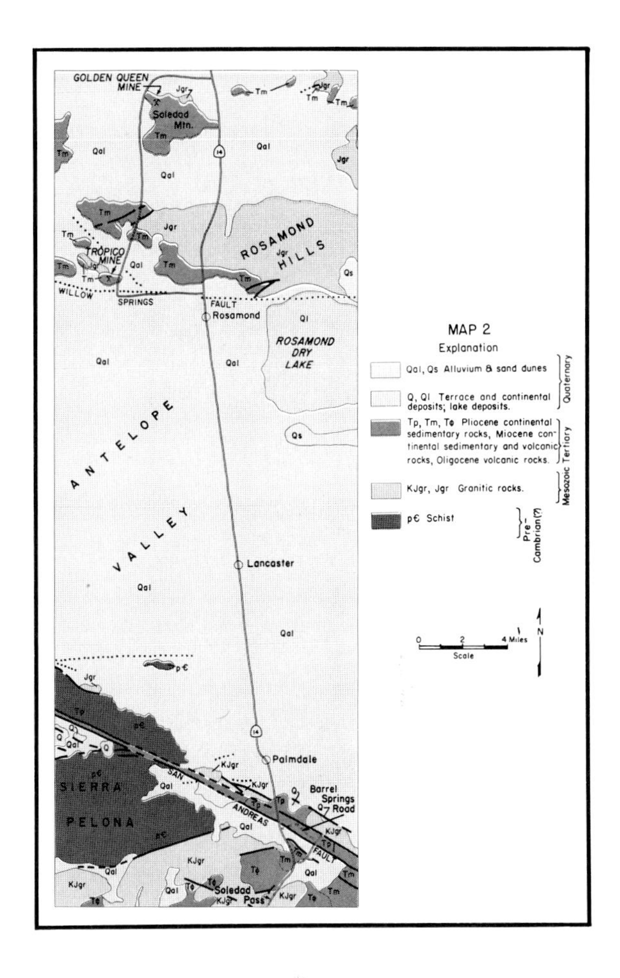


**FIGURE 17-8** *Map 2 showing geology along State Highway 14 from the San Andreas fault at Palmdale across Antelope Valley to Soledad Mountain.*

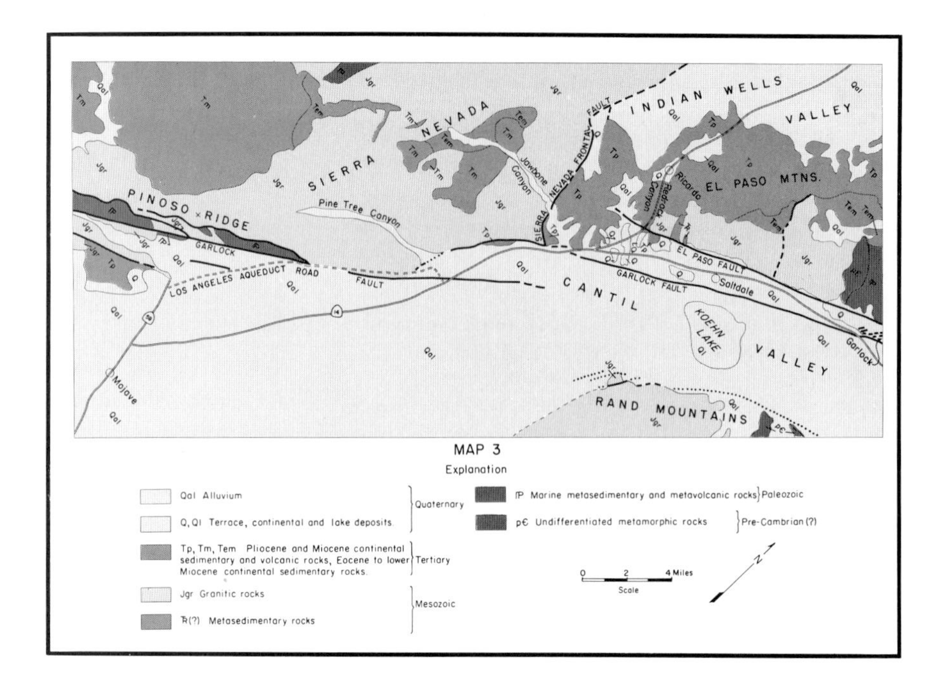


**FIGURE 17-9** *Map 3 showing geology along State Highway 14 from Mojave along the Garlock fault zone.*

**17-10** *Red Rock Canyon on Highway 14 (visible in photo), eastern Kern County. The gently dipping stratified rocks are highly colored sandstone, conglomerate, and mudstone of the nonmarine Pliocene Ricardo Formation. (Mary Hill photo.)*

## EL PASO MOUNTAINS TO SEARLES LAKE

Map 4 (Figure 17-12) shows geology along the highways from El Paso Mountains to Searles Lake. In the eastern part of El Paso Mountains, 35,000 feet of Paleozoic chert, limestone, shale, and lesser volcanic rocks (*IP*—the Garlock Series) stand on end and strike north. Permian fusulina in the upper half of the section indicate its age, but strata in the lower half of the section may be much older. Paleocene nonmarine rocks (*Tem*) exposed in these mountains have yielded the oldest mammalian fossils in California.

A secondary road continues on to Randsburg, Johannesburg, and Atolia (the latter two on U.S. Highway 395)—all old mining towns. Randsburg yielded at least $12 million in gold during the period 1895 to 1918. The gold came from quartz veins in Mesozoic granitic rocks. Nearby Red Mountain is famous for silver, and Atolia for exceptionally rich tungsten deposits (scheelite) in quartz monzonite.

Turn off U.S. Highway 395 to the northeast, at Johannesburg, onto a secondary road toward Searles Lake. On the northern side of the low Summit Hills, the road crosses the Garlock fault, which makes the sharp contact between granite and valley alluvium. The road then crosses the low Spangler Hills, which lie at the south end of the Argus Range—one of the north-trending Great Basin ranges.

Approaching the town of Trona, the broad, white, salt-encrusted floor of dry Searles Lake lies on the right. During the several stages of glaciation in the Pleistocene Epoch the cool, humid climates caused freshwater lakes to form in the basins of the Great Basin and Mojave Desert (Figure 17-13). Death Valley—the lowest and largest basin—was the terminal lake for a series of drainage systems including the Owens, Amargosa, and Mojave Rivers and chains of freshwater lakes along their courses.

Searles Lake was the terminus during the last glacial epoch when it was probably a freshwater lake close to 400 feet deep and 16 miles long. Increasingly dry climate caused desiccation; and the lake today consists of a huge body of crystallized salts of sodium, boron, and potassium, with a dry surface and a saline-mud mush beneath. The lake is a major commercial source of

**17-11** *Quarry in stratified Quaternary perlite deposits in the El Paso Mountains of Eastern Kern County. (Charles W. Chesterman photo.)*

borates, lithium salts, potash, soda, trona (sodium carbonate-bicarbonate), common salt, and other salts. Patches of the Pleistocene lake deposits are visible above the present valley floor, as are also well-preserved shore-line terraces. A wonderful group of 100-foot high, calcareous tufa pinnacles can be seen in the lake basin 5 miles south of the highway. Searles Lake deposits yield 30 percent of the annual production of borates in California and all the potash produced in the state.

## SEARLES LAKE TO THE CREST OF THE PANAMINT RANGE

From Searles Lake to the Panamint Range the geologic features may be followed on Map 5 (Figure 17-15). About 12 miles north of Searles Lake, the road crosses over a low pass between the Argus and Slate Ranges, typical Great Basin ranges. In the pass itself are exposures of stratified Paleozoic rocks intruded by granite. Straight ahead is a marvelous panorama across the Panamint Valley to the great fault scarp that marks the western base of the Panamint Mountains. This is an active fault scarp, along which the valley was downdropped and the range elevated in very late geologic time. The low benches seen are Pleistocene terraces and shore lines representing fluctuating levels of the freshwater lake that once occupied Panamint Valley.

Exposed along the western side of the Panamint Range are 8,000 feet of Cambrian, Ordovician, Silurian, and Devonian strata (Illustration 7-2). These strata lie on Precambrian metasedimentary rocks seen in the lower part of the range.

Evidence of the recency of movement in the Panamint fault zone is a mile-wide, downfaulted trench in alluvium—the Wildrose graben at the mouth of Wildrose Canyon. The road crosses the graben with its two bounding faults 200 feet high. A couple of miles beyond the graben are

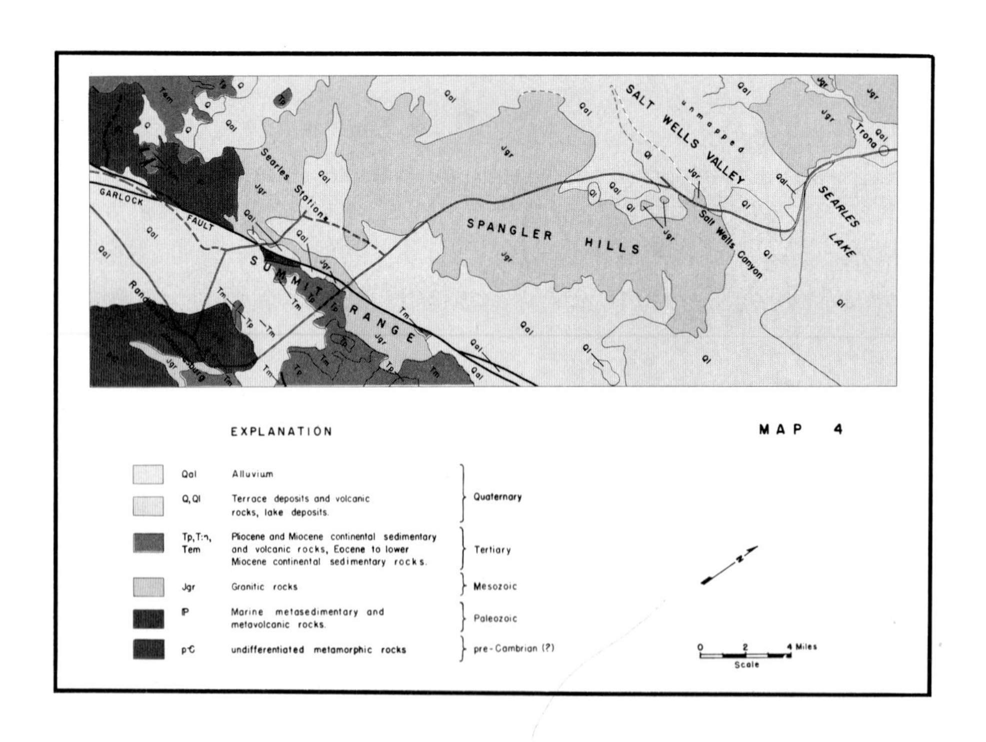


**FIGURE 17-12** *Map 4 showing geology along paved secondary roads from Randsburg to Trona on Searles Lake.*

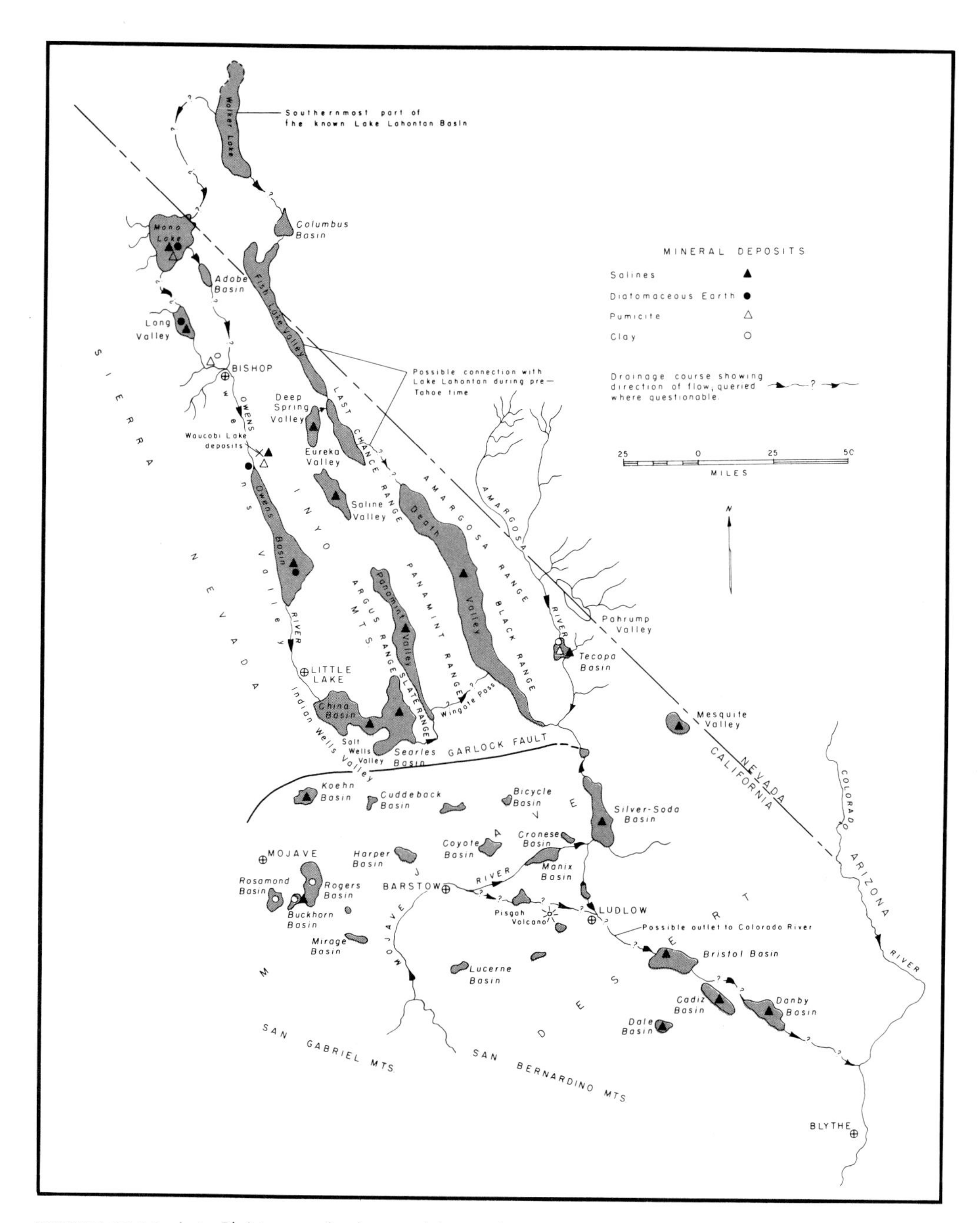


**FIGURE 17-13** *Late Pleistocene freshwater lakes and rivers in the Great Basin and Mojave Desert. (California Division of Mines and Geology,* Mineral Information Service, *April, 1961.)*

**17-14** *Looking east at Paleozoic strata in the Slate Range (see Figure 17-15). Slightly tilted shore lines of late Pleistocene Lake Searles are near center of the photo. Just below the shore lines is a curved fault scarp so recently formed that it has cut deposits of the alluvial fans. (B. W. Troxel photo.)*

exposures of tilted beds of a coarse fanglomerate—the late Tertiary Nova Formation (*T*)—which is seen lying unconformably on folded older Tertiary gravels and Precambrian rocks.

## WILDROSE TO THE FLOOR OF DEATH VALLEY

Map 6 (Figure 17-16) shows roadside geology from Wildrose Canyon to Death Valley. Up Wildrose Canyon the road passes through earlier Precambrian (Archean) granite gneisses, overlain by later Precambrian (Algonkian) Noonday dolomite (Photograph 6-10). Six miles past Wildrose ranger station (entrance to Death Valley National Monument) are several beehive charcoal kilns that were built in the 1880s to make charcoal for a smelter at the Modoc lead-silver mine in the Argus Range. Strata of the Noonday dolomite cross the canyon at the kilns. The flat here offers the first spectacular view of Death Valley and the Funeral and Black Mountains across the valley (Photograph 3-12).

North of Emigrant Pass, the road crosses Harrisburg Flat, an old, uplifted erosion surface. A well-marked turnoff leads 8 miles east to Aguereberry Point, noted for its spectacular view of Death Valley. This great structural and topographic, fault-bounded trough stretches north-south for 150 miles and east-west for only about 10. Stratified rocks on the road are part of the lower Cambrian section. Looking eastward from the point, successively higher formations in the Paleozoic section are seen.

Below Emigrant Spring, after the road joins State Highway 190, the steep western face of Tucki Mountain exposes fanglomerate beds of the Nova Formation and cliff-forming interlayered basalt flows. Tucki Mountain consists generally of east-dipping later Precambrian and Paleozoic strata that form an unusually complete section.

Stovepipe Wells hotel and resort were named after an early-day well, concealed by sand but kept open by a stovepipe driven through the sand. Downhill from Stovepipe Wells and extending

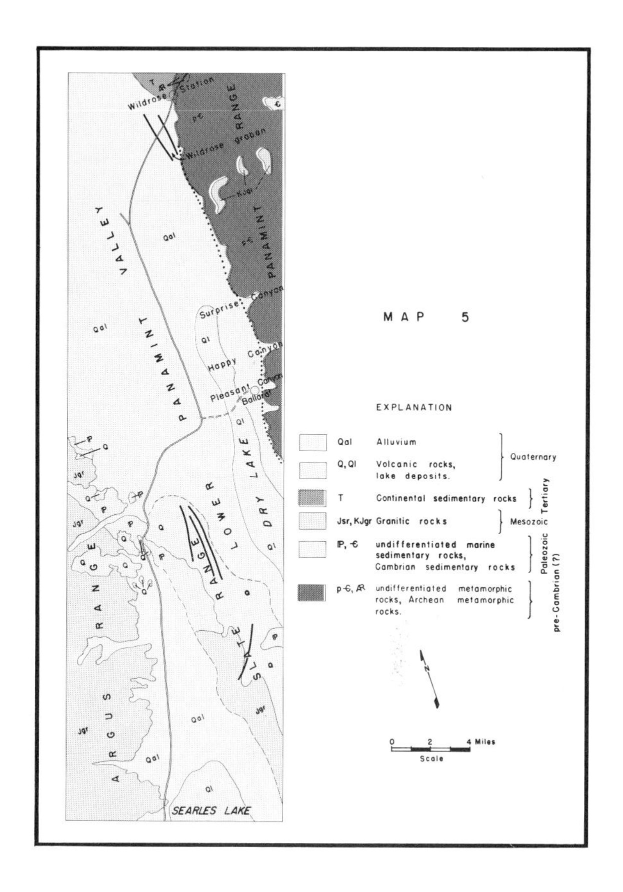


**FIGURE 17-15** *Map 5 showing geology along paved secondary road from Searles Lake across the Argus and Slate Ranges to Wildrose Station in the Panamint Range.*

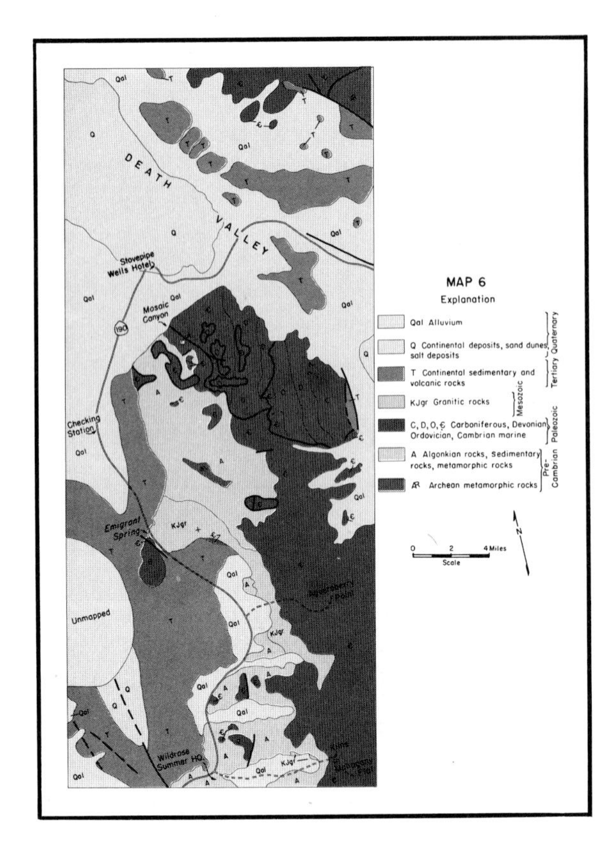


**FIGURE 17-16** *Map 6 showing geology along secondary road and State Highway 190 into Death Valley by way of Stovepipe Wells.*

onto the floor of Death Valley are the famous, much-photographed Death Valley sand dunes. A marvelous view of the valley may be seen looking eastward from Stovepipe Wells; southward are seen the faulted later Precambrian, Cambrian, and Ordovician strata on Tucki Mountain. An interesting side trip involves walking into Mosaic Canyon where water-polished yellowish-brown dolomite is exposed on the canyon floor. On Highway 190, a couple of miles east of Stovepipe Wells, Cambrian and Ordovician formations can be seen in the face of Tucki Mountain.

## THE FLOOR OF DEATH VALLEY

Geologic features may be followed across the floor of Death Valley on Map 7 (Figure 17-18). Highway 190 extends across the dunes and salt-encrusted valley floor to Furnace Creek, then south along the base of the extremely steep western front of the Funeral and Black Mountains. Branch roads follow the valley floor north to Ubehebe Craters (Photograph 4-2) and famous Scotty's Castle. In the Furnace Creek area are extensive exposures of the Pliocene Furnace Creek Formation and slightly older, highly colored, deformed Tertiary lake beds and gravels (*TQ* in Map 7 and Photograph 17-19). Very recent faults mark the base of the mountains and offset alluvial fans on the valley floor. At Furnace Creek Ranch—a resort area—the old 1882 Harmony borax mill and a steam tractor are on display. Borates were mined from the dry lake beds on the floor of Death Valley before 1830. The white saline crust on the valley floor is made up of common salt and mixtures of other salts, including borates. Alternating layers of salt and mud exist to a depth of more than 1,000 feet. Monument headquarters is about 3 miles north of Furnace Creek Ranch.

A side trip 24 miles southeast of Furnace Creek Inn leads to the top of the Black Mountains and Dante's View. From this vantage point, a marvelous view encompassing Death Valley north to south and due west to the Panamint Mountains and Telescope Peak can be seen (Photograph 1-11). More than 5,000 feet below is Badwater, which is 282 feet below sea level.

**17-17** *Quite similar to the dunes near Stovepipe Wells, these are sand dunes in Eureka Valley—north of Death Valley. The snow-capped high Sierra Nevada is seen on the skyline. (Sarah Ann Davis photo.)*

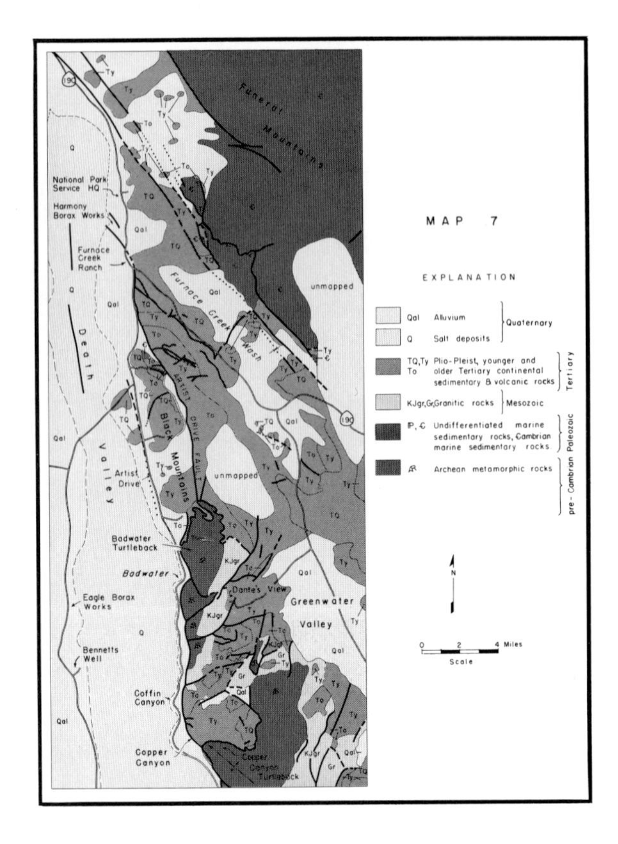


**FIGURE 17-18** *Map 7 showing geology along State Highway 190 on the floor of Death Valley and the western flank of the Black Mountains.*

At Furnace Creek, a paved secondary road diverges from Highway 190 and continues due south along the eastern side of the floor of Death Valley. For the first 15 miles south of Furnace Creek, the road cuts through and lies at the base of highly colored continental sedimentary beds and volcanics of various ages of the Tertiary Period, particularly clearly seen along Artist Drive and in Golden Canyon. Then, for several miles, the road lies at the steeply faulted base of Precambrian rocks of the Black Mountains. This western flank of the mountains is cut by numerous steep-walled, narrow canyons with nearly perfect alluvial fans at their mouths. However, these fans are very much smaller than the huge fans across the valley on the eastern side of the Panamint Range, testimony to the more youthful age of the Black Mountains fault scarp. Many of the branch faults cut the alluvial fans. At Copper Canyon and Coffin Canyon the colored Tertiary sediments and volcanics are again exposed.

## EASTWARD ACROSS THE BLACK MOUNTAINS

Map 8 (Figure 17-20) continues the geologic trip across the Black Mountains. The road extends southward around Mormon Point and then climbs east across the Black Mountains by way of Jubilee and Salsberry Passes. At Mormon Point and Shoreline Butte the shore lines of Pleistocene Lake Manly may be clearly seen on the ancient Precambrian rocks forming the heart of the mountains and cut into the Tertiary rocks. Remnants of old shore lines show that Lake Manly was at least 600 feet deep and 150 miles long.

In Jubilee Pass are exposed masses of chaotically mixed rock fragments that form a highly colored breccia of Tertiary age lying in complexly faulted basins that have been downdropped into older Precambrian gneisses. Three miles east of Jubilee Pass, the great Amargosa thrust fault is exposed on the south side of the road. Blocks of later Precambrian and Cambrian rock formations have been thrust over earlier Precambrian gneiss. Tertiary volcanic rocks are exposed in Salsberry Pass.

## SHOSHONE TO BAKER AND THE SODA MOUNTAINS

This segment of the trip may be followed on Maps 9 & 10 (Figures 17-21 and 17-22). East of Salsberry Pass, the highway drops down into the little desert town of Shoshone between the Dublin Hills on the west and the Resting Spring Range on the east. The valley floor here is covered with silt, mud, and ash beds that were deposited in Pleistocene Lake Tecopa (Figure 17-13). The Resting Springs Range is an eastward tilted fault block of Cambrian strata overlain by Tertiary

**17-19** *Looking across deeply dissected Tertiary sediments in the foreground and Death Valley in the middle ground toward the Panamint Mountains. Highest peak on the horizon is 11,000 feet in elevation; the floor of Death Valley here is more than 200 feet below sea level. (Mary Hill photo.)*

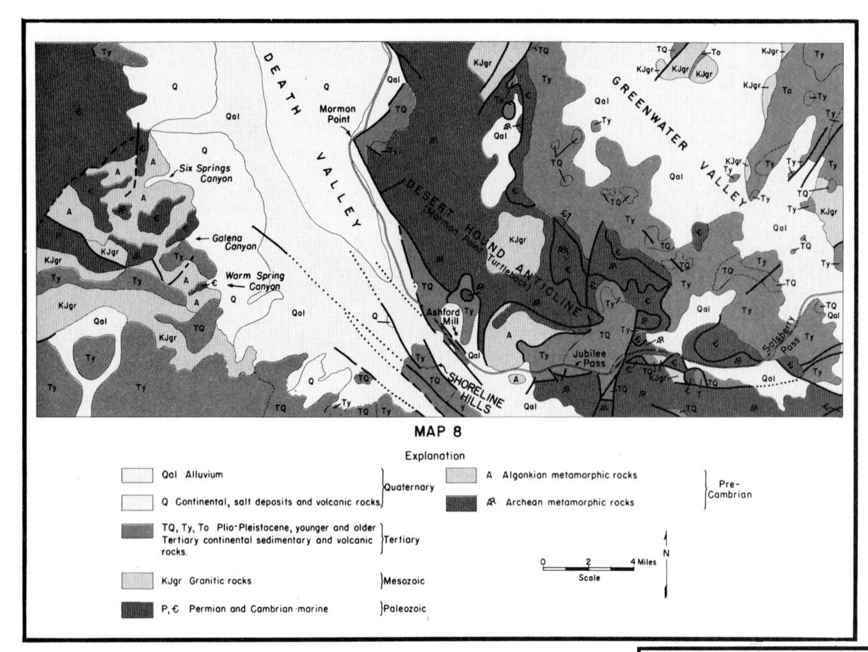


**FIGURE 17-20** *Map 8 showing geology in southern Death Valley.*

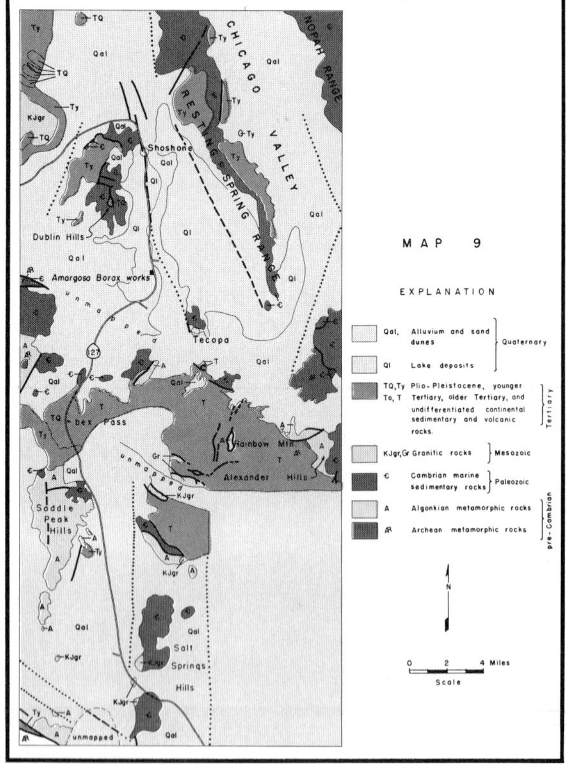


**FIGURE 17-21** *Map 9 showing geology along State Highway 127 from Shoshone to the Salt Spring Hills.*

sediments and volcanics. The strata and structures of the Dublin Hills are similar to those in the Resting Spring Range. State Highway 127 continues south for 56 miles from Shoshone to Baker.

In Ibex Pass are excellent exposures of Tertiary "chaos," consisting of huge slabs of later Precambrian and Cambrian strata. Just south of the pass, the chaos is made up of fragments of granitic rocks. The Saddle Peak Hills consist of later Precambrian (*A* on map 9) Pahrump Group rocks and Noonday dolomite. The Avawatz Mountains appear on the high south skyline to the right of the highway. About 13 miles south of Ibex Pass, Highway 127 turns and passes through a gap in the Salt Springs Hills—the gap followed by the late Pleistocene Mojave River when it once flowed northward into Lake Manly, which occupied Death Valley in the late Pleistocene.

Eight miles south of the Salt Springs Hills gap, Silurian Lake and the Silurian Hills appear just east (left) of the highway (map 10, Figure 17-22). The Silurian Hills are an easily seen, excellent example of complex faulted structures found in the Great Basin. Earlier Precambrian, later Precambrian, Paleozoic strata, and Mesozoic granitic rocks are all exposed. Chaotic later Precambrian rocks of the Pahrump Group lie unconformably on earlier Precambrian gneisses. Thrust-fault plates of steeply dipping Paleozoic dolomite and limestone have been shoved over the Precambrian rocks. The thrusting is older than the Mesozoic granite, for granite has been intruded across the thrust faults.

Ten miles south of Silurian Lake is Silver Lake, and beyond it is Soda Lake—others in the string of remnants of the large freshwater lakes of the late Pleistocene (Silver-Soda Basin on Figure 17-13). Silver Lake has occasionally been reached by the Mojave River in its greatest flood periods in historic times. The Pleistocene lake that filled the Silver-Soda Basin has been called Lake Mojave. The low hills west of the highway are foothills of the Soda Mountains, seen on the right from Silver Lake, Baker (junction with Interstate Highway 15), and for 17 miles along Interstate 15. The Soda

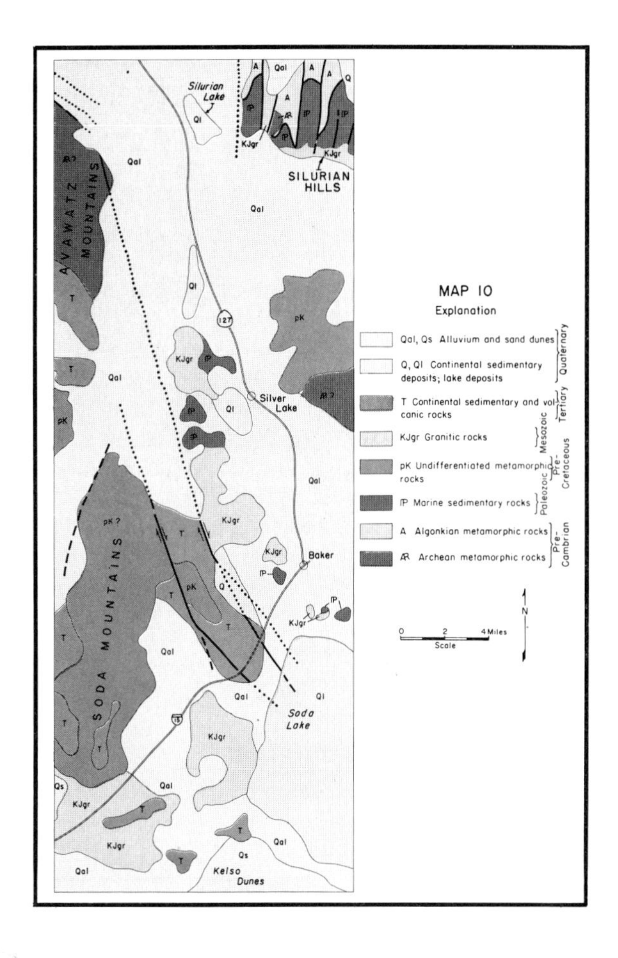


**FIGURE 17-22** *Map 10 showing geology along State Highway 127 to Baker, thence along Interstate Highway 15 across the Soda Mountains.*

Mountains consist mostly of a complex of Jurassic-Triassic metavolcanic and metasedimentary rocks, which occur as roof pendants in the more extensive Late Mesozoic granitic rocks. The mountains are flanked by Tertiary volcanic and sedimentary strata. Ancient thrust faulting and younger normal faulting have contributed to structural complexity.

## SODA LAKE TO THE SAN ANDREAS FAULT ZONE

The hills and ranges that border Interstate 15 from Baker to the San Andreas fault zone at Cajon Pass (Maps 11, 12, and 13—Figures 17-23, 17-24, and 17-25) are generally made up of Late Mesozoic granitic rocks, which contain roof pendants of Paleozoic and later Precambrian formations. Tertiary and Quaternary sedimentary and volcanic rocks lie unconformably on the older rock formations or are faulted against them. Tertiary formations are also folded.

**FIGURE 17-23** *Map 11 showing geology along Interstate Highway 15 from Cave Mountain to the Calico Mountains.*

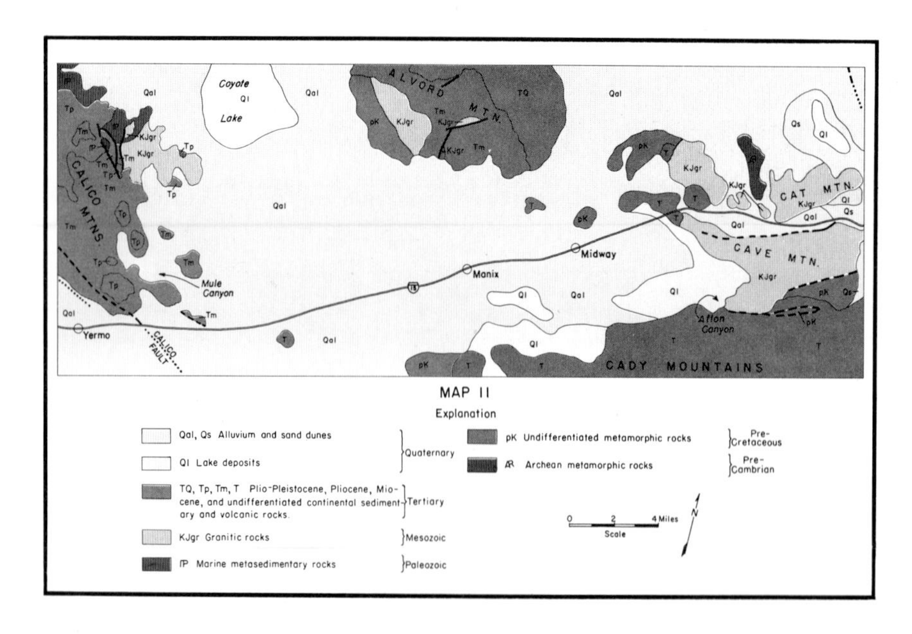


The Kelso dunes, up to 500 feet high, may be seen south of Soda Lake. They have been formed by prevailing west winds across the desert picking up sand from the usually dry bed of the Mojave River. About 15 miles southwest of Baker, granitic Cat Mountain is on the right (map 11, Figure 17-23). It gets its name from a large cat-shaped sand dune high on its eastern slope.

A few miles beyond Cat Mountain is Manix Basin, covered by sediments deposited in the floor of the former 200-square-mile Lake Manix (Figure 17-13). Fossil vertebrates show that these lake beds were formed in early Pleistocene to late Pleistocene time. The beds were displaced several inches in a horizontal, left-lateral direction along the Manix fault in 1947, during a minor earthquake.

From a few miles beyond Manix to Barstow, the Calico Mountains are seen north of the highway. They consist of brilliantly colored, folded, Tertiary volcanic and sedimentary strata which lie unconformably on Paleozoic sedimentary and Mesozoic granitic rocks. Very rich silver deposits were mined in the Calico Mountains from 1882 to 1896. For the years 1884 to 1897, colemanite beds in the Tertiary formations were the world's major source of borates. The southwestern side of the

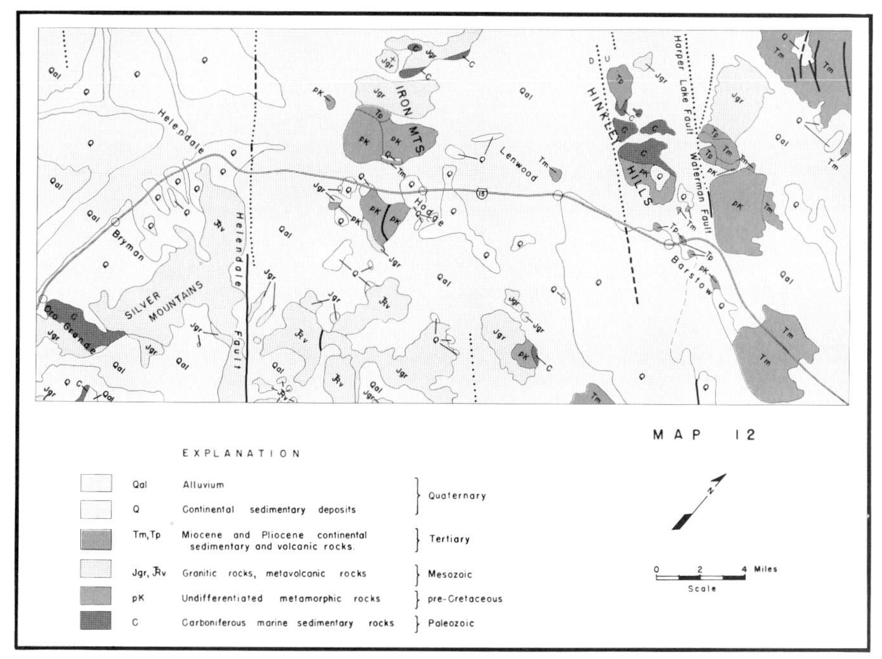


**FIGURE 17-24** *Map 12 showing geology along Interstate Highway 15 from Barstow to Oro Grande.*

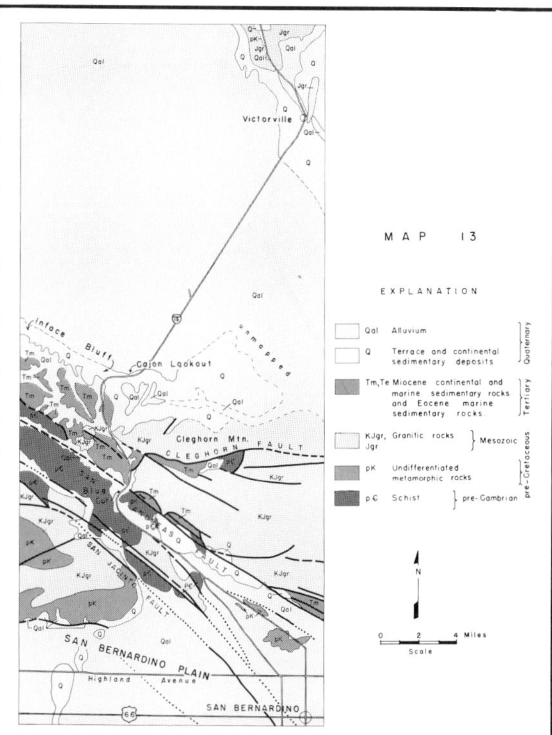


**FIGURE 17-25** *Map 13 showing geology along Interstate Highway 15 from Victorville to the San Andreas fault zone in Cajon Pass.*

Calico Mountains is marked by the northwest-striking Calico fault, which extends for a length of about 100 miles. It is not readily seen from the highway but is one of four similar major faults in a zone 35 miles wide, clearly shown on the San Bernardino map sheet of the *Geologic Map.*

From Barstow to Victorville for 31 miles (Figures 17-24, 17-25), Interstate 15 follows a chord across a broad loop of the Mojave River; the older U.S. Highway 66 follows the course of the river itself. The desert along this stretch of highway consists of groups of isolated hills that are "islands" of complex associations of Mesozoic granite, late Paleozoic strata, and Jura-Triassic metavolcanic rocks in a "sea" of alluvium. The northwest-striking faults have had very recent movement, at least partly right-lateral; and Pleistocene alluvium has been gently domed between them. In the Victorville–Oro Grande district, limestone is quarried from the Late Paleozoic Oro Grande Series and the overlying Permian Fairview Valley Formation and is processed for cement at a nearby plant.

From Victorville to the San Andreas fault zone in Cajon Pass (Figure 17-25), Interstate 15 follows a straight course for 20 miles across a nearly level, huge alluvial fan made up of coarse fragments derived from the San Gabriel and San Bernardino Mountains. A broad sweeping panorama of the desert can be seen from Cajon Summit (elevation 4,301 feet).

As the highway approaches the summit, magnificent outcrops of Tertiary and Quaternary sandstone and conglomerate may be seen. The upper Pleistocene Inface gravel seen in deep highway cuts dips 25 to 30 degrees north; it lies on top of thick Pleistocene strata of land-laid gravel and clay, which, in turn, lie unconformably on granite and the upper Miocene Cajon Formation. The Cajon Formation consists of nonmarine sandstone and conglomerate forming huge, steeply dipping, slablike outcrops, which extend for miles northwest of the highway. Small areas of the early Miocene marine Vaqueros Formation and Paleocene sandstone and conglomerate are exposed along the highway north of the San Andreas fault in the Blue Cut.

Having reached the Blue Cut, we have now completed the desert circuit! Interstate 15 follows fractured and brecciated Pelona schist for 12 miles within the great San Andreas fault zone. Offset Cajon Creek, a series of parallel ridges, and fault scarps in Late Quaternary terrace deposits show the recency of fault movements. Surface displacement no doubt extended along this segment of the fault during the great Fort Tejon earthquake in 1857.

*The active San Jacinto fault lies at the southwestern base of the San Jacinto Mountains block northeast of Hemet.*

# 18 rocks and structures in the peninsular ranges and salton trough

■ Geography and geologic setting of the Peninsular Ranges and Salton Trough were discussed briefly in Chapter 1. Outlines of these natural provinces and names of the principal ranges, basins, and rivers may be seen in Figure 16-1. Major rock units and fault zones are indicated on the 1:2,500,000-scale colored geologic map of California (available from the U.S. Geological Survey or the California Division of Mines and Geology). For more detailed geology, see the Salton Sea, San Diego-El Centro, and Santa Ana map sheets of the *Geologic Map of California* (index, Figure 18-1).

Generally bounded on the north and east by the Transverse Ranges and the Chocolate Mountains of the Mojave Desert, the Peninsular Ranges extend more than 30 miles offshore to form the largely submerged continental borderland. Principal island eminences of the borderland are Santa Catalina, San Clemente (Illustration 10-4), Santa Barbara, and San Nicolas. The San Andreas fault zone sharply bounds the Salton Trough on the east and northeast; the Raymond-Malibu fault zone, which extends far westward out to sea, possibly into the deep-sea Murray fracture zone, forms an equally sharp northern boundary to the Peninsular Ranges province.

Resemblance of the Peninsular Ranges province to the Sierra Nevada is striking. Both are northwest-trending mountain blocks, several hundred miles long, elevated on the east along faults and tilted gently westward. Like the Sierra, geologic history of the Peninsular Ranges began with the thick accumulation of Paleozoic marine sedimentary rocks and continued with marine sedimentary and volcanic deposition into Jurassic time. A few fossils of probable late Paleozoic age and some Jurassic fossils have been found, but the long pregranitic history is not so well known as that of the Sierra Nevada.

A second major episode in geologic history of the province was a Mid-Cretaceous period of mountain building during which the older rocks were intensely deformed and metamorphosed. Granitic rocks of the southern California batholith represent this episode.

Postbatholithic rocks show intermittent shallow-marine and continental deposition and deep erosion of the batholith and remnants of its roof

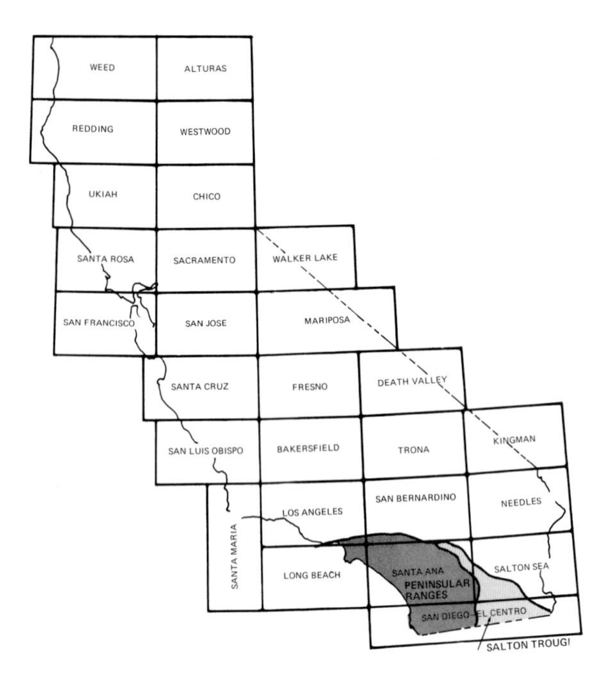


**FIGURE 18-1** *Index of map sheets* (Geologic Map of California) *covering the Peninsular Ranges and Salton Trough.* (*California Division of Mines and Geology.*)

of older rocks. Cenozoic history includes minor volcanic activity. Uplift, warping, and faulting, particularly active in late Tertiary and Quaternary time, formed the rugged ranges seen today.

## SOUTHERN CALIFORNIA BATHOLITH

The great mass of granitic rocks that constitutes the heart of the Peninsular Ranges province dominates rock outcrops from the Santa Ana Mountains in the north for more than 1,000 miles to the southern tip of Baja California (Figure 18-2). The granitic province is only about 70 miles wide in southern California and extends northwesterly for about twice that distance within the state. As in all batholiths—including the Sierra Nevada and Klamath Mountains—the granitic rocks include many types and were intruded in a series of pulses as more or less discrete plutons. Rock types include gabbro (the "black granite" used as dimension stone in San Diego County), diorite, quartz diorite, quartz monzonite, and granite (Table 2-2). Quartz diorite is by far the most abundant. Radiometric ages show dates of 90 million to 100 million years ago for consolidation of the batholith; some older and younger dates have been measured but most are in the Cretaceous age range.

As in the Sierra Nevada and the Klamath Mountains, mineralization accompanied the intrusions. Gold, copper, nickel, silver, and tungsten have all been mined in the granitic and older rocks, but the ores have not been rich. The granitic rocks themselves have been more valuable, used as dimension stone and crushed stone.

Most famous of the mineral deposits are the gem minerals of a 25-mile-long belt in San Diego County. There the granitic rocks are cut by a series of elongate pegmatite dikes from a fraction of an inch to 100 feet in thickness. Besides coarse crystals of quartz, feldspar, and mica, the pegmatites have yielded pink, black, and green tourmaline; pink spodumene; beryl; topaz; and garnet. Three principal districts—Pala, Rincon, and Mesa Grande—have produced a total of $2 million in gems. The Ramona district has been a source of topaz. Tourmaline has been the major mineral in the pegmatities of San Diego County. These three districts, the principal mines, and references to the state's publications are shown in Figure 18-4.

Rocks of the southern California batholith are easily seen from any of the numerous fine highways that crisscross the Peninsular Ranges. State Highway 76 (Santa Ana map sheet of *Geologic Map of California*) is one of the best; it extends from the coast at Oceanside eastward across the gem belt—with side roads to Pala and Mesa Grande—past Lake Henshaw in the great Elsinore fault zone and joins State Highway 78 to continue through the old Julian nickel district. It is the principal access road to the huge Anza Borrego Desert State Park. The state park preserves and protects several hundred square miles of desert mountains and valleys in the Peninsular Ranges province. State 78 continues eastward through Borrego Valley—across the very active San Jacinto fault zone—and into the southern end of the Salton Trough. Interstate 8 crosses the Laguna and Jacumba Mountains of the Peninsular Ranges (summit on this highway is at elevation 4,118 feet) and descends into the Salton Trough near the southern border of the state.

The more northerly State Highway 74 is one of the most interesting. Starting at Capistrano Beach, it crosses the Santa Ana Mountains, first through several miles of Tertiary formations and then, successively, marine Upper Cretaceous rocks, Jurassic sedimentary and volcanic formations, and the granitic batholith. The road descends into the rift valley of the Elsinore fault zone—in which Lake Elsinore lies in a graben—and goes eastward across more batholithic and older rocks into the San Jacinto rift valley through Hemet. A few miles off this highway at Idyllwild is Mount San Jacinto State Park on the western slope of the San Jacinto Mountains. The road follows the San Jacinto fault zone for 25 miles and then crosses between the San Jacinto and Santa Rosa Mountains over a 5,000-foot pass and drops down the very steep eastern escarpment of the Peninsular Ranges into Coachella Valley at Palm Desert. This route exposes miles of the rocks of the southern California batholith and the earlier Mesozoic rocks. In the San Jacinto fault zone are thick beds of Pleistocene gravels.

**FIGURE 18-2** *Geologic section across the Peninsular Ranges province from Del Mar, on the coast north of San Diego, to the Salton Sea. The dominant role of the southern California batholith in the Peninsular Ranges is apparent. Remnants, or roof pendants, of prebatholitic rocks are represented by the Jurassic Santiago Peak Volcanics near the coast, the Late Jurassic (?) Julian schist in the heart of the province, and Paleozoic (?) and pre-Cretaceous metamorphic rocks in the eastern San Ysidro and Santa Rosa Mountains. Tertiary beds in the coastal area occur as superficial deposits. The eastern flank of the Santa Rosa Mountains drops steeply down beneath Late Cenozoic sediments of the Salton Trough. Two great active fault zones slice through the eastern half of the ranges: Elsinore fault zone, zone, more than 12 miles wide; and San Jacinto fault zone, about 8 miles across.*

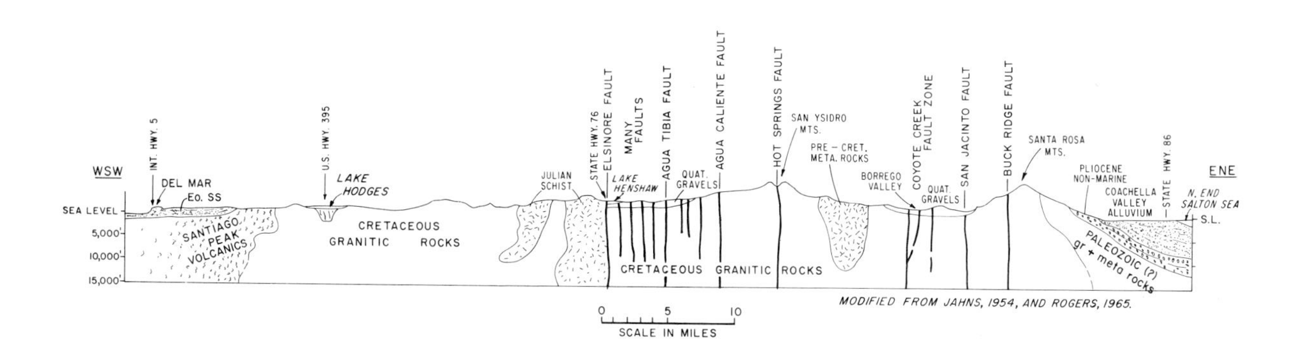

State Highway 71 follows southeasterly from Pomona in the Los Angeles Basin across upper Miocene shale and sandstone in the Puente Hills. Crossing the Santa Ana River, it follows the Elsinore fault zone for 45 miles. Here are remnants of the shallow-marine and brackish-water Paleocene Martinez Formation and occasional exposures of granite and prebatholithic rocks as slivers in the fault zone. Coarse granitic sediments of Quaternary age occupy much of the fault trough.

Highest mountain in the Peninsular Ranges is 10,831-foot San Jacinto Peak at the crest of the San Jacinto Mountains. An aerial tramway from Palm Springs takes the visitor high up the seemingly near-vertical eastern slope of San Jacinto Peak to about the 8,500-foot contour. Not only are there very close views of the complex gneisses and granitic rocks marginal to the batholith, but there are more distant spectacular views of the Salton Trough and the Little San Bernardino Mountains beyond.

**18-3** *Light-colored granitic rocks of the southern California batholith intruding older, dark, folded, metamorphic rocks in the Oriflamme Mountains, San Diego County. (F. H. Weber photo.)*

## PREBATHOLITHIC ROCKS

Rocks older than those of the granite batholith are widespread as roof pendants in the central part of the batholith and in larger areas flanking the Peninsular Ranges (Photograph 18-3). An older group of such rocks consists of highly metamorphosed schists, gneisses, quartzite, and crystalline limestone. Direct evidence is lacking but, by analogy with similar rocks in the nearby Transverse Ranges, they are usually considered late Paleozoic and early Mesozoic in age.

A somewhat later group of rocks—less metamorphosed—includes the fossil-bearing Late Jurassic Bedford Canyon Formation and the Santiago Peak Volcanics in a belt a few miles wide from the Santa Ana Mountains to the San Ysidro

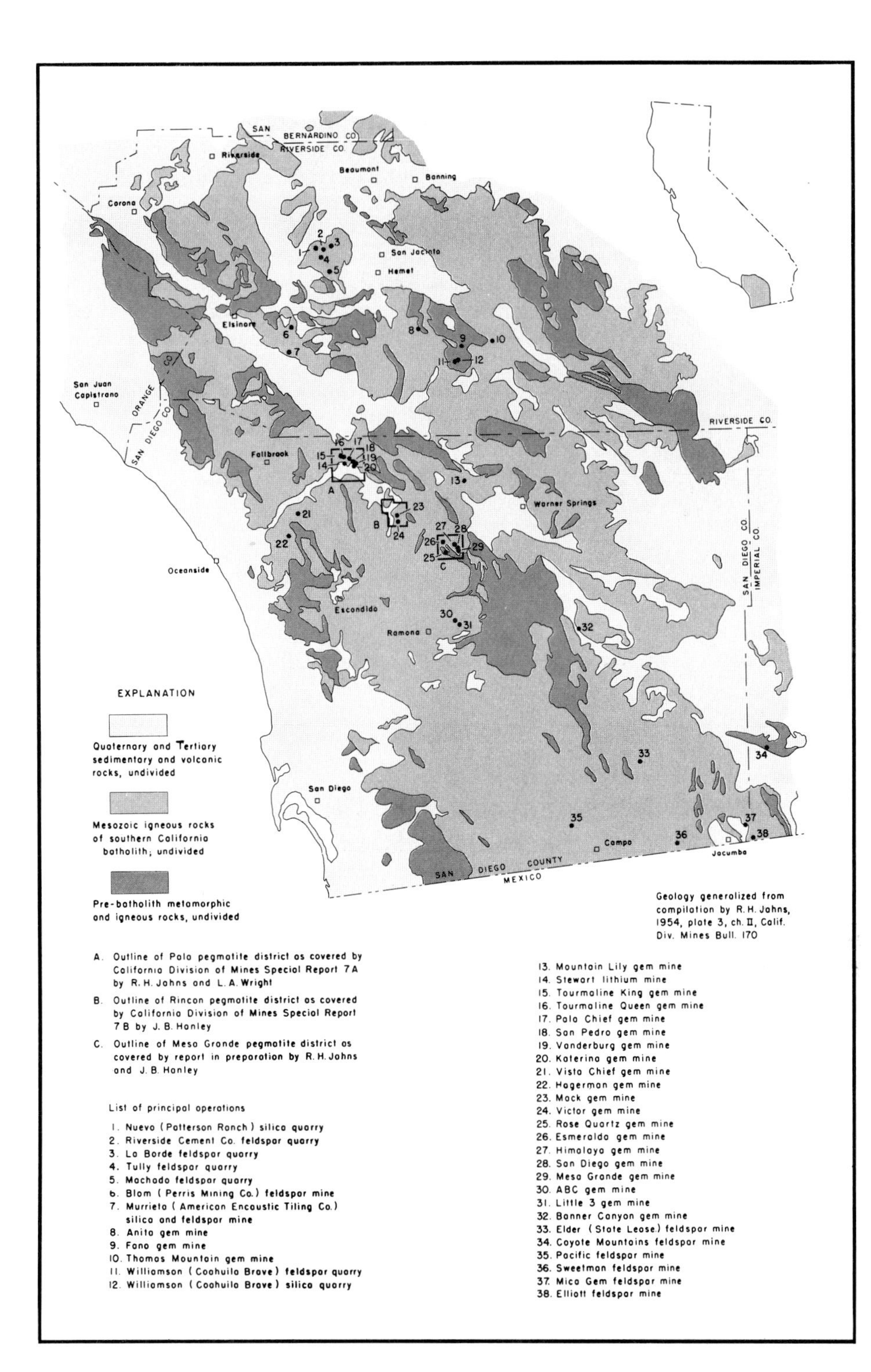


**FIGURE 18-4** *Gem districts and mines in the Peninsular Ranges in San Diego County. (California Division of Mines and Geology Bull. 170.)*

**18-5** *Group of crystals from Little Three Gem mine (No. 31, Figure 18-4) in the Ramona district. Dark crystals are smoky quartz, light are topaz coated with the rare pink lithium mineral cookeite. (F. H. Weber photo.)*

Mountains just north of the Mexican border. The Julian schist (Highway 78) has been considered Triassic, but may be a more metamorphosed equivalent of the Bedford Canyon Formation. The western belt of Jurassic-Triassic formations is readily seen on most roads that extend a few miles inland from the coast, particularly from San Diego, Oceanside, and Capistrano Beach.

The islands and submarine rock samples of the continental borderland (Long Beach map sheet of the *Geologic Map of California*) all consist of Tertiary sedimentary and volcanic rocks, except that several square miles of rocks very similar to the Franciscan Formation of the Coast Ranges crop out on Santa Catalina. A tiny outcrop of similar rocks appears in the Palos Verdes Hills. It is interesting to note that middle Miocene coarse sedimentary breccias in the Los Angeles Basin, which are also exposed at San Onofre on Coast Highway 1 (Interstate 5) about 9 miles southeast of Capistrano Beach, are made up of fragments of Franciscan rocks. This strongly suggests that a portion of the continental borderland now below sea level was exposed to erosion in the middle Miocene Epoch.

The mid-Cretaceous granitic rocks and older metamorphic rocks lie deeply buried beneath late Cenozoic sediments in the Salton Trough. Studies of the types and behavior of earthquake waves—both natural and artificial—show that sediments in the Salton Trough are about 2.3 miles deep in California and perhaps 4 miles deep near the Mexican border.

## LATE CRETACEOUS AND CENOZOIC ROCK FORMATIONS

The Peninsular Ranges–Salton Trough area was probably land undergoing erosion during early Cretaceous time, since no sedimentary rocks of this age have been found. The reddish land-laid Trabuco conglomerate in the Santa Ana Mountains is the oldest sediment of Late Cretaceous time.

Following deposition of the Trabuco, seas advanced across the western part of the Peninsular Ranges province several times. These seas, now represented by exposures of marine, dark-gray sandstone, shale, and conglomerate lying unconformably on granitic rocks along the San Diego coast and in the western Santa Ana Mountains, advanced from the west. In the same area, thin marine and interbedded continental Paleocene and Eocene shale, sandstone, and conglomerate are present. Upper Cretaceous rocks are best exposed in the sea cliffs of San Diego County, from Point Loma northward. State Highway 1 from Long Beach (Interstate 5 from Capistrano Beach south) follows the coastal terrace and exposes the Cenozoic formations.

Exposed Eocene strata appear near the coast: the Rose Canyon shale (formerly mined for brick making) in northwestern San Diego City; the white to light-brown Torrey sand a few miles north at Torrey Pines State Beach; and the Poway conglomerate, including sandstone and mudstone, extending northward into the Santa Ana Mountains. There is no evidence that late Cretaceous, Paleocene, and Eocene seas extended more than a few miles inland.

**18-6** *Looking north at the Pala district* (A *on Figure 18-4*). *The large numbers of pegmatite dikes—some gem-bearing—appear as light-colored bands.* (*F. H. Weber photo.*)

**18-7** *Lake Hodges near Escondido, U.S. Highway 395, in San Diego County. This landscape is typical of the granitic-metamorphic terrain a few miles inland from the coast. (Mary Hill photo.)*

The belt of postgranite sediments widens markedly toward the north into the Los Angeles Basin to the Transverse Ranges. The Los Angeles Basin was persistently downwarped in late Tertiary times and received many thousands of feet of marine middle Miocene to early Pleistocene sediments. These sediments have produced the enormous amounts of petroleum that rank the Los Angeles Basin among the most prolific producers of all time. Marine middle and upper Miocene rocks extend south along the coast about as far as Oceanside.

Oldest Tertiary rocks in the Salton Trough proper are coarse, nonmarine sedimentary rocks of Miocene age. They are overlain by coarse, oyster-bearing, shallow-marine sediments of the early Pliocene Imperial Formation. The seas in which the Imperial sediments were deposited probably advanced northward from the Gulf of California; not from the west. Many thousands of feet of coarse, nonmarine sediments, with minor thicknesses of lake beds and volcanics, make up the late Tertiary and Quaternary section of the Salton Trough.

Marine middle and upper Pliocene sandstone and conglomerate (San Diego Formation) extend along the coast and are particularly well exposed in the sea cliffs and in gullies cut through the flat overlying terrace deposits in western San Diego County. Thin marine Pleistocene deposits are extensive in the terraces of the San Diego area and crop out also in the western Los Angeles Basin.

## STRUCTURE AND LATE HISTORY OF THE SALTON TROUGH

The Salton Trough is a land-locked basin that represents the direct northward extension of the Gulf of California, separated from the Gulf only by a low divide consisting of sediments of the Colorado River delta. The present Salton Sea—about the salinity of sea water—was formed in 1905-1906 when the lower Colorado River burst through its dikes and poured into the basin, filling it to an elevation of about 200 feet below sea level. Bringing the river back into control before floodwater rose high enough to threaten the town of Indio (at sea level), and other basin towns, was a tremendous task! Level of the lake is now kept about constant by excess irrigation waters from Imperial and Coachella Valleys. Wave-cut shore lines, freshwater shells, and occasional deposits of travertine show that the basin was occupied intermittently during the Pleistocene Epoch, and up to a few hundred years ago, by freshwater lakes. The largest of these Ice-Age lakes, whose level rose slightly above present sea level, has been called Lake Coahuila. Shore lines of ancient Lake Coahuila are readily seen from State Highway 111—along the eastern side of the sea—and, on the west, from State 86 which follows the western shore.

The Salton Trough—and doubtless its vastly greater extension to the south, the Gulf of California—is bounded and cut internally by active northwest-striking faults. The straight faults of the San Andreas fault zone form the northeastern margin of the trough and the sea itself and extend southward into the gulf. The straight southwestern margin of the Algodones Dunes is controlled by the San Andreas fault (Photograph 3-15). The southwestern fault margin is not quite so simply and clearly marked. Interrupted faults extend along the steep eastern scarps of the San Jacinto and Santa Rosa Mountains, however, and the great active San Jacinto and Elsinore fault zones slice southeasterly across the Peninsular Ranges and across the southwestern Salton Trough into the gulf.

Gravity and seismic-wave studies show that the continental crust of the earth is lacking below the Gulf of California. The crust thickens gradually northward, but is still only 12 to 15 miles thick beneath the Salton Sea—6 to 10 miles less thick

**18-8** *Looking north across La Jolla, northwestern San Diego, at the dissected surface of the broad Pleistocene marine terrace. Nearly flat-lying Eocene formations are exposed in the canyons. (F. H. Weber photo.)*

than the crust beneath the bordering Peninsular and Transverse Ranges.

All three of these major fault zones are marked by rift valleys, prominent scarps, warm springs, strips of vegetation, and linear valleys that are deeply filled with Quaternary sediments. Fault traces are generally straight and fault planes nearly vertical. Historic displacements have been largely right-lateral, but the steepness of bounding scarps on both sides of the Salton Trough and the tremendous thicknesses of late Tertiary and Quaternary sediments which fill the trough give convincing evidence of great vertical displacements in late geologic history. Folding of the sedimentary strata within the basin, along with both strike-slip and vertical fault movements, suggests the complexity of the geologic forces that have been operating. Displacement within the Salton Trough is going on very rapidly; precise surveys show that the southwestern margin of the trough moved several feet northwestward, relative to the northeastern margin, between 1941 and 1954, without noticeable surface faulting. Displacement along faults of the Salton Trough has averaged about 1 inch per year in historical times.

Scientific evidence is strong that the Gulf of California developed as a great rift trough as the Peninsula of Baja California pulled obliquely northwestly away from the mainland of Mexico. The earth's crust was thinned as it was stretched and split across the trough. High heat flow through the thin crust, hot springs, and volcanism (Photograph 18-13) reinforce evidence that the Gulf of California and the Salton Trough lie on an active portion of the East Pacific rise. Here, sea-floor spreading is in action.

**18-9** *Torrey Pines, just north of San Diego. Coarse, dark, Late Quaternary terrace deposits lying uncomformably on white Eocene Torrey sand.* (*F. H. Weber photo.*)

**18-10** *Looking north at Dana Point, near Capistrano Beach. The formation exposed in the sea cliff at the point is dark, coarse, middle Miocene San Onofre breccia. A fault through the cove has exposed light-colored, thin-bedded, later-Miocene Capistrano shale and sandstone dipping toward the east. The San Onofre breccia consists of rock fragments typical of the Franciscan Formation now exposed only on Santa Catalina Island and in the Palos Verdes Hills. The inference is strong that the San Onofre sediments were derived from a middle Miocene offshore land mass now below sea level. A marine terrace truncates both formations; higher and older terraces are on the skyline. (Mary R. Hill photo.)*

## EARTHQUAKES OF THE SALTON TROUGH AND PENINSULAR RANGES

San Diego and vicinity in the southern Peninsular Ranges in California is one of the more stable—less seismic—areas in the state. In contrast, the northern and eastern Peninsular Ranges and the Salton Trough constitute an area of extremely high earthquake frequency. Immediate explanation, of course, is that this is the 60-mile-wide belt of active faults making up the San Andreas fault system (see the 1:2,500,000 colored geologic map of California, available from the U.S. Geological Survey or the California Division of Mines and Geology).

The first earthquake of record in California was experienced in 1769 by the Portolá expedition camped on the Santa Ana River near the present town of Olive. Perhaps it originated on the Elsinore fault, only 5 miles to the northeast. Records are very scant, but in 1775 the Anza expedition felt a sharp earthquake 25 miles southeast of Hemet, evidently originating on the San Jacinto fault; and in 1812 Mission San Juan Capistrano was severely damaged by a strong earthquake. Displacement on the Newport-Inglewood fault, just 2 or 3 miles offshore from

**FIGURE 18-11** *Salton view, northwest from the Santa Rosa to the Orocopia Mountains. The complex San Andreas fault zone follows the base of the mountains. (Scratchboard drawing by Peter Oakeshott.)*

**18-12** *View east across Carrizo Wash and Coyote Mountains; Superstition Mountains in the background and the Chocolate Mountains on the distant skyline across the Salton Trough. This typical desert landscape was photographed about 25 miles west of El Centro in Imperial County. (F. H. Weber photo.)*

Capistrano Beach, may well have caused the latter. Movement on this fault offshore from Newport Beach caused the moderately-strong (Richter magnitude 6.3) Long Beach earthquake of 1933, which cost 115 lives and $65 million in property damage.

Perhaps California's greatest earthquake was that of 1857, originating on the San Andreas fault north of Fort Tejon in the Tehachapi. Surface ruptures on the fault extended at least as far south as San Bernardino and perhaps to the Salton Sea. Magnitude of this earthquake is unknown, of course, since there were no seismographs to record the wave motion, but it probably was on the order of 8 on the Richter scale, judging from the area over which it was felt.

None of the many later earthquakes in the Peninsular Ranges-Salton Trough region has been much over magnitude 7, but there have been many in the range of magnitude 5 to 7. Among the larger were the 1899 San Jacinto earthquake, and others along this fault zone in 1918, 1934, 1942, 1951, and 1954. Along the Elsinore fault zone and extending into Baja California were strong earthquakes in 1910, 1915, 1919, 1934, 1935, and 1936, convincing evidence of the activity of that fault.

The 1940 Imperial Valley (or El Centro) earthquake, magnitude 7.1, which originated on the Imperial fault—in the San Jacinto fault zone—southeast of El Centro, resulted in deaths of nine people and caused $6 million in damages to

**18-13** *Mud volcano, near Niland, southeastern margin of Salton Sea. The gas-inflated mud bubble is about a foot across. Hot water springs, carbon dioxide gas, very recent volcanic activity, and the active San Andreas fault nearby are all associated with the rifting process that formed the Salton Trough and Gulf of California. (Norman A. Moore photo, courtesy* Desert Magazine, *State Division of Mines and Geology,* Mineral Information Service, *July, 1967.)*

**18-14** *Looking west at the bare granitic rocks of Borrego Mountain, southeastern corner of Santa Ana map sheet of* Geologic Map of California. *Land-laid Pliocene gravels lie near the base of the mountains. The Coyote Creek fault, which caused the 1968 Borrego Mountain earthquake, is in the margin of the valley along the foot of the mountains. (F. H. Weber photo.)*

irrigation works. Surface fault displacement reached a maximum of 19 feet in a right-lateral sense; the All-American Canal at the Mexican border was offset 14 feet. It is interesting that even very small earthquakes are accompanied by surface displacement on the Imperial fault; some creep-type movements also occur without earthquakes.

On April 8, 1968, accompanying the Borrego Mountain earthquake, a maximum right-lateral displacement of 15 inches occurred on the Coyote Creek fault—in the San Jacinto fault zone—at the hamlet of Ocotillo Wells. Magnitude was 6.5 on the Richter scale and the temblor was discernible over 60,000 square miles. Most interesting feature of this earthquake was that it triggered surface movements up to 1 inch on the Superstition Hills, Imperial, and Banning-Mission Creek faults. Implications for extension of damage as earthquakes on one fault trigger others are obvious!

The *Geologic Map of California* and our journeys along mountain and desert highways across the Peninsular Ranges and Salton Trough have shown us something of the complex geologic history of these two strikingly different, yet adjacent, geologic provinces. Admittedly, in California, we see only the northern fraction of the Peninsular Ranges and of the Salton Trough. These great features extend far south into Baja California and the Gulf of California, respectively.

Why are these contrasting provinces adjacent? Why is the broad geologic setting similar to the Sierra Nevada and its adjacent basins and ranges to the east? How are the Salton Sea, Gulf of California, and the Peninsular Ranges related to the great East Pacific Rise?

As geologists and geophysicists extend their studies southward into the far greater area of these provinces in Mexico, and as we map and study more intensively our California side, some of the answers we are seeking in attempting to reconstruct geologic history may develop. But not before there are more and deeper questions!

□ What of the geologic future? What great changes will take place in California landscapes—and indeed throughout the Earth—in the limitless aeons to come?

*Change*—evolution—has been the keynote of this history of California's landscapes, affecting the solid earth, its atmosphere and waters, and the living parts of landscape as well. Only the basic geologic principles and processes appear to remain constant throughout time, but certainly man's understanding of natural processes may be expected to continue to grow—probably ever more rapidly as he acquires an expanding base for understanding.

The present uneasy crust and changing land forms will prevail in this geologically dynamic state for millions of years to come. Fundamental elements of the geologic environment of man—and his evolving descendants—will continue to be slipping faults, earthquakes, active volcanoes, landslides and mudflows, and uplift and subsidence of the land. Man continues to rapidly develop the knowledge of how to control and adjust to his geologic setting. There are also increasingly hopeful signs that he will be applying this knowledge to a life supported by clean air and clear water in harmony with the beautiful, ever-changing mountains, valleys, and coasts of California!

# epilogue

# guide to supplemental reading and sources of information on the geology of california

■ Principal sources of information on the geology of California are the California Division of Mines and Geology and the United States Geological Survey. The Division issues a large number of publications, maintains a comprehensive library open to the public in the Ferry Building in San Francisco, and answers inquiries on all facets of California geology and mineral resources orally and by mail.

Many publications are available from the U.S. Geological Survey. It maintains sales and information offices in San Francisco and Los Angeles.

The most complete mineral collections in the state are open to the public in the Mineral Exhibit of the California Division of Mines and Geology, Ferry Building, San Francisco.

The University of California, various state and federal agencies, a number of book publishers, and about 300 periodicals publish material on California geology at irregular intervals. Specific reference to some of these appears below.

**PUBLICATIONS OF THE CALIFORNIA DIVISION OF MINES AND GEOLOGY**

*List of Available Publications* (list includes only those publications of the Division that are in print); obtainable from the Division without charge.

**how to order**

We recommend that the serious student using this book have at hand the highly colored, small-scale *Geologic Map of California* on a 16 × 20 inch sheet, available for $1.75 from:

1. Superintendent of Documents
   Government Printing Office
   Washington, DC 20402
   Refer to: Colored geologic map of California, scale 1:2,500,000

and for $1.00 from:

2. California Division of Mines and Geology
   P. O. Box 2980
   Sacramento, California 95812
   Refer to: State Bulletin #190, colored geologic map of California, scale 1:2,500,000

Other publications may be purchased over-the-counter from the Division of Mines and Geology at any of its three offices:

Ferry Building, Room 2022, San Francisco 94111, Phone: (415) 557-0633

2815 "O" Street, Sacramento 95816, Phone (916) 445-5716

Junipero Serra Building, 107 S. Broadway, Room 1065, Los Angeles 90012, Phone: (213) 620-3560

Mail orders should be sent to the Sacramento headquarters office. All publications are sent postpaid. Send checks or money orders (no stamps, please), rather than cash; make them payable to the California Division of Mines and Geology. California residents should include 5 percent tax. *California Geology* is not taxable.

If you order Special Reports, Bulletins, or Map Sheets, give the number and title (for example, Bulletin 144, *Copper in California*); if you order issues of the *California Journal of Mines and Geology* or *Reports of the State Geologist,* give the month and year (for example, *California Journal of Mines and Geology,* April 1958).

Address orders to:

California Division of Mines and Geology,
P. O. Box 2980, Sacramento, California 95812

Include your name, address, and ZIP code.

Following is a selection of Division publications and their prices. All are in print (January 1977) with the exception of Bulletin 170. Formats of the state publications vary widely, and all are issued at irregular intervals except the monthly *California Geology*. The *List of Available Publications, 1977* is obtainable by request and is free.

**bulletins**

Bulletin 141, *Geologic Guidebook along Highway 49—The Sierran Gold Belt—The Mother Lode Country.* By Olaf P. Jenkins et al. 1948. 164 pp., 2pls., 221 figs., 10 maps. $2.50

Bulletin 154, *Geologic Guidebook of the San Francisco Bay Counties: History, Landscape, Geology, Fossils, Minerals, Industry, and Routes to Travel.* 1951. 392 pp., 1 pl., 7 maps, 280 figs. $2.50

Bulletin 170, *Geology of Southern California.* Richard H. Jahns, editor. Out of print, but available in most California libraries. Chap. I, General Features; Chap. II, Geology of the Natural Provinces; Chap. III, Historical Geology; Chap. IV, Structural Features; Chap. V. Geomorphology; Chap. VI, Hydrology; Chap. VII, Mineralogy and Petrology; Chap. VIII, Mineral Deposits and Mineral Industry; Chap. IX, Oil and Gas; Chap. X, Engineering Aspects of Geology; Geologic Guides 1 to 5; Map Sheets 1 to 34.

Bulletin 171, *Earthquakes in Kern County, California, during 1952.* Gordon B. Oakeshott, editor. 1955. 283 pp., 2 pls., 263 figs. $4.00. (A symposium on the stratigraphy, structural geology, and origin of the earthquakes; their geologic effects; seismologic measurements, application of seismology to petroleum exploration; structural damage and design of earthquake-resistant structures). Part I, Geology; Part II, Seismology; Part III, Structural Damage.

Bulletin 182, *Geologic Guide to the Merced Canyon and Yosemite Valley, California.* Gordon B. Oakeshott, editor. 1962. 68 pp., 2 pl., illus. $1.50

Bulletin 183, *Franciscan and Related Rocks, and Their Significance in the Geology of Western California.* By Edgar H. Bailey et al. 1964. 177 pp., 2 pls., 35 figs., 82 photos. $5.00

Bulletin 189, *Minerals of California.* By J. W. Murdoch and R. W. Webb. 1966. 559 pp., 2 figs., 27 photos. $5.00

Bulletin 190, *Geology of Northern California.* Edgar H. Bailey, editor. 1966. 508 pp., 1 pl., over 400 photos and diagrams, illus. $6.00

Bulletin 191, *Mineral Resources of California.* 1966. 450 pp., 1 pl., 88 figs., 49 tables. $2.00

Bulletin 193, *Gold Districts in California.* By William B. Clark. 1970. 186 pp., 1 pl., 30 figs., 89 photos, 9 tables. $4.00

Bulletin 196, *San Fernando, California, Earthquakes of 1971.* Gordon B. Oakeshott, editor. 1975. Numerous illust., 5 pls. in envelope. $13.00

Bulletin 198, *Urban Geology, Master Plan for California.* By John T. Alfors et al. 1973. 112 pp., 43 illus. $2.50

**special reports**

Special Reports 10A, 10B, 10C on jade. Reprinted as Special Publication 49, *California Jade* (See under Miscellaneous Publications).

Special Report 11, *Guide to the Geology of Pfeiffer Big Sur State Park, Monterey County, California.* By Gordon B. Oakeshott. 1951. 16 pp., 1 pl., 28 figs. $0.25

Special Report 14, *Geology of the Massive Sulfide Deposits at Iron Mountain, Shasta County, California.* By A. R. Kinkel, Jr., and J. P. Albers. 1951. 19 pp., 6 pls., 6 figs. $0.75

Special Report 52, *Index to Geologic Maps of California to December 31, 1956.* By R. G. Strand, J. B. Koenig, and C. W. Jennings. 1958. 128 pp., 1 illus., 57 maps. $1.00

Special Report 52A, *Index to Geologic Maps of California 1957-60.* A supplement to Special Report 52. By James B. Koenig. 1962. 60 pp., 1 illus., 40 maps. $1.00

Special Report 52B, *Index to Geologic Maps of California 1961-64.* A supplement to Special Report 52. By James B. Koenig and Edmund W. Kiessling. 1968. 1 illust., 48 maps. $1.75

Special Report 73, *Economic Geology of the Panamint Butte Quadrangle and Modoc District, Inyo County, California.* By Wayne E. Hall and Hal G. Stephens. 1963. 39 pp., 12 pls., 2 figs., 11 photos, 8 tables. $2.00

Special Report 74, *Index to Graduate Theses on California Geology to Dec. 31, 1961.* By Charles W. Jennings and Rudolph G. Strand. $1.00

Special Report 76, *Recent Volcanism at Amboy Crater, San Bernardino County, California.* By Ronald B. Parker. 1963. 23 pp., 1 pl., frontis., 6 figs., 12 photos. $1.00

Special Report 80, *Geology and Mineral Resources of Mt. Diablo, Contra Costa County, California.* By Earl H. Pampeyan. 1964. 31 pp., 5 pls., 10 figs., 4 tables. $1.50

Special Report 83, *A Geologic Reconnaissance in the Southeastern Mojave Desert.* By Allen M. Bassett and Donald H. Kupfer. 1964. 43 pp., 1 pl., 4 figs., 6 photos, 1 table. $1.50

Special Report 85, *Economic Geology of the French Gulch Quadrangle, Shasta and Trinity Counties, California.* By John P. Albers. 1965. 43 pp., 8 pls., 11 figs. $2.00

Special Report 91, *Short Contributions to California Geology.* 1967. 60 pp., illus. Containing "Petrography of Six Granitic Intrusive Units in the Yosemite Valley Area, California," by Arthur R. Smith; "The Plutonic and Metamorphic Rocks of the Ben Lomond Mountain Area, Santa Cruz County, California," by Gerhard W. Leo; "The Origin of Tuscan Buttes and the Volume of the Tuscan Formation in Northern California," by Philip A. Lydon; "Chittenden, California, Earthquake of September 14, 1963," by Earl E. Brabb; "Notes on the Types of California Species of the Foraminiferal Genus *Orthokarstenia* Dietrich, 1935," by Joseph J. Graham and Dana K. Clark. $1.50

Special Report 92, *Short Contributions to California Geology.* 1967. 55 pp., illus. Containing "Mineralogy of the Kalkar Quarry, Santa Cruz, California," by Eugene B. Gross et al; "A Test of Chemical Variability and Field Sampling Methods, Lakeview Mountain Tonalite, Lakeview Mountains, Southern California Batholith," by A. K. Baird et al; "The Effects of Provenance and Basin-edge Topography on Sedimentation in the Basal Castaic Formation (Upper Miocene, Marine), Los Angeles County, California," By Robert J. Stanton, Jr.; "Sedimentary Rocks of Late Precambrian and Cambrian Age in the Southern Salt Spring Hills, Southeastern Death Valley, California," by Bennie W. Troxel; "Reconnaissance Geology of the Helena Quadrangle, Trinity County, California," by Dennis P. Cox. $1.50

Special Report 94, *Geology of the Desert Hot Springs-Upper Coachella Valley Area, California.* By Richard J. Proctor. 1968. 50 pp., 1 pl., 13 figs., 10 photos. $2.50

Special Report 97, *Geologic and Engineering Aspects of San Francisco Bay Fill.* Harold B. Goldman, editor. 1969. 130 pp., 4 pl., 6 figs., 3 tables. $3.00

Special Report 98, *Natural Slope Stability as Related to Geology, San Clemente Area, Orange and San Diego Counties, California.* By Robert P. Blanc and George B. Cleveland. 1968. 19 pp., 4 pl., 4 figs., 8 photos. $2.00

Special Report 99, *Geology of the Dry Mountain Quadrangle, Inyo County, California.* By B. Clark Burchfiel. 1969. 19 pp., 1 pl., 15 figs. $2.00

Special Report 100, *Short Contributions to California Geology.* 1969. 68 pp., illus. Containing "Clay mobility, Portuguese Bend, California," by Paul F. Kerr and Isabella M. Drew; "The chemical 'fingerprinting' of acid volcanic ocks," by R. N. Jack and I. S. E. Carmichael; "Cretaceous and Eocene coccoliths at San Diego, California," by David Bukry and Michael P. Kennedy; "Stratigraphy and petrology of the Lost Burro Formation, Panamint Range, California," by Donald H. Zenger and Eugene F. Pearson; "Rapid method of sampling diatomaceous earth," by George B. Cleveland. $2.00

Special Report 101, *Geology of the Elysian Park-Repetto Hills Area, Los Angeles County, California.* By Donald L. Lamar. 1970. 45 pp., 2 pls., 27 figs. $2.50

Special Report 102, *Index to Geologic Maps of California 1965–68.* By Edmund W. Kiessling. 1972. 78 pp., 26 maps. $2.75

Special Report 104, *Upper Cretaceous Stratigraphy on the West Side of the San Joaquin Valley, Stanislaus and San Joaquin Counties, California.* By Charles C. Bishop. 1971. 29 pp., 2 pls., 17 figs. $5.00

Special Report 111, *Geology and Engineering Geologic Aspects of the Southern Half Canada Gobernadora (7½′) Quadrangle, Orange County, California.* By Paul K. Morton. 1974. 30 pp., 1 pl., 2 figs. $3.00

Special Report 112, *Geology and Engineering Geologic Aspects of the San Juan Capistrano (7½′) Quadrangle, Orange County, California.* By Paul K. Morton, William J. Edgington, Donald L. Fife. 1974. 64 pp., 1 pl., 2 figs., 2 tables. $3.00

Special Report 115, *Index to Graduate Theses and Dissertations on California Geology 1962 through 1972.* By Gary C. Taylor. 1974. 89 pps., 1 pl. *See Special Report 74 for theses from 1892 through 1961.* $3.00

Special Report 118, *San Andreas Fault in Southern California.* John C. Crowell, editor. 1975. 272 pp., numerous illus., folded regional map in pocket. $5.00; $4.00 without map

Special Report 122, *Engineering Geology of the Geysers Geothermal Resource Area, Lake, Mendocino, and Sonoma Counties, California.* By C. Forrest Bacon et al. 1976. 35 pp., 18 photos, 4 figs. $3.50

Special Report 123, *Character and Recency of Faulting, San Diego Metropolitan Area, California.* By Michael P. Kennedy et al. 1975. 33 pp., 2 pls., 4 figs. $4.00

Special Report 124, *Oroville, California, Earthquake, 1 August 1975.* Roger W. Sherburne and Carl J. Hauge, editors. 1975. 151 pp., numerous photos and figs. $3.50

Special Report 125, *Mines and Mineral Deposits in Death Valley National Monument, California.* By James R. Evans, Gary C. Taylor, and John S. Rapp. 1976. 61 pp., 6 maps, 2 figs., 24 photos, 11 tables. $3.50

**california geology (formerly mineral) information service)**

*California Geology,* a monthly publication, is designed to report on the progress of earth science in California, and to inform the public of discoveries in geology and allied earth sciences of interest and concern to their lives and livelihood. It also serves as a news release on mineral discoveries, mining operations, statistics of the minerals industry, and new publications.

Subscription price, January through December, $3.00. Back issues are generally available.

The following list refers to some of the topics covered in issues of the past few years.

The Long Beach-Compton earthquake of Mar. 10, 1933 Mar. 1973
The disposal of nuclear waste Apr. 1973
Earthquake activity, Monterey Bay to Halfmoon Bay May 1973
Mudflows at Big Sur June 1973
Pleistocene Lake San Benito July 1973
Mining activity in California Dec. 1973
Model for the origin of Sierran granites Jan. 1974
Glaciers—a picture story Feb. 1974
The earthquake trail at Point Reyes National Seashore Apr. 1974
Chemical stabilization of landslides May 1974
What price gold? June 1974
Major structural features in the Mountain Pass area July 1974
Origin of Lake Merced, San Francisco Aug. 1974
Diversion of Furnace Creek Wash, Death Valley, California Oct. 1974
Geology and land-use development in Ventura County Nov. 1974
Surprise Valley fault Dec. 1974
The Stanislaus River—a study in Sierra Nevada geology Jan. 1975
Geological hazards along the coast south of San Francisco Feb. 1975
Geologic guide, Sierra Nevada—Basin and Range May 1975
The Santa Barbara earthquake of June 1925 June 1975
Acoustic emission along San Andreas fault July 1975
Living glaciers of California Aug. 1975
The geology of Moaning Cave, Calaveras County, California Sept. 1975
Seismic hazards and urbanization in Santa Clara County Oct. 1975
Vertebrate fossils Nov. 1975
Oroville earthquake Dec. 1975
Pygmy forest Jan. 1976
Sandstone caves in the central Santa Cruz Mountains Mar. 1976
Landsat 1 imagery Apr. 1976
Roadcut geology in the San Andreas fault zone May 1976
Pleistocene glaciation, Trinity Alps May 1976
Off-road vehicle effects on California's Mojave Desert June 1976
Landslides—the descent of man July 1976
Palmdale bulge Aug. 1976
Lime industry in California Oct. 1976
Seismic safety for California dams Nov. 1976
Pueblo water rights of the City of Los Angeles Dec. 1976
Economic evaluation of Pine Mountain phosphate deposit Dec. 1976

**county reports**

NEW SERIES

1. *Mines and Mineral Resources of Kern County, California.* By Bennie W. Troxel and Paul K. Morton. 1962. 370 pp. 10 pls., 120 figs., 25 tables. $5.00

2. *Mines and Mineral Resources of Calaveras County, California.* By William B. Clark and Philip A. Lydon. 1962. 217 pp., 4 pls., 20 figs., frontis., 47 photos. $4.00

3. *Mines and Mineral Resources of San Diego County, California.* By F. Harold Weber, Jr. 1963. 709 pp., 10 pls., 59 figs., 89 photos, 3 tables. $8.00

4. *Mines and Mineral Resources of Trinity County, California.* By J. C. O'Brien. 1965. 2 pls., 3 figs., 18 photos, frontis. $3.50

5. *Mines and Mineral Resources of Monterey County, California.* By Earl W. Hart. 1966. $5.00

6. *Shasta County, California, Mines and Mineral Resources.* By Philip A. Lydon and J. C. O'Brien, 1974. 154 pp., 2 pls., 7 figs., 44 photos, 3 tables. $7.50

**map sheet series**

8. *Geology of the Palo Alto Quadrangle, Santa Clara County, California.* By T. W. Dibblee. 1966. $1.50

**9.** *Geology of the Kelseyville Quadrangle, Sonoma, Lake and Mendocino Counties, California.* By James R. McNitt. 1968. $1.50

**10.** *Geology of the Lakeport Quadrangle, Lake County, California.* By James R. McNitt. 1968. $1.50

**11.** *Geology of a Portion of Western Marin County, California.* By Harold J. Gluskoter. 1969. $1.50

**12.** *Geology of the Southeast Quarter of Trinity Lake Quadrangle, Trinity County, California.* By Philip A. Lydon and Ira E. Klein. 1969. $1.50

**13.** *Preliminary Reconnaissance Map of Major Landslides, San Gabriel Mountains, California.* By D. M. Morton and R. Streitz. 1969. $1.00

**14.** *Geology of the Furnace Creek Borate area, Death Valley, Inyo County, California.* By James F. McAllister. 1970.

**15.** *Preliminary Reconnaissance Map of Major Landslides, San Gabriel Mountains (Los Angeles County), California.* By D. M. Morton and R. Streitz. 1969. $1.00

**16.** *Geology and Slope Stability of the Southwest Quarter Walnut Creek 7½′ Quadrangle, Contra Costa County, California.* By Richard B. Saul. 1973. $3.50

**17.** *Geology and Mineral Deposits of the Mescal Range (15′) Quadrangle, San Bernardino County, California.* By James R. Evans. 1971. $1.50

**18.** *Geology of the Northeast Quarter Shoshone 15-minute Quadrangle, Inyo County, California.* By Charles W. Chesterman. 1973. $2.00

**19.** *Geology of the Lakeview-Perris (7½′) Quadrangles, Riverside County, California.* By D. M. Morton. 1972. $3.50

**20.** *Geology of the Southeast Quarter Tecopa (15′) Quadrangle, San Bernardino and Inyo Counties, California.* By Lauren A. Wright. 1974. $2.50

**21.** *Geology of the Bodie Quadrangle, Mono County, California.* By Charles W. Chesterman and C. H. Gray, Jr. $4.50

**23.** *Maximum Credible Rock Acceleration from Earthquakes in California.* By Roger W. Greensfelder. 1974. $1.50

**24.** *Geology of the Arroyo Grande (15′) Quadrangle, San Luis Obispo County, California.* By Clarence A. Hall, Jr. 1974. $4.75

**26.** *Offshore Surficial Geology of California.* By Edward E. Welday and John W. Williams. 1975. $7.50

**28.** *Geology of the West Central Part of the Mt. Wilson 7½′ Quadrangle, San Gabriel Mountains, Los Angeles, County, California.* By Richard B. Saul. 1976. $2.50

## geologic map of california

The state is covered by a series of twenty-seven geologic map sheets comprising a geologic atlas on a scale of 1:250,000, or about 1 inch equals 4 miles. These sheets are lithographed in full color and are printed on a topographic base at $2.50 each. Accompanying each map is an explanatory data sheet that contains an index chart to the selected references used in the compilation of the map, a table showing stratigraphic nomenclature, an index map indicating the available U.S. Geological Survey topographic quadrangle maps of the region, and one or more photographs showing some of the salient geologic features of the area.

In 1967, the Division of Mines and Geology issued the first in its series, *Bouguer Gravity Map of California.* The map consists of sheets of the *Geologic Map of California,* scale 1:250,000, lithographed in color. The sheets have geologic contacts and faults subdued. Bouguer gravity contours are shown over the entire area of each sheet. They are sold with a brief descriptive text and with a stratigraphic data sheet, all packaged in a manila envelope. Only part of the state gravity contour maps had been completed by 1977.

## miscellaneous publications

*Sixty-eighth Report of the State Geologist* (Programs and accomplishments of the California Division of Mines and Geology during fiscal year 1974–75). 1975. Free

*Legal guide for prospectors and miners.* 1962, 1970, 1973. By Charles L. Gilmore and Richard M. Stewart. $1.00

*Directory of mineral producers in California during 1973.* $2.00

*De Argento Vivo: Historic documents on Quicksilver and Its Recovery in California Prior to 1860.* By Elisabeth L. Egenhoff. 1953. 144 pp., illus. $1.00

*Fabricas: A Collection of Contemporary Pictures and Statements on the Mineral Materials Used in Building in California Prior to 1850.* By Elisabeth L. Egenhoff. 1952. 189 pp., 268 illus. $1.00

*The Elephant as They Saw It: A Collection of Contemporary Pictures and Statements on Gold Mining in California.* By Elisabeth L. Egenhoff. 1949. 128 pp., 66 figs. Out of print.

*A Walker's Guide to the Geology of San Francisco.* A supplement to *Mineral Information Service,* November 1966. $0.50. ($0.25 each for 10 or more).

*Field Trip A: Point Reyes Peninsula and San Andreas Fault Zone.* By Alan J. Galloway. $0.50

*Field Trip B: San Francisco Peninsula.* By M. G. Bonilla and Julius Schlocker. $0.50

*Field Trip C: San Andreas Fault from San Francisco to Hollister.* By Earl E. Brabb, Marshall E. Maddock, and Robert E. Wallace. $0.50

*Field Trip D: Hydrology Field Trip to East Bay Area and Northern Santa Clara Valley.* By S. N. Davis. $0.50

*Field Trip E: Sacramento Valley and Northern Coast Ranges.* By D. O. Emerson and E. I. Rich. $0.50

*Field Trip F: Yosemite Valley and Sierra Nevada Batholith.* By Dallas L. Peck, Clyde Wahrhaftig, and Lorin D. Clark. $0.50

*Field Trip G: Mineralogy of the Laytonville Quarry, Mendocino County, California.* By Charles W. Chesterman. $0.50

Set of seven guides (A to G) $1.50
(Guidebooks A to G are also included as a part of Bulletin 190, *Geology of Northern California.*)

*Earthquake and Geologic Hazards Conference, December 7 and 8, 1964, San Francisco, California.* Record of proceedings of conference sponsored by the Resources Agency of California. 154 pp., 5 illus. $1.00

*Landslides and Subsidence Geologic Hazards Conference, May 26 and 27, 1965, Los Angeles, California.* Record of proceedings of conference sponsored by the Resources Agency of California. 190 pp., 31 illus. $1.00

Special Publication 42, *Fault Hazard Zones in California.* By Earl W. Hart. 1976. Indexes to maps of special studies zone. 27 pp. $1.00

Special Publication 47, *Active Fault Mapping and Evaluation Program.* By Division of Mines & Geology. 1976. Fault-map of California, scale approx. 1 in = 45 mi. 42 pp. $2.00

Special Publication 48, *Second Report on the Strong-Motion Instrumentation Program.* Thomas E. Gay, Jr., editor. 1976. 39 pp. $1.00

## PUBLICATIONS OF THE U.S. GEOLOGICAL SURVEY

The United States Geological Survey maintains seven public inquiries offices to provide oral and written information and to make over-the-counter sales of geologic maps, reports, and topographic maps.

Mail orders for maps west of the Mississippi River should be addressed to:

Distribution Section, Denver Federal Center, Denver, Colorado 80225.

Mail orders for book publications should be sent to:

Superintendent of Documents, Government Printing Office, Washington, D.C. 20402.

Public inquiries offices of the U.S. Geological Survey in California are:

7638 Federal Building, 300 N. Los Angeles Street, Los Angeles, Calif. 90012.

504 Custom House, 555 Battery Street, San Francisco, Calif. 94111.

### popular publications

Popular publications of the U.S. Geological Survey are intended to answer public inquiries concerning the earth sciences and activities of the federal Geological Survey. Some of these are booklets, such as *Exploration Assistance, Earthquakes,* and *Suggestions for Prospecting.* Another series consists of leaflets on such subjects as Topographic Maps, Volcanoes, San Andreas Fault, Astrogeology, Geologic Time, and Gold. Each booklet or leaflet is about 16 to 24 pages. All are free from:

Information Office, Geological Survey, Washington, D.C. 20242.

### scientific and technical publications

The Geological Survey has published a large amount of technical and scientific materials on the geology of California. Publications are chiefly in the form of bulletins, professional papers, and a number of map series. The USGS is the publisher of topographic maps that cover the state on various scales.

### photographs

Photographic Library, U.S. Geological Survey, Room 2274, Building 25, Denver Federal Center, Denver, Colorado 80225, maintains a collection of 140,000 photographs representing geologic studies of the United States. These may be viewed by the public at the Denver Federal Center and prints may be obtained for nominal prices. The photographs are catalogued by subject, geographic location, and author or photographer. Some color transparencies are available also. Permission for the use of USGS photos is not required, but credit should be given to the photographer and to the Geological Survey.

USGS Professional Paper 590, 1968, *A Descriptive Catalog of Selected Aerial Photographs of Geologic Features of the United States* is available for a nominal price from the Superintendent of Documents, Washington, D.C. 20402.

### sampling of u.s. geological survey publications on california geology

Bulletin 1398, *Recent Landslides in Alameda County, California (1940–71): An Estimate of Economic Losses and Correlations with Slope, Rainfall, and Ancient Landslide Deposits.* By T. H. Nilsen, F. A. Taylor, and E. E. Brabb. 1976. 21 pp., 1 pl. $1.65

Circular 525, *Tectonic Creep in the Hayward Fault Zone, California.* By Dorothy H. Radbruch, M. G. Bonilla, and others. 1966. 13 pp. Free

Circular 537, *Effects of the Truckee, California, Earthquake of September 12, 1966.* By Reuben Kachadoorian, R. F. Yerkes, and A. O. Waananen. 1967. 14 pp., plate in pocket. Free

Circular 566, *Tertiary Gold-bearing Channel Gravel in Northern Nevada County, California.* By D. W. Peterson, W. E. Yeend, H. W. Oliver, and R. E. Mattick. 1968. 22 pp. Free

Circular 714, *Selected Geologic Literature on the California Continental Borderland and Adjacent Areas, to January 1, 1975.* By A. E. Roberts. 1975. 116 pp. Free

Circular 729, *Earthquake Prediction—Opportunity to Avert Disaster.* U.S. Geological Survey, 1200 South Eads St., Arlington, Va. 22202. 1976. Free

Circular 730, *Geologic Appraisal of Offshore Southern California: The Borderland Compared to Onshore Coastal Basins.* By J. D. Taylor. 1976. Free

Circular 740, *Earthquakes and Related Catastrophic Events, Island of Hawaii.* By R. I. Tillings, et al. 1975. 33 pp. Free

Earthquake Information Bulletin. Publ. bimonthly by U.S. Geological Survey. Per year, $3.00

GQ-437, *Geologic Map of the Devils Postpile Quadrangle, Sierra Nevada, California*. By N. K. Huber and C. D. Rinehart. 1965, Map sheet 28 by 41 inches. $1.00

GQ-462, *Geologic Map of the Mono Craters Quadrangle, Mono and Tuolumne Counties, California*. By R. W. Kistler. 1966. Map sheet 31 by 39 inches. $1.00

GQ-769, *Areal and Engineering Geology of the Oakland East Quadrangle, California*. By Dorothy H. Radbruch. 1969. Scale 1:24,000. $1.00

I-416, *Pleistocene Lakes in the Great Basin*. By C. T. Snyder, George Hardman, and F. F. Zdenek. 1964. Map sheet 36 by 56 inches. $0.50

I-522, *Approximate Location of Fault Traces and Historic Ruptures within the Hayward Fault Zone between San Pablo and Warm Springs, California*. By Dorothy H. Radbruch. 1967. Map sheet 30 by 48 inches. $1.00

Prof. Paper 160, *Geologic History of the Yosemite Valley*. By Francois E. Matthes. 1930. 137 pp., photos, maps, and figs. Out of print.

Prof. Paper 369, *Geology of San Nicolas Island, California*. By J. G. Vedder and R. M. Norris. 1963. 65 pp. $2.75

Prof. Paper 385, *Geology and Mineral Deposits of the Mount Morrison Quadrangle, Sierra Nevada, California*. By C. D. Rinehart and D. C. Ross, with a section, "A Gravity Study of Long Valley," by L. C. Pakiser. 1964. 106 pp. $2.00

Prof. Paper 437-H, *Land Subsidence in the San Joaquin Valley, California, as of 1972*. By J. F. Poland, B. E. Lofgren, R. L. Ireland, and R. G. Pugh. 1975. 78 pp. $2.20

Prof. Paper 438, *Structural Geology and Volcanism of Owens Valley Region, California: A Geophysical Study*. By L. C. Pakiser, M. F. Kane, and W. H. Jackson. 1964. 68 pp. $2.75

Prof. Paper 494A, *Stratigraphy and Structure, Death Valley, California*. By Charles B. Hunt and Don R. Mabey. 1966. 162 pp., 123 pls. & figs. Out of print.

Prof. Paper 504A, *Glacial Reconnaissance of Sequoia National Park, California*. By F. E. Matthes. 1965. 58 pp., 2 pl. $1.25

Prof. Paper 554-D, *Cenozoic Volcanic Rocks of the Devils Postpile Quadrangle, Eastern Sierra Nevada, California*. By N. K. Huber and C. D. Rinehart. 1967. P. D1–D21, plate in pocket. $0.70

Prof. Paper 579, *The Parkfield-Cholame, California, Earthquakes of June–August 1966: Surface Geologic Effects, Water Resources Aspects, and Preliminary Seismic Data*. By R. D. Brown, Jr., J. G. Vedder, R. E. Wallace, E. F. Roth, R. F. Yerkes, R. O. Castle, A. O. Waananen, R. W. Page, and J. P. Eaton. 1967. 66 pp. $0.50

Prof. Paper 684-B, *Plutonic Rocks of the Klamath Mountains, California and Oregon*. By P. E. Hotz. 1971. 20 pp. $0.35

Prof. Paper 827, *Geology of the Sierra Foothills Melange and Adjacent Areas, Amador County, California*. By W. A. Duffield and R. V. Sharp. 1975. 30 pp., 1 pl. $2.35

Prof. Paper 912, *Petroleum Geology of Naval Petroleum Reserve No. 1, Elk Hills, Kern County, California*. By J. C. Maher, R. D. Carter, and R. J. Lantz. 1975. 109 pp., 25 pls. $15.75

Prof. Paper 941-A, *Studies for Seismic Zonation of the San Francisco Bay Region*. (Basis for reduction of earthquake hazards, San Francisco Bay region, California). By R. D. Borcherdt, ed. 1975. $2.80

Prof. Paper 929, *A New Window on Our Planet*. Richard S. Williams, and William D. Carter, eds. 1976, ERTS-1. (Numerous colored images and figures.) 362 pp. $13.00

*Atlas of Volcanic Phenomena*. 1971. 20 sheets, each 16 by 21½ inches. Per set, $4.25

*Earthquake Information Bulletin,* publ. bimonthly by U.S. Geological Survey, Superintendent of Documents, Washington, D. C. 20402. Per year, $3.00

*New Publications of the Geological Survey,* publ. monthly by U.S. Geological Survey, 329 National Center, Reston, Va. 22092. Free.

## SELECTED GENERAL REFERENCES


Allen, C. R., P. St. Amand, C. F. Richter, and J. M. Nordquist, 1965, Relationship between Seismicity and Geologic Structure in the Southern California Region. *Seismological Society of America Bull.,* vol. 55, no. 4, pp. 753-797. $8.00

American Geological Institute, Falls Church, Va. 22041. *Geology: Science and Profession.* $1.00

American Geological Institute, 1972, *Glossary of Geology.* Falls Church, Va. 22041. 857 pp. Paperback (abridged) $3.50

Beck, Warren A., and Ynez Haase, 1975, *Historical Atlas of California.* University of Oklahoma Press, Norman, 110 pp. Paperback (large size) $4.95

Blackwelder, Eliot, 1931, Pleistocene Glaciation in the Sierra Nevada and Basin Ranges. *Geological Soc. Amer. Bull.,* vol. 42, pp. 865-922. (Out of print but may be available from Johnson Reprint Corp., 111 Fifth Ave., New York, N.Y. 10003.)

Bolt, B. A., W. L. Horn, G. A. Macdonald, and R. F. Scott, 1975, *Geological Hazards.* Springer-Verlag, N.Y., 328 pp. 116 figs.

Bulletin of the Association of Engineering Geologists, Publ. quarterly. AEG Executive Director, 8310 San Fernando Way, Dallas, Tx 75218. Per year, $14.50

Calder, Nigel, 1972, *The Restless Earth—A Report on the New Geology.* Viking, N.Y., 152 pp.

Colbert, Edwin Harris, 1951, *The Dinosaur Book.* McGraw-Hill, N.Y., 156 pp. $5.95

Communications Research Machines, Inc., Del Mar, CA 92014. *Geology Today,* 1973, 530 pp. (A profusely illustrated, multi-authored, beginning-geology textbook built around development of the theory of plate tectonics. New and different!)

Cowen, Richard, 1976, *History of Life,* McGraw-Hill, N.Y., 145 pp. Paperback, $2.95

Davies, T. A., 1976, 573 Holes in the Bottom of the Sea—Some Results from 7 Years of Deep-sea Drilling. *Journal of Geological Education,* vol. 24, no. 5, Nov., pp. 143-155.

Davis, Stanley N., Paul H. Reitan, Raymond Pestrong, *Geology—Our Physical Environment.* McGraw-Hill, N.Y., 470 pp.

Dyson, James L., 1962, *The World of Ice.* Knopf, N.Y., 292 pp. $6.95

Earth Science Curriculum Project, 1973, *Investigating the Earth.* Houghton Mifflin, Boston, 529 pp. $9.60

Fenton, Carroll Lane, 1940, *The Rock Book.* Doubleday, Garden City, N.Y., 357 pp. $8.95

———, 1952, *Giants of Geology.* Doubleday, Garden City, N.Y., 333 pp. $4.50; paperback, $0.95

———, 1958, *The Fossil Book: A Record of Prehistoric Life.* Doubleday, Garden City, N.Y., 496 pp. $15.00

*Geotimes:* American Geological Institute, Falls Church, Va. Ten issues per year. Per year, $6.00

Green, Jack, and Nicholas M. Short, 1971, *Volcanic Landforms and Surface Features—A Photographic Atlas and Glossary.* Springer-Verlag, N.Y. 519 pp., quarto format.

Harbaugh, John W., 1974, *Geology Field Guide to Northern California.* Wm. C. Brown Company Publishers,Dubuque, Iowa. 123 pp. Paperback, $2.95

Hargraves, R. B., 1976, Precambrian Geologic History. *Science,* July, pp. 363-371.

Hill, Mary, 1975, *Geology of the Sierra Nevada.* California Natural History Guides: 37. University of California Press, Berkeley, Calif. 232 pp. $8.95; pocket-size paperback, $3.95

Hodgson, J. H., 1964, *Earthquakes and Earth Structure.* Prentice-Hall, Englewood Cliffs, N.J., 166 pp. $3.95

Hurley, Patrick M., 1959, *How Old Is the Earth.* Doubleday, Garden City, N.Y., 160 pp. Paperback, $1.25

______, 1968, The Confirmation of Continental Drift. *Scientific American,* April, pp. 52–64.

Kesler, Stephen E., 1976, *Our Finite Mineral Resources.* McGraw-Hill, N.Y., 120 pp. Paperback, $5.39

King, Philip B., 1959, *The Evolution of North America.* Princeton University Press, Princeton, N.J., pp. 1–22. $8.50

Lawson, A. C., and others, 1908, The California Earthquake of April 18, 1906, Report of the State Earthquake Investigation Committee. *Carnegie Inst. Washington Pub. 87,* vol. 1, pts. 1 and 2, 451 pp. Reprint 1970. $22.50

Legget, Robert F., 1962, *Geology and Engineering.* McGraw-Hill, N.Y., 884 pp.

Long, Leon E., 1974, *Geology.* McGraw-Hill, N.Y. 526 pp. $16.00 (Attractive, well-written and illustrated text generally suitable for the year course in physical and historical geology.)

Macdonald, Gordon A., 1972, *Volcanoes.* Prentice-Hall, Englewood Cliffs, N.J. 510 pp. $18.50 (Extensively illustrated, valuable lists of references, lists of active volcanoes; comprehensive coverage of the earth's volcanoes, for beginner and advanced student.)

Macdonald, Gordon A., and Agatin T. Abbott, 1970, *Volcanoes in the Sea—The Geology of Hawaii.* University of Hawaii Press, Honolulu. 440 pp. $15.00 (The most up-to-date, comprehensive, authoritative coverage of the great Hawaiian volcanoes; beautifully illustrated in black and white; large format.)

Matthews, William H., III, 1973, *A Guide to the National Parks.* Doubleday, Garden City, N.Y., 2 vols., 552 pp. Paperback, $5.95

______, 1968, *A Guide to the National Parks: Their Landscape and Geology.* Natural History Press, N.Y. Two volumes: vol. I, *Western Parks,* $7.95; vol. II, *Eastern Parks,* $6.95

______, 1962, *Fossils: An Introduction to Prehistoric Life.* Barnes and Noble, N.Y., 337 pp. $5.75; paperback, $2.25

______, 1969, *The Story of Volcanoes and Earthquakes.* Harvey House, Inc., Irvington-on-Hudson, N.Y. 122 pp., many illustrations. $4.50 (Very clear and well written; for beginning students of all ages.)

Menard, H. W., 1964, *Marine Geology of the Pacific.* McGraw-Hill, N.Y., 271 pp. $13.50

Moore, J. Robert, ed., 1971, 41 articles from *Scientific American* dealing with marine biology, oceanography, and ocean resources. Freeman, San Francisco, 417 pp. Paperback, $7.00

*National Geographic Magazine,* Jan. 1973, This Changing Earth (45 pp. of very graphic text and colored illustrations on plate tectonics, volcanoes, faults, and earthquakes.)

Oakeshott, Gordon B., 1976, *Volcanoes and Earthquakes.* McGraw-Hill, N.Y., 143 pp. Paperback, $3.95

Park, Charles F., Jr., 1975, *Earthbound: Minerals, Energy, and Man's Future.* Freeman Cooper and Company. San Francisco. 279 pp. Paperback, $3.95

Press, Frank, and Raymond Siener, eds., 1974, *Planet Earth.* Freeman, San Francisco, 303 pp. Paperback, $6.95

Putnam, William C., 1964, *Geology.* Oxford University Press, N.Y., 480 pp. $10.95 (Text ed. $9.00)

*Scientific American,* 1976, Continents Adrift and Continents Aground, Readings from *Scientific American,* Introduction by J. Tuzo Wilson. Freeman, San Francisco, 230 pp. Paperback, $5.95

Seismological Society of America, *Bulletins,* since 1911, issued six times per year. Subscriptions ($45.00 per year) and single issues (From $6.00 to $10.00 per issue) available from the Secretary, William K. Cloud, Seismographic Station, Department of Geology and Geophysics, Univ. of California, Berkeley, Calif. 94720. (Technical papers on seismology and engineering seismology as well as descriptive and technical accounts of most of the earth's significant active faults and earthquakes.)

Shelton, John S., 1966, *Geology Illustrated.* Freeman, San Francisco, 434 pp. $10.00

Shephard, Francis P., 1964, *The Earth Beneath the Sea.* Atheneum, N.Y., 275 pp. Paperbound, $1.65

Tank, Ronald, ed., 1976, *Focus on Environmental Geology.* Oxford University Press, N.Y., 538 pp. $8.95; paperback, $4.95

Townley, S. D., and M. W. Allen, 1939, Descriptive Catalog of Earthquakes of the Pacific Coast of the United States, 1769 to 1928. *Seismological Society of America Bull.,* vol. 29, no. 1, pp. 1-297. Out of print.

United States Geological Survey, 1966, Part II, *Water Resources,* in *Mineral and Water Resources of California.* Committee on Interior and Insular Affairs, U.S. Government Printing Office, Washington, D.C., 650 pp. (Pages 451-650 comprise a comprehensive summary of the state's water resources, problems, and water distribution and use problems. Obtain from California Division of Mines and Geology, Sacramento. $2.00)

Weitz, Joseph L., 1966, *Your Future in Geology.* Richards Rosen Press, N.Y., 192 pp. $3.78

Williams, Howel, 1932, Geology of the Lassen Volcanic National Park, California. *Univ. California Dept. Geol. Sci. Bull.,* UC, Berkeley, vol. 21, pp. 195-385.

——, *Mount Shasta, a Cascade Volcano. Jour. Geol.,* vol. 40, pp. 417-429. 1932.

Wilson, J. Tuzo, ed., 1972, *Continents Adrift.* Freeman, San Francisco, 172 pp. $7.00; paperback, $4.50

# glossary

■ Basic authority is *Glossary of Geology* by the American Geological Institute.

Terms which are defined in groups in the text—including names of fossils, minerals, fauna, flora, rocks, and rock formations, and the names of the units of geologic time—have not been repeated here. For such terms, see index and text.

**Aa** Lava with sharp-pointed, clinkery surfaces.

**Absolute Dating** Dating of rocks directly in years, as by radiometric dating.

**Accretion** Increase or extension, e.g., addition to land by building of a delta.

**Agglomerate** Essentially the same as **volcanic breccia.**

**Aggregate** Sand and gravel used with cement to make concrete. Pumice fragments are sometimes used to make a lightweight aggregate.

**Aggradation** Building up of a basin or surface by deposition of sediments.

**Algae** Primitive, single-celled plant life.

**Alluvial Fan** A fan-shaped deposit built up by stream deposition when the velocity of the stream is reduced by suddenly reaching a lower gradient, as when emerging from a narrow valley. Alluvial *cone* is synonymous.

**Alpine Glaciation** Mountain and valley type of action by ice, as distinguished from continental and piedmont glaciation.

**Angular Unconformity** See **unconformity.**

**Antecedent Stream** A stream that cuts across a ridge or range of hills because it was there before uplift occurred and had enough eroding power to maintain its course.

**Anticline** A fold in which the limbs dip away from each other; a *dome* is a special case of an anticline.

**Artesian** Groundwater confined under pressure so that it flows freely when an opening (well) to the surface is made.

**Asthenosphere** The "weak" layer of the earth below the lithosphere (crust). It is the "low-velocity" zone of the seismologist and is the layer over which crustal plates move.

**Ash, Volcanic** Explosively formed fragments of volcanic pyroclastic material smaller than 4 millimeters in diameter.

**Atom** The unit of reference for any element; the smallest particle that maintains all the physical and chemical properties of the chemical element.

**Augen Gneiss** A coarse-grained, banded metamorphic rock with eye-shaped structures or minerals in it.

**Avalanche** A snow slide, but the term is often used for slides of rock, mud, ice, or mixtures of these; usually, a sudden fall.

**Badlands** Very rough, narrow ridge-and-valley topography, steeply gullied. Usually due to occasional rapid runoff in arid or semiarid climates.

**Barchan** A crescent-shaped sand dune.

**Batholith** A large mass of granitic rock—measurable in terms of hundreds of miles in length and up to a hundred miles or more in width—formed from the slow cooling of magma several miles below the surface.

**Bed** A layer or stratum of rock.

**Bedding, Graded** See **graded bedding.**

**Bedrock** Solid rock which generally underlies softer, looser, more weathered rock or soil.

**Blister, Volcanic** A swelling on an extrusion of lava, due to gas pressure, usually several yards in diameter.

**Blue Schist** Schistose metamorphic rock of the glaucoplane-schist facies.

**Breccia** Any coarse-grained rock made up of angular fragments, e.g., volcanic breccia.

**Caldera** A very large, roughly circular volcanic depression, often due to collapse, e.g., Crater Lake, Oregon or Kilauea Caldera.

**Carbon 14** An isotope of carbon which is incorporated in all living things and is used as a basis for dating dead organic matter (accurate up to 20,000 to 30,000 years).

**Cinder Cone** A cone-shaped peak composed almost wholly of volcanic fragments formed by explosive action.

**Cinders** Fragments of volcanic origin, roughly 4 to 32 millimeters in diameter.

**Cirque** A bowl-shaped depression at the head of a glacial valley, formed by glacial erosion.

**Clastic** Made up of particles or fragments, like many sedimentary rocks.

**Clay** Plastic material made up of particles which will pass through a U.S. Standard no. 200 sieve; or, a clay mineral—one of a complex of hydrous aluminum silicates.

**Cleavage** (1) The breaking of a mineral along definite crystallographic planes; (2) the tendency of a rock to split along planar structures.

**Coast Range Thrust** A great thrust fault, best seen on the eastern flank of the Coast Ranges, which may have brought the Late Jurassic-Cretaceous Great Valley Series many miles westward over Franciscan rocks of the same general age.

**Coal** A combustible rock consisting of over 50 percent carbon, formed from compaction and induration of vegetation through geologic time.

**Columnar Jointing** The development of planar structures in rocks, leaving masses of columns; commonest as an effect of cooling in volcanic rocks, e.g., Devil's Post Pile.

**Composite Volcano** Buildup of alternate lava flows and the deposits of violent eruption.

**Compound** Two or more chemical elements that are chemically bonded together.

**Concretion** A hard, compact, rounded mass of mineral matter precipitated from water solution, formed in place in sedimentary rocks, e.g., limey concretions in shale.

**Conformable** Referring to strata in an unbroken sequence. See **unconformity.**

**Contact** Junction between two different kinds of rock or rock formations—a surface.

**Continental Drift** Movement of continental blocks over the asthenosphere. See **plate tectonics.**

**Continental Platform** The continent plus its shelves and slopes.

**Continental Rise** The gentle slope of sediments built up between the base of the continental slope and the abyssal plain.

**Continental Shelf** The gentle slope between the shoreline an the continental slope; usually the margin out to 200 meters deep.

**Continental Slope** The part of the continental margin between the shelf and the continental rise or abyssal plain. The slope is usually 3 or 4 degrees—thus, it is steeper than the shelf.

**Convection** Movement in a fluid (air, water, magma, asthenosphere) due to differential heating. May be the mechanism for sea floor spreading.

**Core of the Earth** The central nucleus of the earth, beginning at a depth of nearly 2,900 kilometers. There is an outer liquid core and a dense nickel-iron (?) inner core.

**Correlation** The process of connecting and relating rock units to show that they are approximately equal in age.

**Country Rock** The wall rock into which an intrusion may be emplaced or a vein deposited.

**Crater** The surface basin of a volcano, formed by explosive action and settling; usually at the summit of a volcanic cone or along a rift.

**Cross-bedding** An arrangement of layers in stratified rock such that minor layers are curved and at various angles to the major bedding surfaces. Characteristic of sand dunes, sandy deltas, and stream channels.

**Crust of the Earth** Lithosphere; the outermost layer of the earth, above the mantle, which is from 3 to 35 miles thick and is thinner under the oceans than under the continents.

**Daughter Elements** Intermediate members of a radioactive series between *parent* and *end product.*

**Delta** A fan-shaped deposit of alluvium at the mouth of a stream, where the stream enters quiet water; named after the Nile delta which looks like the Greek letter.

**Dendritic Drainage** A stream drainage pattern in which streams branch in several directions like an espaliered tree.

**Differentiation, Magmatic** Separation of a magma into liquids and solids of various compositions, both chemical and mineral; developing more than one rock type from a single magma.

**Dip** Angle of inclination of a stratum of rock, or fault, or other plane surface, measured with respect to the horizontal plane. See **strike.**

**Dome, Volcanic** A rounded extrusion of lava or welded tuff; low-angle if basaltic lava, steep if viscous or rhyolitic rock. See **plug** and **plug dome.**

**Earthquake** A sudden movement or trembling of the earth caused by the abrupt release of strain.

**Elastic Rebound** The theory of the origin of earthquakes which states that movement along a fault is the result of an abrupt release of accumulated elastic strain; theory proposed by H. F. Reid in 1911 as a result of the San Francisco 1906 earthquake.

**Electron** The negative building unit within an atom.

**Element** The simplest particle of matter into which a substance can be divided by ordinary physical and chemical means. Over 100 elements have been discovered, but only about 92 are stable.

**Epicenter** The point, in an earthquake, on the surface of the earth vertically above the focus or hypocenter.

**Epoch** The division in the geological time scale within a *period,* e.g., Pliocene Epoch.

**Era** A major unit in the geological time scale, e.g., Mesozoic Era.

**Erosion** Tearing down and wasting away of the land surface by such agents as running water and gravity.

**Escarpment** A steep or vertical step or cliff in the land surface, above or below sea level.

**Eugeosyncline** A geosyncline characterized by both volcanism and deposition of clastic sediments.

**Eustatic** Refers to worldwide changes of sea level, such as those associated with continental glaciation.

**Evolution** The theory that life has developed gradually, generally from the simple to the more complex forms; progressive change.

**External Drainage** Drainage with normal outlets to the sea.

**Facet** A plane surface or face produced by abrasion or erosion; may apply to a desert wind-faceted pebble or a cut face on a mountain spur.

**Fan, Deep Sea** Continental rise; the great fan of sediment that develops at the base of the continental slope, extending out onto the abyssal plain, e.g., Monterey Fan, Delgada Fan.

**Fanglomerate** A conglomerate made up of heterogeneous, coarse, angular fragments such as might occur in the upper part of a desert alluvial fan.

**Fault** A break in the earth's crust along which movement has taken place.

**Fault Creep** Fault slip; the slow, spasmodic displacement that occurs along some active faults.

**Fault Scarp** A step or cliff in the land surface caused by faulting. If eroded back, it is called a "fault-line scarp."

**Fault Slip** See **fault creep.**

**Fault System** A large group of related faults.

**Fault Zone** A fault made up of numerous minor fault traces. Hazardous fault zones are delineated—by law—in California.

**Faults, Lateral** Faults in which the displacement is predominantly horizontal, to the right or to the left.

**Fault, Normal** A tensional fault, with fault surface dipping toward the downthrown block.

**Fault, Reverse** A compressional fault, with fault surface dipping toward the upthrown block.

**Faults, Vertical** Faults in which the displacement is predominantly vertical.

**Fauna** The entire animal population of a given time, place, or environment.

**Flora** The entire plant population of a given time, place, or environment.

**Focus, Earthquake** The point at which an earthquake shock appears to originate: also called **hypocenter.**

**Foliation** A planar arrangement of textural or structural features in rocks, e.g., as in a schist.

**Foreset Beds** The cross-bedded strata on the underwater front of a delta.

**Formation** A group of rocks of similar age and having certain features in common; usually a mappable unit.

**Fossils** Any remains of ancient plant or animal life. A *marker* fossil is one which is characteristic of a certain geologic horizon or formation; also called *guide* and *index* fossils.

**Fracture Zones, Sea Floor** Huge zones of faulting on the sea floor, related to oceanic-crust breakup and movements, e.g., Mendocino fracture zone.

**Fumarole** A vent, usually volcanic, emitting gases and condensed water vapor; characteristic of the late stages of active volcanism.

**Geomorphology** The scientific study of landforms.

**Geologic Column** A diagram which shows the geologic units and their relative ages in a column on a page analogous to their position in the field.

**Geologic Time Scale** An orderly chronologic arrangement of era, periods, and epochs, and sometimes lesser divisions.

**Geology** The scientific study of the earth.

**Geosyncline** A long downwarping strip of the earth's crust subsiding as sediments are added; up to hundreds of kilometers long.

**Geothermal** Pertaining to heat—steam—within the earth, e.g., The Geysers geothermal area.

**Geyser** An eruptive hot spring. Most of the material erupted is water (steam, water vapor).

**Glacial Flour** Very fine rock particles ground by glacial erosion and abrasion, which are little weathered chemically.

**Glacial Till** Unsorted, unstratified glacial drift or rock debris deposited by glaciers.

**Glaciation** The process of modifying a land surface by the action of glaciers.

**Glacier** A mass of ice on land formed from the accumulation of snow and ice. *Alpine* glaciers develop in mountain valleys; *continental* glaciers cover a broad land surface. Glaciers move slowly downslope.

**Global Tectonics** See **plate tectonics.**

**Gneiss** A coarsely banded, coarse-grained metamorphic rock.

**Graben** A downfaulted block of the earth's crust.

**Graded bedding** Stratification in which a bed may change grain size upward and downward, e.g., sand grading upward through clay into shale.

**Gravel** Loose accumulation of rounded rock particles larger than 2 millimeters in diameter.

**Graywacke** Dark sandstone containing feldspars and rock fragments, which are poorly sorted and consisting mostly of angular grains.

**Great Valley Series** The tremendously thick series of marine stratified sedimentary rocks of Late Jurassic to Late Cretaceous age best developed in a deep geosyncline forming the eastern flank of the Coast Ranges and western part of the Great Valley of California.

**Greenschist** A schistose metamorphic rock including green minerals, such as chlorite, actinolite, and epidote; a facies of low-grade metamorphism.

**Greenstone** Informal field name for a greenish, metamorphosed, basic-to-ultrabasic igneous rock.

**Groundmass** The finer-grained interstitial mineral material in a porphyritic rock.

**Half-life** The time it takes in years for half the atomic nuclei in an atom to break down by radioactive decay.

**Heat Flow** Usually refers to the rate of heat transfer within rocks; the result of thermal conductivity and thermal gradient.

**Horst** An uplifted block of the earth's crust between faults.

**Hypocenter** See **focus.**

**Ice Age** A time in earth history when continental glaciers were particularly widespread; usually refers to the Pleistocene glaciation.

**Igneous Rock** Rock consolidated from once-molten magma. *Extrusive* if it reaches the surface of the earth: *intrusive* if the magma consolidates below the surface.

**Interior Drainage** Stream drainage in which there is no outlet to the sea, usually because there is insufficient water in a desert climate.

**Island Arc** A chain of volcanic islands rising from the deep-sea floor, convex toward the ocean—like the Aleutians; associated with deep-sea *trenches*.

**Isostasy** A condition of balance between high-standing, low-density masses of rock and low-standing denser bodies, e.g., less-dense continental crust "floating" higher on the asthenosphere than high-density oceanic crust.

**Isotope** One or more species of an atom of the same chemical element differing only in the number of neutrons, which causes a slight difference in atomic weight. Some isotopes are stable and others break down through radioactive decay; e.g., carbon 12 and carbon 14 are isotopes of carbon.

**Jointing** The formation of more-or-less regular fracture systems in rocks; often planar surfaces but may be curved, as in the formation of granite domes like Half Dome, Yosemite.

**Landform** Any feature of the earth's surface having a characteristic shape produced by natural geologic processes. Mountains, valleys, volcanic cones, granite domes, and sand dunes are all landforms.

**Landscape** The association of landforms, vegetation, etc., which makes up the surface of the earth and is visible in a single view or characteristic of a certain area or time.

**Landslide** A more-or-less abrupt downslope movement of loose and weathered soil and rock, moving in a mass. See **mass wasting.**

**Lapilli** Products of explosive volcanic action—pyroclastics—with particles 1 to 64 millimeters in diameter; term often used for more-or-less rounded, pea-sized to walnut-sized particles which cooled in the air during explosive eruptions.

**Lateral Moraine** A ridge-shaped accumulation of glacial drift or till along the sides of valley or alpine glaciers; best seen after melting of the ice.

**Lava** Magma which has reached the surface of the earth in a still-molten condition.

**Lava Tube** A tubular cavern left when a lava flow has crusted over while the inner part of the lava has flowed on. If the surface collapses, a *lava trench* is left.

**Lithosphere** The crust of the earth.

**Low-velocity Zone** A zone or layer in the earth, usually in the upper mantle (asthenosphere), where seismic-wave velocities are relatively low because the rock is in a meltng or near-melting condition.

**Mafic** Igneous rocks high in magnesium and iron, like peridotite.

**Magma** Molten rock below the surface of the earth.

**Magnetic Reversals** Refers to natural changes in the earth's magnetic field in which the earth's polarity reverses itself.

**Magnitude, Richter** A measure of the strength of an earthquake, on a scale devised by California seismologist Charles F. Richter. The maximum trace amplitude of an earthquake wave is measured in thousandths of a millimeter, from seismograms of a certain standard seismograph at a standard distance from the epicenter. The logarithm of the amplitude is then determined. One number on the scale (e.g., from $M = 5$ to $M = 6$) is equivalent to a change in total energy generated of about 32 times.

**Mantle of Earth** Zone or layer of the earth below the crust and above the core; about 2,880 kilometers thick.

**Mass Wasting** The general processes of movement of soil and weathered rock downslope under the influence of gravity and all forms of weathering and erosion.

**Meander** A wandering stream moving over a low gradient; may also be applied to a single loop of such a stream.

**Melange** Heterogeneous mixture of irregular blocks of rocks of diverse origin, e.g., Franciscan melange; a French term meaning *mixture*.

**Metamorphic Rock** A rock altered by heat, pressure, and chemical action, often developing new metamorphic minerals.

**Metasedimentary Rock** A metamorphosed sedimentary rock.

**Mineral** A naturally occurring, inorganic chemical compound (sometimes an element, e.g., gold, sulfur); usually has a characteristic crystal form.

**Mineralization** The process of natural introduction of mineral matter into rock substance; usually implies a possible commercial value.

**Miogeosyncline** A nonvolcanic type of geosyncline, composed of stratified sedimentary rocks (see **geosyncline** and **eugeosyncline**).

**Moat, Volcanic** A circular valley around a volcanic neck or dome, formed when an upwelling igneous plug did not quite fill the crater.

**Mohorovicic Discontinuity—Moho** The base of the earth's crust where a change in seismic-wave velocities occurs.

**Moraine** A ridgelike deposit of glacial drift or till left (1) along the sides of a valley or alpine glacier—*lateral* moraine; (2) a the front of a melting glacier—*terminal* and *recessional* moraine; or (3) on the land surface—*ground* moraine.

**Mud Cracks** Cracks in earth or sediment due to shrinking during desiccation, often in polygonal patterns.

**Mudflow** Downslope movement of loose, weathered rock and soil with a lot of water included.

**Natural Provinces** Basic geologic regions that have developed distinctive characteristcs of trend, structure, relief, drainage, and landscape. We may also speak of climatic, faunal, floral, and geomorphic provinces.

**Neutron** One of the elementary particles of an atom; they have approximately the same masses as protons but lack the positive charge.

**Nevadan Orogeny** Mountain-building epoch in the Late Jurassic in the Sierra Nevada and Coast Range regions.

**Obsidian** Volcanic glass, usually high-silica in composition, formed by quick cooling.

**Ocean Basin** The bowl in which the ocean lies; it does not include the continental shelves and slopes.

**Ophiolite** Usually applied to rocks formed in the upper mantle, including serpentine and associated pillow basalt, chert, and shale.

**Orbicular Structure** A spheroidal structure developed in some igneous rocks. The "orbs" may be a separate mineral or merely a structural form.

**Orogenics** Mountain-building processes.

**P waves** Longitudinal, compressional, or "push-pull" type of earthquake waves. They move the fastest of the earthquake waves.

**Pa hoe hoe** Twisted, ropy, smooth structure of a lava flow; contrasts with *aa* lava.

**Pangaea** A hypothetical supercontinent supposed to have existed in the Late Paleozoic or earlier. Pangaea included all known continents.

**Pegmatite** An extremely coarse grained igneous rock, most commonly the composition of granite.

**Period, Geologic** Time unit in the geologic time scale within the *eras*. A period includes *epochs*.

**Perlite** A glassy rhyolite in which concentrically layered spherules or "pearls" are formed.

**Petrology** The scientific study of rocks.

**Phenocrysts** The larger crystals in a porphyritic igneous rock.

**Photosynthesis** Synthesis of chemical compounds in the presence of light, as affects plants in sunlight.

**Piercement** A structure in which older rock is thrust up through younger rock as in a neck or dome, e.g., salt domes.

**Pillow Lavas** Basaltic lava which has been extruded under water and cooled to form pillow-shaped and pillow-sized structures. (Pillow structure is regarded as evidence of extrusion under water.)

**Plain** A large, relatively low lying area developed by any of various causes.

**Plate Tectonics** Global tectonics theory based on the separation and movement of huge continental and oceanic plates which move over the asthenosphere and interact with each other.

**Plateau** A relatively high, nearly flat summit area; but may be quite mountainous, e.g., Modoc Plateau.

**Playa** A desert pond or lake which is dry most of the time.

**Plug** A near-vertical, pipelike igneous rock column, such as in a *plug dome volcano,* a *volcanic neck* or *spine,* or a *salt plug* (dome).

**Plug Dome** A volcanic dome formed by magma welling up and cooling to fill a crater or by being shoved up from gas pressure below, e.g., Lassen Peak.

**Pluton** An irregular igneous intrusion which may be miles across but is a small part of a batholith; often of a distinctive petrologic type.

**Plutonic** Of deep-seated, igneous origin.

**Porphyritic** An igneous rock texture in which larger mineral crystals appear embedded in a finer groundmass.

**Proton** One of the units within the atom; it carries a unit positive charge equal and opposite to the negative charge of an electron, but the proton has a mass over 1,800 times that of an electron.

**Pumice** Light-colored, lightweight, vesicular, glassy volcanic rock; usually near rhyolite in composition and characteristically explosive in origin.

**Pyroclastic** Explosive, fragmental volcanic rocks; tephra.

**Radioactive Decay** The process of the spontaneous breakdown of the nuclei of certain elements; the basis for radiometric dating.

**Radiometric Dating** The dating of minerals in rocks by using the principle of *radioactive decay* and the known *half-life* of isotopes of certain elements.

**Relief, Topographic** Difference in elevation between higher and lower points on the surface of the earth.

**Relative Time** Comparison of ages—as for example in a stratigraphic succession—without knowing ages in absolute terms.

**Ridges (Plate Tectonics)** Great oceanic ridge and fault zones from which eruptions of upper-mantle basalt and mafic rocks from centers of sea floor spreading take place; also called *rises.* Oceanic plates move in both directions away from the ridges, e.g., mid-Atlantic ridge, East Pacific rise.

**Rift Valley** A valley primarily resulting from faulting.

**Right Lateral** See **fault, lateral.**

**Ripple Marks** Small-scale ridges and hollows consisting of sediment—most often sand or silt—produced by wave and current action; elongated at right angles to water currents or wind.

**Rise** (1) In *plate tectonics,* an oceanic ridge, e.g., East Pacific rise; (2) in oceanography, a deep-sea fan near the base of the continental slope, e.g., a *continental rise.*

**Rock** A natural aggregate of two or more minerals, although some rocks consist essentially of only one mineral; the solid materials which make up the fundamental units of the earth's crust.

**Rock Cycle** A sequence of events leading to the formation, partial destruction, or modification of rocks; often expressed in a diagram showing the relationships between various types of rocks.

**Rock Formation** A group of rocks of similar age with certain features in common; usually a mappable unit.

**Rock Glacier** A rock slide, usually made up of angular fragments showing little chemical weathering, which may move en masse in steep mountain areas.

**Roof Pendant** Part of the older roof rock of a batholith, which has been exposed by erosion.

**Root** A downward "bulge" or extension of lower-density crustal material, as in "root" of the Sierra Nevada.

**S waves** Transverse or shear earthquake waves; wave motion is transmitted transverse to the line of propagation. S waves move slower than *P waves* and are not transmitted through liquids.

**Sag Pond** Fault-caused depression in uneven ground in a fault zone.

**Salt Pan** An undrained, small, natural hollow where water evaporates and leaves salts. See **playa.**

**Sand** Rock fragments or particles having diameters in the range of $\frac{1}{16}$ to 2 millimeters; unconsolidated.

**Scarp** See *fault scarp.*

**Schist** A foliated, crystalline, metamorphic rock. A schist is finely granular or platy.

**Scoria** Vesicular, cindery, blocky crust on a lava flow; heavier, darker, and more crystalline than pumice.

**Sea Cliff** The cliff along a sea coast primarily due to undercutting by wave and current action.

**Sea Floor Spreading** In *plate tectonics,* refers to movement of oceanic plates away from each other and away from spreading centers along the mid-oceanic ridges; presumably convection in the upper mantle is the source of material and energy.

**Seamount** A hill or mountain on the sea floor, either peaked or flat-topped; of volcanic origin.

**Seastack** Nearshore island separated from the mainland by wave and current action.

**Sedimentary Rock** A rock formed from the consolidation of loose sediment; normally, but not always, stratified; also applies to rock formed by chemical and organic means.

**Seismic Zonation** Marking zones or areas of differing seismic history; may be related to the expected frequency of earthquakes in different zones or belts.

**Seismogram** The record of an earthquake on a chart.

**Seismograph** The instrument used to record earthquake wave motion.

**Seismology** The scientific study of earthquakes, including study of the interior of the earth from earthquake waves.

**Serrated Ridge** A saw-toothed ridge, particularly the result of headward erosion by glaciers.

**Shield Volcano** A volcano made up of the consolidation of highly fluid lava flows, hence having a very low conical profile, e.g., Mauna Loa, Hawaii.

**Sial** Magma very high in silica and alumina; therefore, the primary material of the continents, like granite.

**Sill** A sheetlike intrusion parallel to stratification, schistosity, or structures in rocks.

**Silt** A term for soil texture or fine-grained sediment made up of particles in the range 0.002 to 0.05 millimeter in diameter; unconsolidated.

**Sima** Magma very high in magnesia and iron; the primary material of the upper mantle and the oceanic plates.

**Soil** (1) The weathered rock and organic matter which is the natural medium for plant growth; (2) in engineering, refers to all unconsolidated earthy material above bedrock.

**Spatter Cone** Small cones built up by the splashing and spattering of lava spouting out of lava tubes and flows.

**Spilite** An altered basalt or andesite in which the feldspar has been albitized; often occurs as pillow basalt, slightly metamorphosed to greenstone.

**Spine** See **volcanic neck** and **plug.**

**Stratification** Layering or bedding in sedimentary rocks due to deposition. Layering in other types of rocks should not be called stratification.

**Strato-volcano** See *composite volcano.*

**Strike** The direction of the line of intersection of a plane of stratification, fault surface, or other natural plane with the horizontal plane. See also *dip*.

**Strike slip** Used to describe a fault in which the principal displacement has been horizontal in the direction of the strike of the fault.

**Subduction Zone** The zone where one crustal block descends beneath, or is overridden by, another—e.g., Nazca plate (Pacific plate) being subducted beneath the South American crustal plate.

**Submarine Canyon** A steep-sided, V-shaped valley on a continental shelf or on a continental slope, similar to a stream valley on land. See **turbidity current.**

**Syncline** A trough-shaped fold. In stratified rocks, the youngest rocks appear in the axial region of the fold.

**Synclinorium** Huge, complex syncline.

**Tactite** A rock formed by contact metamorphism, often of complex mineral composition including garnet and epidote.

**Tailings, Mine** Waste rock left after mining and milling.

**Talus** Angular rock fragments of any size which accumulate near the foot of a steep slope. Talus cones and aprons are common in steep mountain areas.

**Tarn** A small lake, usually of glacial origin.

**Tectonics** The study of the earth's structure and the forces and processes that produce structural changes.

**Tephra** See **pyroclastic,** a synonym.

**Terrace** A natural, raised, near-flat platform; a bench; often indicating the former level of a water surface, but formed in a variety of ways.

**Texture** Size, shape, and arrangement of minerals in rocks.

**Thrust Fault** A low-angle reverse fault in which one block overrides another.

**Till** Unsorted and unstratified glacial drift.

**Topography** The general configuration of the land surface.

**Transform Fault** In plate tectonics, a *strike-slip* fault characteristic of oceanic ridges and along which the ridges are offset; may form the boundary between two plates.

**Tree Rings** Growth rings which may be used to determine the approximate age of a tree.

**Trench** A narrow, elongated depression of the deep-sea floor between the continental margin and the abyssal deep of the ocean.

**Tufa** A chemical sediment consisting of calcium carbonate formed by evaporation around a spring. Tufa may take on weird forms.

**Tuff, Volcanic** A compacted pyroclastic deposit, mainly volcanic ash.

**Turbidity Current** A density current, usually in water and due to a heavy load of suspended sediment. An important mode of erosion of submarine canyons, and a mode of transportation and deposition used to build shelf deposits and deep-sea fans.

**Turbidity Deposition** Deposition of sediments primarily by the action of *turbidity currents.*

**Ultramafic Rocks** *Peridotites* and *serpentine;* igneous rocks high in iron and magnesium.

**Unconformity** A break in the sedimentary record due to uplift, erosion, and other causes. If the older rocks have been folded, so that the younger strata lie at an angle to them, it is called an *angular* unconformity. See also **conformable.**

**Varves** Seasonally banded sediments deposited in glacial meltwater. One varve is a pair representing summer and winter glaciolacustrine sediments, or one year.

**Vesicular Structure** The structure produced in volcanic rocks by the escape of gases. The holes left are *vesicles;* if later filled with mineral matter, they become *amygdules.*

**Volcanic Bombs** Head-sized, melon-shaped, smooth fragments twisted at the ends, which represent masses of lava cooled as they fell through the air.

**Volcanic Dome** See **Dome, volcanic.**

**Volcanic Neck** A volcanic plug or spine.

**Volcanic Rock** That rock of igneous origin which has been extruded on the surface of the earth, or in the near-surface as in a volcano.

**Wall Rock** See **Country rock.**

**Waves, Earthquake** Seismic waves; the wave motion caused by abrupt movement of blocks of the earth's crust along a fault, although earthquake waves may originate in other ways. See **P wave** and **S wave.**

**Weathering** The natural destructive process by which rock is broken down by chemical and mechanical means. Weathering occurs near the surface of the earth and prepares rock for its removal by erosion.

# index

## A



## B

# C

# E



# F

# G

# H



# I

# J

# L

# N

# P

# R

# T

## Z